E. Wyndham (Edward Wyndham) Tarn

The Mechanics of Architecture

E. Wyndham (Edward Wyndham) Tarn

The Mechanics of Architecture

ISBN/EAN: 9783743686816

Printed in Europe, USA, Canada, Australia, Japan

Cover: Foto ©berggeist007 / pixelio.de

More available books at **www.hansebooks.com**

THE MECHANICS

OF

ARCHITECTURE

THE MECHANICS

OF

ARCHITECTURE

A

*TREATISE ON APPLIED MECHANICS ESPECIALLY
ADAPTED TO THE USE OF ARCHITECTS*

By E. WYNDHAM TARN, M.A.

Architect

AUTHOR OF "THE SCIENCE OF BUILDING," "PRACTICAL GEOMETRY FOR THE ARCHITECT"
ETC, ETC,

Illustrated with 125 Diagrams

Capio Lumen

LONDON

CROSBY LOCKWOOD AND SON

7, STATIONERS'-HALL COURT, LUDGATE HILL

1892

" Eruditus geometria . . . philoscphos diligenter audierit."

" Præterea de rerum natura quæ græce φυσιολογια dicitur philosophia explicat, quam necesse est studiosins novisse, quod habet multas et varias naturales quæstiones."—Vitruvius on the Education of the Architect.

" Architecti est scientia pluribus disciplinis et variis eruditionibus ornata . . . . ea nascitur e fabrica et ratiocinatione . . . . at qui utrumque perdidicerunt, uti omnibus armis ornati citius cum auctoritate quod fuit propositum sunt adsecuti."
—Vitruvius, Book I.

PREFACE.

THE numerous works that have been written upon the subject of Applied Mechanics seem to be generally intended more for the use of the Engineer than the Architect. The training of the Engineer being essentially a scientific one, he is compelled to devote much time to the study of mechanics; while the Architect, whose chief aim is to become an Artist, has to give most of his hours of study to the arts of drawing and designing; so that he has but little time, and often less inclination, to make acquaintance with the science of his profession, being generally contented to take his formulæ from some "pocket-book," without caring to enquire into the principles upon which those formulæ are based.

In the following pages the Author has endéavoured to supply a want which is felt by many Architects, by bringing together, in a small compass, all that it is essential for the Architect to know upon this subject, and to give in as simple a form as possible an outline of the principles upon which all good construction should be based. In doing so it has been assumed that the reader possesses a moderate acquaintance with algebra and trigonometry; and although it has been necessary to the solution of some problems to invoke the powerful aid of the Calculus, yet by far the greater part of the work can be understood and the examples worked out by those who have no knowledge of the higher mathematics.

The Author would call particular attention to the theories and examples of Roofs, Arches, Vaulting, and Domes, to which he has given great prominence in this work, as being those subjects in which the Architect is especially interested. He has also worked out numerous examples showing the practical application of every theory and formula, in order that the reader may never be at a loss to understand how to use them.

Those who wish to pursue the various subjects more deeply are referred for information to the following Authors, to all of whom the present writer desires to acknowledge his deep obligations :—

Bow. Economics of Construction, by R. H. Bow.
Clarke. Graphic Statics, by G. S. Clarke.
Fenwick. Mechanics of Construction, by S. Fenwick.
Graham. Graphic and Analytical Statics, by R. H. Graham.
Moseley. Mechanical Principles of Engineering and Architecture, by Henry Moseley.
Rankine. Manual of Applied Mechanics, by W. J. M. Rankine.
Stoney. Theory of Strains, by B. B. Stoney.
Weisbach. Mechanics of Engineering, by Julius Weisbach.
Wray. Application of Theory to Practice, by H. Wray.

CONTENTS.

CHAPTER I.

FORCES IN EQUILIBRIUM.

CHAPTER II.

MOMENTS OF FORCES.

CHAPTER III.

CENTRE OF GRAVITY.

CHAPTER IV.

RESISTANCE OF MATERIALS TO STRESS.

CHAPTER V.

DEFLEXION OF BEAMS.

CHAPTER VI.

STRENGTH OF PILLARS.

CHAPTER VII.

ROOFS, TRUSSES.

CHAPTER VIII.

ARCHES.

CHAPTER IX.

DOMES, SPIRES.

CHAPTER X.

BUTTRESSES, SHORING, RETAINING WALLS, FOUNDATIONS.

CHAPTER XI.

EFFECT OF WIND ON BUILDINGS.

CHAPTER XII.

MISCELLANEOUS EXAMPLES AND SOLUTIONS

MECHANICS OF ARCHITECTURE.

CHAPTER I.

FORCES IN EQUILIBRIUM.

1. MEASURE OF FORCE.—The term "Force" is applied to a certain action of one body upon another which causes a change in the circumstances of one or other, or both of the bodies. It is the object of the science called "Mechanics" to examine into and determine the laws of force and the results which different forces produce. By the term "Applied Mechanics" (as far as the present treatise is concerned) we mean the application of the laws of force to Architecture.

It is usual to represent mechanical forces by means of straight lines whose directions indicate the directions in which the forces act upon a body; and a length measured to scale upon any of these lines is taken to represent the magnitude of the force in tons, hundredweights or pounds avoirdupois.

When a force acts at any point of a *rigid* body, that is, of a body which does not yield or change its shape under the action of the force, an exactly similar and equal force is set up in the body, acting however in a diametrically opposite direction to the impressed force;

this opposing force is called the "reaction" of the body, and is always equal in magnitude and opposite in direction to the impressed force. Thus, if a load is placed on hard ground, a force exactly equal to the weight of that load presses upwards upon it and prevents it from sinking into the ground. A familiar example of *reaction* is seen when a man in a boat pushes himself away from the bank of a river by means of a pole; the *reaction* of the bank being equal to the pressure exerted by him upon the pole. It is by *reaction* that the paddles or screw of a steam-vessel, or the oars of a boat, propel the vessel through the water.

The only *impressed force* that we have to take account of in building construction is the force of "gravity," which is the term applied to the attraction of the earth upon every particle of matter which comes within its influence, and which is commonly called weight, being proportional to the "mass" of the body attracted, and measured as before stated in tons, hundredweights and pounds. The force of gravity is always considered, as far as buildings are concerned, to act in parallel lines, and in a direction perpendicular to the surface of still water. When we speak of the "load" which any structure has to sustain, we mean the force which the earth exerts upon every particle of the mass which is placed upon that structure, by pulling it towards its surface.

There is another kind of force which we shall have to consider, and which is altogether independent of "gravity," namely, the "resistance" which all solid bodies offer to any impressed force tending to crush them or tear them asunder. This resistance is called the "force of cohesion," and varies greatly in different bodies, as we shall see hereafter. When the force pushes against the body it is called a "compressive" or "crushing stress;" when the force tends to pull it asunder, it is called a "tensile" or "extending stress." When it tends to cause the body to bend or break across in the direction of the force, it is termed a "transverse stress;" and when it acts transversely as a cutting force tending to cause the body to separate into slices, and the slices to slide off each other, it is termed a "shearing stress."

The term "stress" is applied to the action of an external force upon the fibres or particles of a body when applied to its extremities either as a pushing or pulling force; while the term "strain" is applied to the effect of the stress in changing the shape of the body. Thus the change of length arising from a tensile or compressive force is termed "longitudinal strain." By "unit-stress" is meant the stress upon a unit of sectional area; "inch-stress" being the stress where the unit is one square inch, which is the unit usually employed in England.

2. RESULTANT OF FORCES. — When two or more forces act either in the same or in different directions on a body, if there is some one force, P, that will exactly balance them or keep the body in "equili-brium," so that no movement takes place; then a force R equal and opposite to P will evidently balance the force P, and will consequently produce the same effect upon the body as the original forces themselves would produce. This single force, R, is called the

"resultant" of these forces, while the original forces
themselves are called the "components" of the single
force R. Thus, for example, if we place a body, A,
upon a table and attach to the body three cords having
weights P, Q and S, at their ends, and acting in the
directions AP, AQ and AS (fig. 1), which keep the

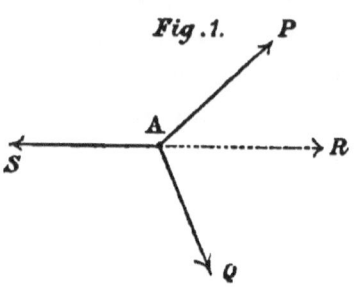

Fig. 1.

body in equilibrium; then
the force S is equal and
opposite to the resultant
of P and Q, since it ex-
actly balances them; the
force P is also equal and
opposite to the resultant
of Q and S, for the same
reason; and the force Q
is equal and opposite to the resultant of P and S. If
we produce the line SA in the direction AR, and take
a force R equal to S but acting in the opposite direction; R will be the *resultant* of P and Q, while P and
Q will be the *components* of R. And the same for the
other forces.

3. RESOLUTION OF FORCES.—We have now to determine the magnitude and direction of the resultant, R,
of the forces P and Q which act at the point A (fig. 1), or
that force which exactly balances the force S. Let the
magnitude of the forces P and Q be represented on
scale by the lengths AB and AC (fig. 2) and draw BD
parallel to AC, CD parallel to AB. Then it is proved
in treatises on elementary mechanics that the diagonal
AD of the parallelogram ABDC represents the
resultant, R, of the forces P and Q, both in magnitude
and direction, and may be measured on the same scale

as was used for AB and AC which represent P and Q.
In order that R and S may balance each other, they
must be equal in magnitude and opposite in direction;
so that the line AE, which represents S, must be a
continuation of the line DA, and must be equal in
length to DA. The force S, represented by AE, will
then balance the forces P and Q. Similarly, to find
the resultant, T, of S and Q, draw CF parallel to AE,
and EF parallel to AC; then the diagonal AF will be
their resultant T, both in magnitude and direction, and
must be equal and opposite to the other force P, in

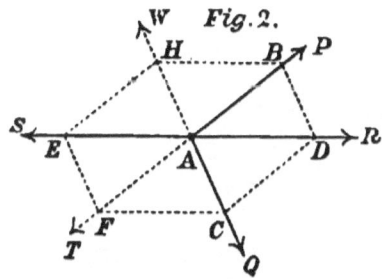

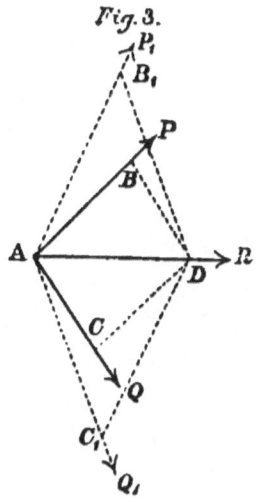

order that the forces may be in
equilibrium. In the same way
by drawing the parallelogram
AEHB we find the force W, represented by the diagonal AH,
which is the resultant of P and
S, and is equal and opposite to the force Q, when there
is equilibrium between the forces.

Conversely, if the force R, represented by the line
AD (fig. 3) acts at a point A, and we desire to find
two forces P and Q equivalent to R, and which would
act in the directions AB and AC; we draw DB

parallel to AC, and DC parallel to AB; then AB will represent the force P, and AC the force Q, both in magnitude and direction; all three forces being measured by the same scale. The forces P and Q are called the "Components" of the force R, which is said to be "compounded" of these two forces. When P and Q are used in place of R, the latter is said to be "resolved" into the forces P and Q.

The component forces may be taken to act in *any* directions, as AB_1, and AC_1, so long as the angle between them, ($B_1 AC_1$) is less than two right angles.

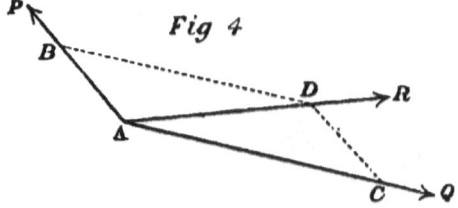

Fig 4

Thus, where the angle BAC (fig. 4) is very obtuse, P and Q will still be the components of R, represented by the diagonal AD of the parallelogram ABDC; but if AB comes to be in the same straight line as AC, then the line DC becomes parallel to AC, or C goes off to an infinite distance, while B coincides with A; so that one component vanishes while the other becomes infinite, and the resultant, R, coincides with the greater force.

When the forces P and Q, represented by AB and AC (fig. 5,) are equal, then the direction AD of their resultant, R, will evidently bisect the angle BAC. If the angle BAC equals 120°, then the triangles ABD and ACD will be equilateral, and AD is equal to AB or to AC; or the resultant, R, is equal to each of its components, P or Q. If the angle BAC is greater than 120°, then each component will be greater than the resultant. If BAC is less than 120°, each component will be less than the resultant; but in *all* cases the sum, P + Q, of the components is greater than the resultant, since the sum of any two sides of a triangle is always greater than the third side.

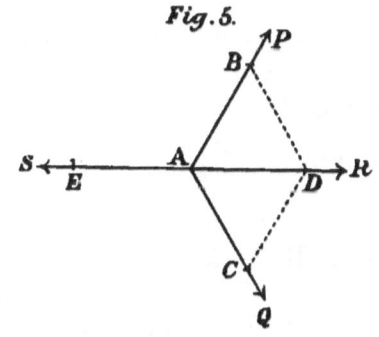

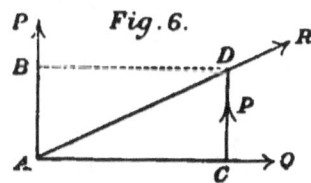

In practical applications of this theorem we often have the directions of the component forces perpendicular to one another, or the force R has to be resolved into two others at right-angles to one another. Let AD (fig. 6) represent R, AB and AC its components P and Q at right angles to one another. Complete the parallelogram ABDC; then AC represents the component Q, and CD the component P.

$$P = CD = AD \times \frac{CD}{AD}; \quad Q = AC = AD \times \frac{AC}{AD}.$$

Now the ratio $\frac{CD}{AD}$ is called the *sine* of the angle DAC,

and the ratio $\frac{AC}{AD}$ is called the *cosine* of the angle DAC.

And since the values of the *sines* and *cosines* of all

angles are to be found in tables, it follows that when the angle DAC and the length AD, which represents R, are known, we can determine the components P and Q from the equations

$$P = R . \sin. DAC; \quad Q = R . \cos. DAC.$$

If the angle DAC is 30°, its *sine* is $\frac{1}{2}$ or ·5, and its *cosine* is ·866; in which case P $= \frac{1}{2}$ R, and Q = ·866 R. If the angle DAC is 60°, its *sine* is ·866, and its *cosine* is ·5; then P = ·866 R, and Q $= \frac{1}{2}$ R. When the angle is 45°, the *sine* and *cosine* are equal, each being $\frac{1}{2} \sqrt{2}$ or ·707; so that P = Q = ·707 R.

When the two components P and Q are known, their ratio $\frac{P}{Q} = \frac{DC}{CA}$ is called the *tangent* of the angle DAC, and by referring to a table of *tangents* the angle DAC is found, from which the value of R can be ascertained.

For, $R = AD = \frac{AD}{AC} \times AC = \frac{AC}{\frac{AC}{AD}} = AC$ divided by the *cosine* of DAC.

The length AD can also be calculated directly when AC and DC are known, by help of the 47th proposition of 1st book of Euclid; for in a right-handed triangle

$$AD^2 = AC^2 + DC^2$$

so that R is the square root of the sum of the squares of P and Q; or $R = \sqrt{P^2 + Q^2}$. For example, if P = 5 and Q = 7, we have $R = \sqrt{25 + 49} = 8·6$; then sin. DAC $= \frac{P}{R} = \frac{50}{86} = ·5814$, which we find in the tables to be the *sine* of 35° 33′, as the value of the angle DAC for the above values of P and Q.

If the angle DAC is given, it will suffice to enable us to determine the relations between R, P, and Q; since we can obtain the values of sin. DAC and cos. DAC from the tables. Thus, if DAC = 35° 33′, we find sin. DAC = ·5814, and cos. DAC = ·8136; or

$$\frac{P}{R} = ·5814, \frac{Q}{R} = ·8136, \text{ and } \frac{P}{Q} = \frac{5814}{8136} = ·7146 = \text{tan.}$$

DAC.

4. TRIANGLE OF FORCES.—It will be evident from the last article that if the three forces P, Q and S (fig. 5) are represented in direction and magnitude by the three sides of the triangle ABD, taken in order; that is, AB, BD, DA; they will balance each other at the point A of their application.

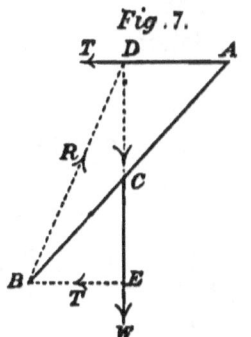

Fig. 7.

For AB is the direction of the force P, BD parallel to that of the force Q, DA that of the force S which acts in the opposite direction to the resultant R of P and Q. Conversely, if three forces acting on a body keep it in equilibrium, their directions must meet in one point. For example: Suppose a rigid rod ACB (fig. 7) to be kept in equilibrium by the action of a horizontal force T, at A, a vertical force W at its middle point C, and by a third force R acting at B. Then since the directions of W and T meet in the point D, it follows, from the above theorem, that DB must be the direction of the force R acting at B, and meeting the other two in the point D. Draw the horizontal line BE meeting DC in E; then the sides of the triangle BDE will represent the three forces taken in order. That is to say, DE represents W, EB represents T, and BD represents R; and they will act in the directions shown by the arrows. If then any one of the three forces, as W, is known, the other two can be determined, as follows:

$$W = DE = AB \times \sin. ABE, \text{ or } AB = \frac{W}{\sin. ABE}.$$

$$T = BE = \tfrac{1}{2} AB \times \cos. ABE$$

$$= \tfrac{1}{2} AB \times \sin. ABE \times \frac{\cos. ABE}{\sin. ABE}$$

$$= \tfrac{1}{2} W \times \cotan. ABE \quad . \quad . \quad . \quad (1)$$

$$R = BD = \sqrt{DE^2 + BE^2} = \sqrt{W^2 + T^2} \quad . \quad (2)$$

The angle DBE which R makes with the horizontal, is determined from the equation

$$R . \sin. DBE = DE = W;$$

$$\text{or, } \sin. DBE = \frac{W}{R} \quad . \quad . \quad . \quad (3)$$

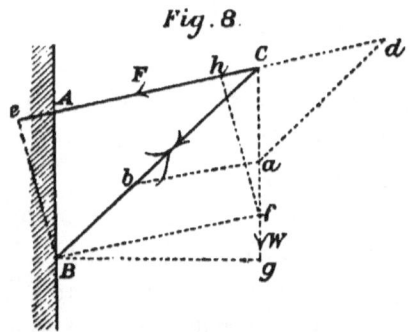

Fig. 8.

A familiar example of the triangle of forces will be found in the bracket or crane, where an inclined strut BC (fig. 8) presses against a wall at B, and also against the end of another beam AC at C where a load W is supported. Let the vertical line C*a* represent the load W, and draw *ad* parallel to BC, *ab* parallel to AC, producing AC to *d*. Then the three lines C*a*, *ab*, *b*C represent the force W, the stress,

F, in AC, and the stress in CB; and these forces are in equilibrium if taken in order. Therefore the force F, in AC which balances the stress produced by W at C, must act from C towards A, and is a pulling one, so that AC might be a cord or chain; while the force acting in BC to balance the stress produced by W must act from C towards B, or the beam BC is subjected to a pushing force.

5. TRIANGULAR FRAME.—Suppose a triangular frame to be formed of three rigid straight bars jointed together at their ends, as ABC (fig. 9); and to be acted

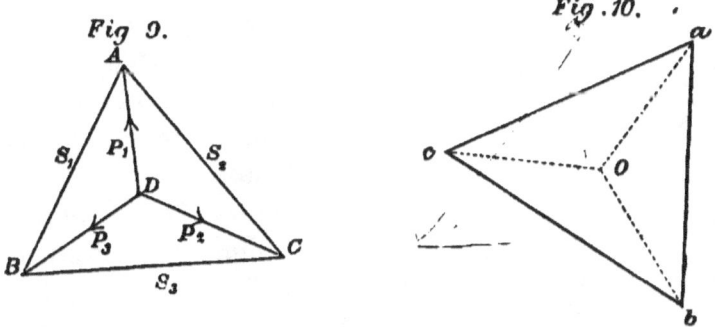

upon at the vertices A, B, C, by the forces P_1, P_2, and P_3, which keep the frame in equilibrium. From what has been just stated (4) the direction of these three forces must meet in a point D, and the forces themselves will be proportional to the sides of the triangle abc (fig. 10) which are respectively parallel to their directions. Suppose these forces to produce in the bars of the frame the stresses S_1, S_2 and S_3. From the vertices of the triangle abc draw lines ao, bo, co, parallel respectively to the sides of the triangle ABC. Then ao being parallel to AB will represent S_1, the stress in the bar AB; *bo* being parallel to AC will represent S_2, the stress in the bar AC; *co* being parallel to BC will represent S_3, the stress in the bar BC. The first triangle ABC (fig. 9) is called the "frame-diagram," and the second triangle *abc* (fig. 10) is called the "stress-diagram." The lines which in the former make a closed figure are represented in the latter by lines meeting in a point; and the lines which in the latter make a closed figure are represented in the former by lines meeting in a point. Consequently the two triangles are termed "reciprocal figures."

6. POLYGON OF FORCES. — The principle of the "triangle of forces" can be extended to any number of forces which are in equilibrium at a point; the forces being represented in magnitude and direction by the sides of a polygon respectively parallel to their directions, the polygon having as many sides as there are forces. Suppose the five forces P, Q, R, S and T, to be in equilibrium at the point A (fig. 11), and to be represented in magnitude and direction by the lines AB, AC, AD, AE, and AK. First, describe a parallelogram ABGC on the lines AB and AC; then the diagonal AG represents the resultant of P and Q.

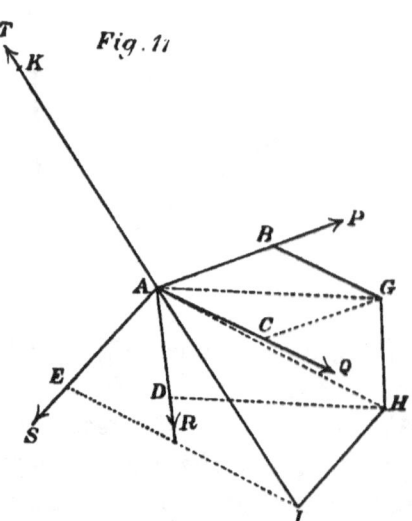

Fig. 11

Next, draw a parallelogram AGHD on the lines AG and AD ; then the diagonal AH represents the resultant of AG and R, or of P, Q and R. Thirdly, draw the parallelogram AHIE, upon the lines AH and AE ; then the diagonal AI represents the resultant of AH and S, or of P, Q, R and S. The force T, represented by AK, must be equal in magnitude and opposite in direction to the last resultant AI, in order that the forces may be in equilibrium at the point A.

The resultant AI *closes* the polygon ABGHIA, the sides of which represent the magnitude and direction of the five forces, P, Q, R, S and T.

7. POLYGONAL FRAME. — As the principle of the "Triangle of forces" can be extended to any number of forces acting at a point, so the principle of the "triangular frame" (5) can be extended to a frame having any number of sides. Suppose a polygonal frame formed of more

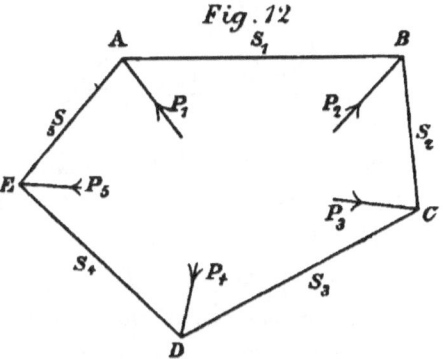

Fig. 12

than three rigid straight rods jointed at their ends, as ABCDEA (fig. 12) to be kept in equilibrium by the forces P_1, P_2, &c. acting at the joints A, B, &c. and to produce in the bars the stresses S_1, S_2 &c. Then each joint must be in equilibrium under the forces which act upon it. The forces P_1, S_1 and S_5, acting at A must be represented by the three sides of a triangle which are parallel respectively to their directions (4). First, let the triangle *Oae* (fig. 13) have the side *Oa* parallel to AB representing S_1 (fig. 12); the side *Oe* parallel to AE representing S_5; and the side *ae* parallel to the direction of P_1. Next, let the triangle *Oab* have the side *Oa* representing S_1, as before; the side *Ob* parallel to BC representing S_2; the side *ab* parallel to the direction of P_2. Thirdly, let

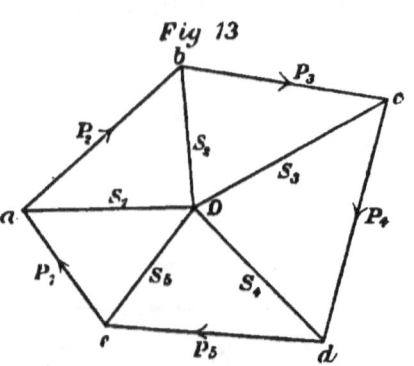

Fig 13

the triangle *Obc* have the side *Ob* representing S_2, as before; the side *Oc* parallel to CD representing S_3; the side *bc* parallel to the direction of P_3. Fourthly, let the triangle *Ocd* have the side *Oc* representing S_3, as above; the side *Od* parallel to DE representing S_4; the side *cd* parallel to the direction of P_4. Lastly, we have left the triangle *Ode*, of which the side *Od* represents S_4, as above; the side *Oe* parallel to EA represents S_5; then the side *de* gives the direction and magnitude of P_5, in order that the forces may balance.

Thus each joint of the frame (fig. 12) furnishes a triangle (fig. 13), and each triangle has one side common to only two other triangles; and all these triangles put together make up the closed polygon *abcdea* (fig. 13), the outer lines of which represent the magnitude and direction of the several exterior forces P_1, P_2 ... P_5; while the inner lines *Oa*, *Ob*, ... *Oe*, represent in magnitude and direction the several stresses S_1, S_2, . . . S_5, in the bars of the frame (fig. 12). Thus ca represents P_1, ab represents P_2, bc represents P_3, cd represents P_4, and de represents P_5. Oa represents S_1, Ob stands for S_2, Oc for S_3, Od for S_4, Oe for S_5.

It will be seen from the above that the fifth force P_5 can only be determined both in magnitude and direction by the line de which closes the stress-diagram, so that if P_1, P_2, P_3 and P_4 are given, or known, beforehand; the force P_5 which balances them, remains to be determined, as in the theorem of the polygon of forces previously given (6). The line de representing P_5 is said to *close* the polygon.

CHAPTER II

8. DEFINITION OF A MOMENT.—Whenever there is a fixed point in a body which is acted upon by external forces, those forces will tend to turn the body round, or cause it to rotate, about the fixed point, provided that the directions of the external forces do not pass through the fixed point. Let O be a fixed point in a body which is acted on at A by the forces P, Q and R (fig. 14). Draw OB perpendicular to the direction of the force P; OC perpendicular to that of force Q; and

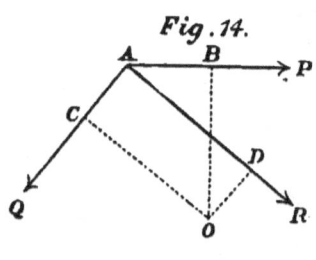

Fig. 14.

OD perpendicular to that of the force R. Then P × OB is called the *moment* of P about O; Q × OC the *moment* of Q about O; and R × OD the *moment* of R about O. The *moment* of a force about a point measures the tendency of that force to cause the body to rotate about an axis through the fixed point, such axis being at right-angles to the plane in which the force acts, or to the plane of rotation.

If two forces, as P and Q, acting at a point A of a body in which there is a fixed point O, keep it in equilibrium: then their *moments* about O must be equal in magnitude, but tending to turn the body in opposite directions; or we must have,

$$P \times OB = Q \times OC;$$

OB and OC being respectively perpendicular to the directions of P and Q, as in (fig. 14).

If R is the resultant (3) of the forces P and Q, then it is demonstrated, in elementary treatises on Mechanics, that the *moment* of R about O is equal to the *sum* of the *moments* of P about O and of Q about O; or, drawing OD perpendicular to the direction of R, we have—

$$R \times OD = P \times OB + Q \times OC \qquad . \quad (4)$$

the *moments* being taken about an axis through O at right-angles to the plane in which the directions of the forces lie.

In the diagram (fig. 14) the *moment* of Q tends to rotate the body in an opposite direction to that which P does, consequently one of the moments must be considered *negative* and the other *positive*. When therefore we speak of the *sum* of the moments of the components, we mean their *algebraic* sum, each moment being taken with its *algebraic sign* of *plus* or *minus*. In any case where the components have opposite signs, the moment of the resultant is equal to the *difference* of their two moments.

The example previously given (4) of the *bracket* or *crane* (fig. 8) will also serve to illustrate the application of the theory of *moments*. For if we call F the force acting in AC, and W the load acting vertically at C, we can determine their relations by taking their moments about the fixed point B, at the foot of the strut. Draw Be perpendicular to CA, Bg horizontal or perpendicular to Ca; then in equilibrium we must have—

$$F \times Be = W \times Bg,$$
$$\text{or } F : W = Bg : Be.$$

Draw Bf parallel to AC , meeting Ca in f; fh parallel to Be; then the triangles BAe, fCh, are equal, and are *similar* to the triangle Bqf. Also the triangle Cab is *similar* to the triangle CfB. Therefore we have—

$$F : W = Bg : Be$$
$$= Bf : BA$$
$$= Bf : Cf$$
$$= ab : Ca$$

as shown to be the case by the "triangle of forces" (**4**).

9. THE LEVER.—Suppose ACB to be a rigid straight rod having a fixed point or "fulcrum" at C (fig. 15) and to be acted on by the forces P and Q at its extremities A and B. Draw Ca perpendicular to the direction of

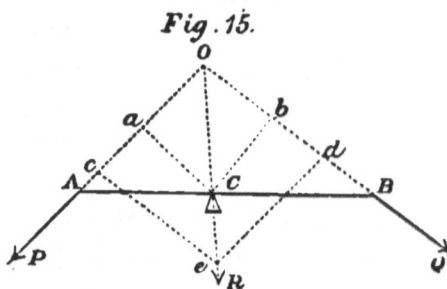

Fig. 15.

P, Cb to that of Q. Then P $\times$ Ca and Q $\times$ Cb are the moments of P and Q about C. If the rod is kept in equilibrium by these forces, then we have (**8**)—

$$P \times Ca = Q \times Cb . \quad . \quad . \quad (5)$$

Suppose that the directions of P and Q meet in the

point O; then since their moments balance about C, the moment of their resultant R must be zero, or its direction passes through the point C. In order to find the value of R, or the pressure which the forces P and Q produce upon the *fulcrum* C, take O*c* to represent P, O*d* to represent Q, and draw the parallelogram O*ced*, making *de* parallel to O*c*, and *ce* parallel to O*d*; then the diagonal O*e* (**3**) represents the resultant R.

If the point O is at a considerable distance from C, the length of the diagonal O*e* is *nearly* equal to the sum of the lines O*c* and O*d*, and the further it is off the more nearly do they approach to equality. Consequently, when the directions of P and Q are parallel to one another, or the point O is at an infinite distance, we must have—

$$R = P + Q$$
$$\text{and also, } P \times AC = Q \times BC \qquad . \quad (6)$$

We have here the principle of the "lever;" and by moving the fulcrum, C, very near to the end, B, of the rod, we see that a very small weight or force acting at A will balance a large one at B; for we get

$$P = \frac{BC}{AC} \times Q. \qquad . \qquad . \quad (7)$$

If Q is 100 lbs., and BC is the hundredth part of AC, then we find from this equation that P = 1 lb. will balance at A the weight Q = 100 lbs. at B.

Also since $\qquad BC = AB - AC$

$$\therefore \frac{BC}{AC} = \frac{AB - AC}{AC} = \frac{AB}{AC} - 1.$$

And we can put the above equation (7) into the form

$$P = \left(\frac{AB}{AC} - 1\right)Q \qquad . \qquad . \quad (8)$$

There are three distinct kinds of lever, which are classed according to the relative positions of the three points A, B, C. In the first kind of lever the fulcrum, C, lies between A and B, as in fig. 15. In levers of the second kind the point B at which the load Q acts lies between A and C (fig. 16). And as the above equation (7) holds good in every case, P must be always less than Q in this kind of lever, since AC is always greater than AB.

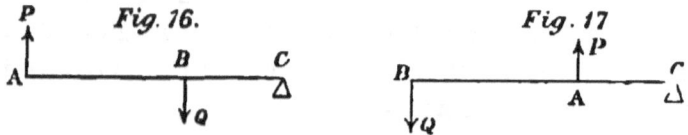

In the third kind of lever (fig. 17) the point A at which P acts lies between B and C, the weight Q acting at one end, B, of the lever, the fulcrum C being at the opposite end. Then since AC is always less than BC, P must always be greater than Q.

10. COUPLES.—In the third kind of lever (fig. 17) the resultant, R, of the forces P and Q, acting at C, will be the difference of P and Q, since they are acting in opposite directions (**8**); or, $R = P - Q$. Then we have—

$$Q \times BC = P \times AC = P (BC - BA).$$

Therefore, $(P - Q) \times BC = P \times AB$

$$\text{or } BC = \frac{P}{P - Q} \times AB = \frac{P}{R} \times AB . \qquad . \quad (9)$$

Now the quantity, or ratio, $\dfrac{P}{P - Q}$, gets greater and greater as $P-Q$ diminishes, or as P and Q approach to equality; so that we may say that when P is equal to Q the point C goes off to an infinite distance, and therefore there is no point of application of the resultant R. The two equal and opposite forces, P, Q, acting at A and B, constitute what is called a "couple;" and the moment $P \times AB$, or $Q \times BA$, is termed the "moment of the couple." The effect of a *couple* is to communicate an angular motion about an axis perpendicular to the plane in which the forces act; and two equal and opposite *couples* acting in the same plane will produce equilibrium in the body on which they act. The effect of a *couple* upon a body is not altered by any change in the forces P, Q, provided that the *moment* remains the same, or that the length of the *arm* of the couple is changed in the inverse ratio to the change in the forces.

11. TRANSPOSITION OF COUPLES. — Suppose a body to have a fixed point, C (fig. 18), and to be acted upon at A and B by two forces, P and Q, whose directions are in the same plane; and that the moments of these two forces about the fixed point C are equal, so that—

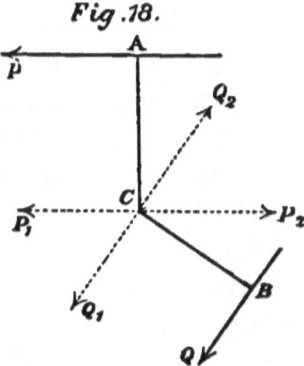

Fig. 18.

$$P \times AC = Q \times BC.$$

At the point C apply the forces P_1 and P_2 acting in opposite directions to one another, but in a direction parallel to that of P at A; and let $P_1 = P_2 = P$.

Then since P_1 and P_2 are equal and opposite, the equilibrium of the body remains undisturbed. Also apply at C the forces Q_1 and Q_2 acting in opposite directions to one another, but in a direction parallel to that of Q at B; and let $Q_1 = Q_2 = Q$. Then the equilibrium will remain undisturbed since Q_1 and Q_2 balance each other.

We have now two equal and opposite *couples*, namely, $P \times AC \times P_2$, and $Q \times BC \times Q_2$, which balance each other and can therefore be removed without affecting the equilibrium of the body. When these *couples* are removed there remain the two forces P_1 and Q_1 acting at C in direction parallel to the original forces P at A and Q at B. Also, by hypothesis, we have $P_1 = P$ and $Q_1 = Q$.

Hence it follows that when two forces act in the same plane upon a body, and their *moments* about a fixed point therein are equal to one another, we can transpose those forces unaltered to that fixed point about which they balance; and they may be considered as acting at that fixed point in directions parallel to their original directions respectively, and producing the same effect upon the body as the original forces did.

This theorem is of importance in determining the stability of arches and domes (Chapters VIII. and IX.)

12. STRESS ON A ROD FIXED AT ONE END.—Suppose a rigid rod AB (fig. 19) to be fixed firmly into a wall at A, and to be loaded at the other end B by a weight W. Then $W \times BA$ is called the "moment of stress" about the point A; and the "moment of stress" about any other point D on the rod is $W \times BD$. Using

the letter "M" to represent the "moment of stress" at any point, D, we have—

$$M = W \times BD \quad . \quad . \quad . \quad (10)$$

or, M is proportional to the distance of D from B. If then we take the vertical line AC to represent on any scale the moment, W × BA, and draw the hypothenuse BC, the vertical line, or ordinate, DE will represent on the same scale the moment W × BD; and similarly for any other point between A and B.

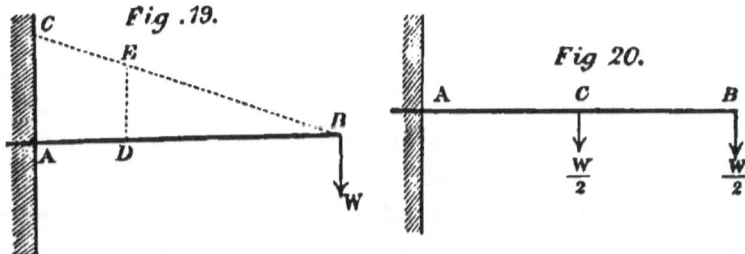

Fig. 19.

Fig 20.

Suppose the load W to be divided into two equal parts, half being placed at the end B of the rod, and half at the middle point C (fig. 20). Then the *moment of stress* at A is,

$$M = \frac{W}{2} \times BA + \frac{W}{2} \times CA$$
$$= \tfrac{3}{4} W \times BA \quad . \quad . \quad (11)$$

or the stress is one-fourth less than when the whole of W is placed at the end B.

Now let W be divided into four equal parts, one-fourth being placed at each of four equidistant points dividing the rod AB into four equal parts; then we have—

$$M = \frac{W}{4} \times AB + \frac{W}{4} \times \tfrac{3}{4} AB + \frac{W}{4} \times \tfrac{1}{2} AB + \frac{W}{4} \times \tfrac{1}{4} AB$$

$$= \tfrac{5}{8} W \times AB \qquad . \qquad . \quad (12)$$

If we divide AB into eight equal parts, and place one-eighth of W at each point, we obtain in the same way—

$$M = \tfrac{9}{16} W \times AB \qquad . \qquad . \qquad . \quad (13)$$

It therefore appears that as we increase the number of points over which W is distributed, the value of M approaches nearer and nearer to

$$M = \tfrac{1}{2} W \times AB \qquad . \qquad . \qquad . \quad (14)$$

and this is its value when the number of points is infinite, or when the load is uniformly distributed over the whole length of the rod; so that the stress at A when the load is distributed is one-half that produced by the same load placed at the further end B.

If we put l for the length of the rod, w for the load per unit of length, then $W = w \times l$, and we have—

$$M = \tfrac{1}{2} w . l^2 \qquad . \qquad . \qquad . \quad (15)$$

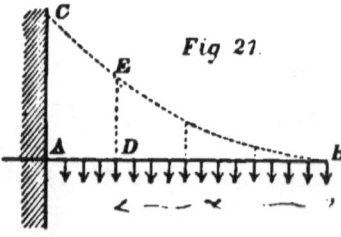

Fig 21.

If x is the distance of any point D from the end B (fig. 21), then putting M for *moment of stress* at D, we have—

$$M = \tfrac{1}{2} w \times x^2 \quad . \quad . \quad (16)$$

Putting $y = \tfrac{1}{2} w \times x^2$, we have the equation of the curve called the *parabola*, the vertex of which is at B. Let the vertical line AC represent on any scale the value of y, or M, when $x = $ BD, as in equation (16); then the vertical DE, measured on the same scale, will represent y when $x = $ BD; and similarly for any other point on AB.

If then we draw a curve through the points C, E . . . B, it will be a parabola. If on the other hand a parabola is drawn through C,E . . . B, its "ordinates" DE, etc., will represent the values of $\frac{1}{2}$ w. x^2. If m is the distance from B of the "focus" of the curve, then by the principle of the parabola, we have—

$$x^2 = 4\ m\ .\ y$$

$$\text{or,}\ \ y = \frac{1}{4\ m}\ .\ x^2$$

$$= \frac{w}{2}\ .\ x^2$$

consequently, $m = \dfrac{1}{2w}$.

For example, if we put $w = 1$ lb. per inch, AB = 12 inches, then AC = $\frac{1}{2} \times 12^2 = 72$, and $m = \frac{1}{2}$, or the distance of the "focus" from B is $\frac{1}{144}$th of the height AC.

13. STRESS ON A ROD SUPPORTED AT EACH END. —Let the rod A B (fig. 22) be supported at A and B and strained at any intermediate point C by a load W. Suppose

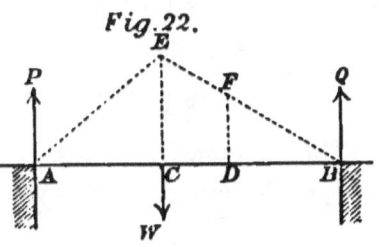

Fig. 22.

P at A and Q at B to represent the vertical reactions of the ends of the rod upon the supports. Then the stress upon the rod will be the same as if we suppose C to be a fixed fulcrum and then take the moments of P and Q about C. Since the forces are supposed to keep the rod in equilibrium, their moments (8) about C must be equal, or

$$Q \times BC = P \times AC,$$

either of which is the *moment of stress* about C; and the rod is subjected to the same straining force as if it were fixed in a wall at A, in which case we found (12) $M = Q \times BC$ by equation (10). Then since the moments of W at C and Q at B may be supposed to balance when taken about the point A, we have—

$$Q \times AB = W \times AC.$$

Therefore,
$$Q = W \frac{AC}{AB}.$$

Substituting this value of Q in the former equation, we get for the moment of stress at C,

$$M = Q \times BC = W \frac{AC \times BC}{AB} \quad . \quad . \quad (17).$$

The value of M is greatest when $AC = BC = \frac{1}{2} AB$, or when C is the centre of the rod; in which case equation (17) becomes,

$$M = \frac{1}{4} W \times AB \quad . \quad . \quad . \quad (18)$$

The moment of stress at any other point, D, on the rod, is the moment of Q with respect to D, namely,

$$M = Q \times BD = W . \frac{AC \times BD}{AB} \quad . \quad . \quad (19)$$

If we take the vertical CE to represent the quantity $W . \frac{AC \times BC}{AB}$, and draw the straight lines AE, BE;

then the ordinate DF will represent the moment of stress at the point D. Also if ordinates are drawn between any other points on AB to the lines BE, AE, they will represent the moments of stress at those points.

Putting AB = l, and AC = x, we have for stress at C, by equation (17),

$$M = W \cdot \frac{x(l-x)}{l} \qquad . \qquad . \quad (20)$$

and when $x = \dfrac{l}{2}$

$$M = \frac{W}{4} \cdot l \qquad . \qquad . \qquad . \qquad . \quad (21)$$

14. DISTRIBUTED LOAD.—Suppose that in the last case the load W is uniformly distributed over the whole length of the rod, w being the load per unit of length; then, W = $w \cdot l$. To find the moment of stress at any point C between A and B (fig.

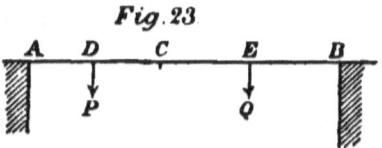

Fig. 23

23), bisect AC and BC in D and E; and let W = P+Q, where P acts along AC, and Q acts along BC. Then we may consider P and Q to act at D and E. Let M_1 be the moment of stress at C produced by P, and M_2 that produced by Q; M the total moment of stress at C. Then $M = M_1 + M_2$; and we have by equation (19)—

$$M_1 = P \cdot \frac{AD \times BC}{AB} = \frac{P}{2} \cdot \frac{AC \times BC}{AB}$$

Also,
$$M_2 = \frac{Q}{2} \cdot \frac{AC \times BC}{AB}$$

Therefore,
$$M = M_1 + M_2 = \frac{P + Q}{2} \cdot \frac{AC \times BC}{AB}$$
$$= \frac{W}{2} \cdot \frac{AC \times BC}{AB} \qquad . \qquad . \qquad . \quad (22)$$

If C is the centre of the rod, then $AC = BC = \frac{AB}{2}$, and

$$M = \frac{W}{2} \cdot \frac{AB}{4} = \frac{W}{8} \cdot AB \qquad . \qquad . \quad (23)$$

Putting $W = w \cdot AB$, the stress at any point C will be found by equation (22) to be,

$$M = \frac{W}{2} \cdot \frac{AC \times BC}{AB} = \frac{w \times AB}{2} \cdot \frac{AC \times BC}{AB}$$
$$= \frac{w}{2} \cdot AC \, (AB - AC) \qquad . \qquad . \quad (24)$$

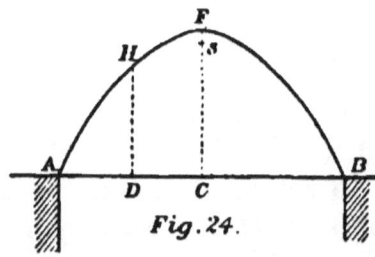

Fig. 24.

and when C is at the centre, $AC = \frac{1}{2} AB = \frac{1}{2} l$; in which case,

$$M = \frac{w}{8} l^2 \quad . \quad . \quad . \quad (25)$$

Suppose C is the centre of the rod AB (fig. 24), and D is any other point at distance x from C, then

$$AD = \frac{l}{2} - x, \quad BD = \frac{l}{2} + x;$$

therefore by equation (24)

$$M = \frac{w}{2} AD \times BD = \frac{w}{2}\left(\frac{l^2}{4} - x^2\right),$$

is the stress upon the point D.

If then we draw a vertical line CF to represent $\frac{1}{8} w . l^2$, and draw a parabola AFB, the ordinate DH will represent on the same scale the moment of stress at D; or the moments of stress at any points along AB are represented by the ordinates of a parabola whose vertex is at F, and whose focus, S, is found by measured $FS = \frac{1}{2w}$, on the same scale that CF represents $\frac{w}{8} . l^2$. If $CD = \frac{l}{4}$, then $DH = \frac{3}{4}$ FC.

15. BEAM LOADED AT TWO OR MORE POINTS.—Suppose a rod AB, supported at each end, to be loaded at D and E (fig. 25) with the weights W_1 and W_2; then **(13)** we have for the stress at the middle point C, caused by the load W_1 at D, from equation (19),

$$M_1 = W_1 . \frac{BC \times AD}{AB} = \frac{W_1}{2} \times AD.$$

The moment of stress at C caused by the load W_2 at E is, by equation (19),

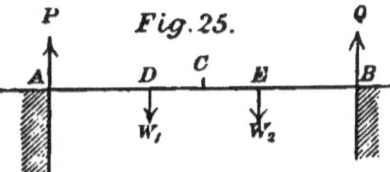

Fig. 25.

$$M_2 = W_2 . \frac{BC \times BE}{AB} = \frac{W_2}{2} BE$$

and

$$M = M_1 + M_2$$

is the total moment of stress at the middle point C.

If we suppose D and E to divide the rod into three equal parts, then $AD = BE = \frac{1}{3}\,l$; and supposing also that $W_1 = W_2 = W$, we have for moment of stress at C,

$$M = M_1 + M_2 = \frac{W}{2} \cdot \frac{l}{3} + \frac{W}{2} \cdot \frac{l}{3}$$

$$= W \times \frac{l}{3} \quad . \quad . \quad . \quad (26)$$

If the rod is divided into four equal parts, and loaded with a weight, W, at each point, then the moment of stress at C will be found in the same manner to be,

$$M = \frac{W}{2}\left(\frac{l}{4} + \frac{l}{2} + \frac{l}{4}\right) = W \times \frac{l}{2} \quad . \quad (27)$$

If there are four such loads placed at equal distances, we find in the same way,

$$M = \frac{W}{2}\left(\frac{2l}{5} + \frac{4l}{5}\right) = W \times \tfrac{3}{5}l \quad . \quad . \quad (28)$$

CHAPTER III.

16. DEFINITION OF CENTRE OF GRAVITY.—The several particles of which any body is composed may be considered as so many weights or forces acting in parallel directions; these forces or weights resulting from the action of "gravity," or the earth's attraction upon the several particles. Such a system of parallel forces must have a "resultant" (2), and the point through which this resultant passes is termed the "centre of gravity" of the body. The centre of gravity may therefore be considered as that point at which the whole weight of the body acts, and where it produces the same effect as the weight of the body would produce.

The centre of gravity of a body may also be considered as that point about which the algebraic sum of the *moments* (8) of all the weights of the particles being taken, is equal to zero; or the moments of the weights of those which are positive exactly balance the moments of the weights of those which are negative, and tend to turn the body in the opposite direction to the former. Since the force of gravity acts equally on all the particles, the position of the centre of gravity is quite independent of the *amount* of the force of gravity, and therefore does not vary with any change in the position of the body.

In order that a heavy body may be supported, the direction of the supporting force must pass through its centre of gravity.   Hence it follows that we may consider the weight of a rigid body to be collected at this point, and that if the centre of gravity is in equilibrium under the forces acting upon it, then the body itself is kept in equilibrium.   From a knowledge of this fact we are enabled readily to find the position of the centre of gravity of any thin flat substance by means of a simple experiment.   Let the body be suspended freely at any point on its perimeter, and a plumb-line dropped from the same point, the direction of the plumb-line being marked upon its flat surface. Do the same at some other point on the perimeter, and the intersection of the two plumb-lines is the centre of gravity, since each of them passes through that point.

The position of this point in a straight rod of uniform density and dimensions throughout, will evidently be at its middle point; and the same will be the case with a rectangular thin plank of uniform thickness, which may be considered as made up of a number of parallel rods.

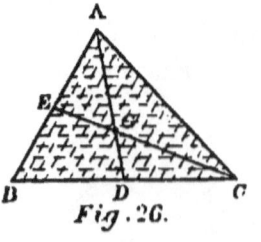

Fig. 26.

17.   Centre of Gravity of a Triangular Plate.—We may consider that a triangular plate as ABC (fig. 26), is made up of a number of straight rods lying parallel to one side or base, BC or AB, and diminishing in length from the base of the vertex.

Bisect BC in D, and draw AD; then the centres of

gravity of the rods lying parallel to BC must be at their intersection with the line AD, and consequently that of the whole triangle must be on that line. Bisect AB in E, and draw CE; then the centres of gravity of the rods lying parallel to AB must be at their intersections with the line CE, and that of the whole triangle must be on that line; and therefore it is at the intersection G of AD and CE. By the geometry of the figure DG is one-third of AD, and EG one-third of CE.

Suppose it is required to find the centre of gravity of a trapezium having two parallel sides, as ABCD (fig. 27), where AD is parallel to BC, but BA and CD will meet if produced in the point E, forming two triangles AED and BEC. Bisect BC in F, and draw EHF. Take Fg equal to one-third of EF,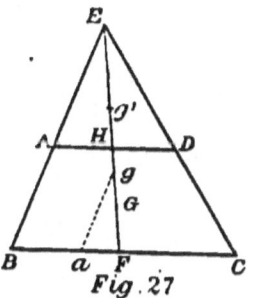

Fig. 27

and Hg' equal one-third of EH; g and g' will be the centres of gravity of the two triangles. Let P represent the area of the triangle AED, Q that of the trapezium ABCD; then we may consider P and Q as forces acting at the centre of gravity of each figure. Let G be the centre of gravity of the trapezium ABCD, the position of which we have to find. Then the moment of P at g' must balance that of Q at G, both being taken about the centre of gravity g of the whole triangle BEC; or we have,

$$P \times gg' = Q \times Gg.$$

Then, $\qquad Gg = \dfrac{P}{Q} \times yy' = \dfrac{P}{Q}(Eg - Eg')$

$$= \dfrac{P}{Q} \times \tfrac{2}{3}(EF - EH).$$

$$= \dfrac{P}{Q} \times \tfrac{2}{3} FH.$$

Now as P and Q represent the areas of the two figures AED and ABCD, we have

$$\dfrac{P}{Q} = \dfrac{EH \times AD}{EF \times BC - EH \times AD};$$

also, $\qquad \dfrac{EH}{AD} = \dfrac{EF}{BC}$, since the triangles are similar.

therefore,

$$\dfrac{P}{Q} = \dfrac{\dfrac{EH}{AD} \times AD^2}{\dfrac{EF}{BC} \times BC^2 - \dfrac{EH}{AD} \times AD^2}.$$

$$= \dfrac{AD^2}{BC^2 - AD^2};$$

therefore, $\qquad Gg = \tfrac{2}{3} \dfrac{AD^2}{BC^2 - AD^2} \times HF \quad . \quad (29)$

We can determine geometrically the position of the point g by taking Ba equal to one-third of BC, and drawing ag parallel to BA and intersecting HF in g. So that the position of G can be found without drawing the triangle EAD, by simply bisecting AD in H, and BC in F, then joining HF and drawing ag in the manner above described. Then the distance Gg can be calculated by the equation (29).

18. Centre of Gravity of any Trapezium. —
Let ABCD (fig. 28) be the given trapezium ; draw the
diagonals AC and BD
intersecting in *b*.
Bisect BD in O and
draw OA, OC. Take
O*a* = ⅓ OA, and O*c*
= ⅓ OC. Then *a* and *c*
are the centres of
gravity of the triangles
BAD and BCD re-
spectively (**17**); and
the centre of gravity,

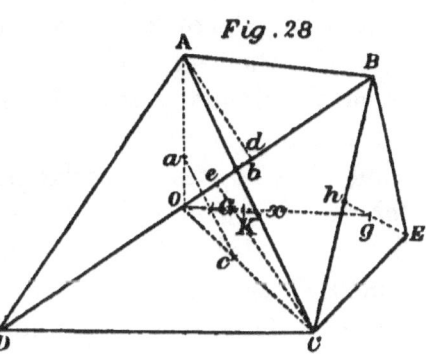

Fig .28

G, of the trapezium will be found on the line *ac* which
is parallel to the diagonal AC. Since the triangles
have a common base BD, their areas must be proportional to the perpendiculars A*d*, C*e*, dropped from the
vertices A and C upon the diagonal BD. And since
C*be*, A*bd* are similar triangles, we have,

$$A b : C b :: A d : C e.$$

On the line AC take C*x* = A*b*, and draw O*x*, cutting
ac in G; then G is the centre of gravity required.
For since O*ac* and OAC are similar triangles, therefore

$$C b \times G c = A b \times G a.$$

In working out this problem it is not necessary to
draw the perpendiculars A*d* and C*e*, which are only
introduced for the purpose of demonstration.

Since either diagonal of a *parallelogram* divides it
into two equal triangles, the centre of gravity will
evidently be at the intersection of the diagonals.

19. Centre of Gravity of any Polygon.—Since a polygon can be divided into triangles, we have first to find the centre of gravity of each triangle **(17)**, and then that of two triangles as above **(18)**, and combine this with the centre of gravity of a third triangle, and so on for as many triangles as the figure is divided into, which is two less than the number of sides. Thus if we suppose the triangle BEC to be added on to the trapezium ABCD (fig. 28), we form a five-sided figure divided into three triangles. Having determined G the centre of gravity of the two triangles ABD, BCD, or of the trapezium ABCD in the manner above described **(18)**, we determine g, the centre of gravity of the triangle BEC, by bisecting BC in h, drawing Eh, and taking $hg = \frac{1}{3}$ Eh. Join Gg, and divide Gg at K in the proportion of the areas of the trapezium ABCD and of the triangle BEC; taking GK to represent the latter and Kg the former. Then K is the centre of gravity of the whole pentagon.

The easiest method, however, in such a case is to cut the figure out in card, and hang it up at two of its vertices with a plumb-line, as before described **(16)**; then the intersection of the two plumb-lines will determine the centre of gravity of the figure.

20. Centre of Gravity of Girder Sections.—The figure BAC (fig. 29) represents a section of "angle-iron," consisting of two parallelograms at right angles to each other, the centres of gravity of which, a and b, can be determined as above **(18)**. Join ab, and divide ab in the inverse proportion of the areas of the two arms by the point G, G*b* being proportional to the arm AC, and G*a* to that of AB; or

$$\frac{G b}{G a} = \frac{\text{Area of AC}}{\text{Area of AB}}.$$

Then G is the centre of gravity of the section.

The centre of gravity of a section of "tee-iron," DABC (fig. 30), can be found by taking *a* the centre of gravity of the vertical parallelogram AD, and A that of the horizontal one BC. Then divide A*a* at G in the inverse proportion of the area of BC to that of AD, AG being to G*a* in the proportion of the area of AD to that of BC. Then G is the required centre of gravity of the section.

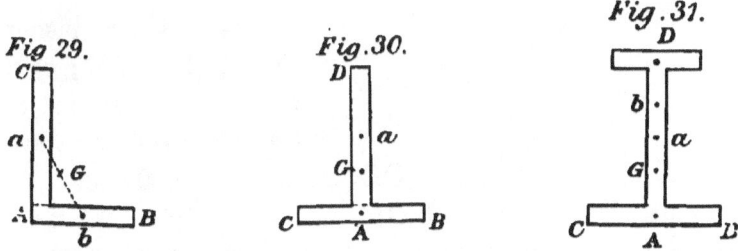

Fig 29.   Fig. 30.   Fig. 31.

To find the centre of gravity of the section of a "double-flanged" beam, as DABC (fig. 31), bisect the "web" AD in *a*, and let A be the centre of gravity of the bottom flange BC, D that of the top flange. By the last method find the centre of gravity *b* of the web and top-flange. Then divide A*b* at G in the inverse proportion of the area of the bottom flange to that of the web and top-flange; AG being proportional to the

latter, and G*b* to the former.   Then G is the required
centre of gravity.

If the top and bottom flanges are equal, it will be
evident that G is at the middle point *a* of the web.

21. CENTRE OF GRAVITY OF CIRCULAR ARCH.—The
circular arch ABCD (fig. 32) consists of the difference
of the two sectors AOB
and DOC; O being
the centre of curva-
ture.   Proceeding as
in the case of a part
of a triangle (fig. 29)
we have to find in the
first place the centres
of gravity, *a* and *b*, of
the two sectors.   Let
the radius OE bisect
the angle AOB; and
when this angle is
less than 60°, we may
take E*a* as very nearly $\frac{1}{3}$ OE, or O*a* = $\frac{2}{3}$ OE; and F*b*
very nearly one-third of OF, or O*b* = $\frac{2}{3}$ OF.   We find
by calculation, if we wish to be more exact, that when
AOB = 40°, then O*a* = ·654 . OE, and O*b* = ·654 . OF;
when AOB = 50°, O*a* = ·646 . OE, and O*b* = ·646 . OF.
When AOB = 60°, O*a* = ·636 . OE, and O*b* = ·636 . OF.
When AOB = 70°, O*a* = ·626 . OE, and O*b* = ·626 . OF.
For AOB = 80°, O*a* = ·614 . OE, O*b* = ·614 . OF.
When AOB = 90°, O*a* = ·6 . OE; O*b* = ·6 . OF.

Then since the areas of the sectors are proportional
to the squares of their radii, we find G, the centre of
gravity of the arch ABCD, from the equation—

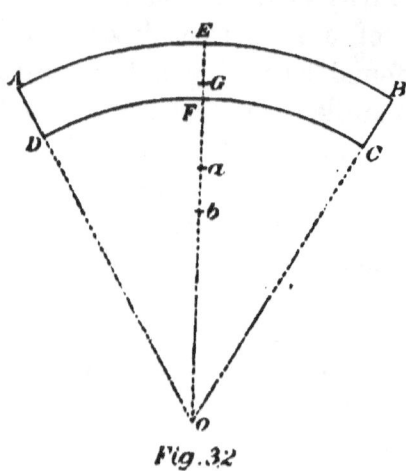

Fig. 32

$$G a = \frac{OF^2}{OE^2 - OF^2} (Oa - Ob) \qquad . \quad (30)$$

Also we have, $\quad OG = Oa + Ga.$

As an example, let the angle AOB = 60°, OE = 12, OF = 10, then $Oa = \cdot 636 \times OE$, $Ob = \cdot 636 \times OF$, $Oa - Ob = \cdot 636 (12 - 10) = 1 \cdot 272$, $Oa = 7 \cdot 632$, $OE^2 - OF^2 = 144 - 100 = 44.$

$Ga = \frac{100}{44} \times 1 \cdot 272 = 2 \cdot 9$, $OG = 7 \cdot 632 + 2 \cdot 9 = 10 \cdot 532.$

When the position of the point G is required with great accuracy, it is found from the equation

$$Ob = \tfrac{4}{3} \frac{R^3 - r^3}{R^2 - r^2} \cdot \frac{\sin. \frac{\theta}{2}}{\theta} \qquad . \quad . \quad (31)$$

Where the angle θ, or AOC, is expressed in "circular measure," that is to say, the angle $57 \cdot 3°$ is the unit of measurement; R and r are put for the radii OE and OF. The following Table gives the values of θ, $\sin. \frac{\theta}{2}$,

and $\frac{\sin. \frac{\theta}{2}}{\theta}$ for angles from 5° to 90°, which will facilitate the calculation; "arc θ" is the length of an arc to a radius unity.

The numbers in the second column of the following table give the *length* of an *arc* of a circle whose radius is unity, corresponding to the angle given in degrees in the first column. To find the *area* of any *sector* of a circle, multiply the length of arc in the second column by half the square of the radius.

θ Degrees.	Arc θ.	Sin. $\frac{\theta}{2}$.	Sin. $\frac{\theta}{2} \cdot \frac{}{\theta}$	θ Degrees.	Arc θ.	Sin. $\frac{\theta}{2}$.	Sin. $\frac{\theta}{2} \cdot \frac{}{\theta}$
5°	·08727	·04362	·49983	48°	·83776	·40674	·48550
6°	·10472	·05234	·49981	49°	·85521	·41469	·48489
7°	·12217	·06105	·49972	50°	·87266	·42262	·48430
8°	·13963	·06976	·49956	51°	·89012	·43051	·48365
9°	·15708	·07846	·49948	52°	·90757	·43837	·48301
10°	·17453	·08716	·49938	53°	·92502	·44620	·48237
11°	·19199	·09585	·49922	54°	·94248	·45399	·48170
12°	·20944	·10453	·49908	55°	·95993	·46175	·48102
13°	·22689	·11320	·49893	56°	·97738	·46947	·48035
14°	·24435	·12187	·49874	57°	·99484	·47716	·47962
15°	·26180	·13053	·49858	58°	1·01229	·48481	·47892
16°	·27925	·13917	·49838	59°	1·02974	·49242	·47820
17°	·29671	·14781	·49816	60°	1·04720	·50000	·47747
18°	·31416	·15643	·49794	61°	1·06465	·50754	·47670
19°	·33161	·16505	·49773	62°	1·08210	·51504	·47596
20°	·34907	·17365	·49746	63°	1·09956	·52250	·47517
21°	·36652	·18224	·49721	64°	1·11701	·52992	·47442
22°	·38397	·19081	·49693	65°	1·13446	·53730	·47362
23°	·40143	·19937	·49665	66°	1·15192	·54464	·47281
24°	·41888	·20791	·49636	67°	1·16937	·55194	·47200
25°	·43633	·21644	·49604	68°	1·18682	·55919	·47117
26°	·45379	·22495	·49572	69°	1·20428	·56641	·47032
27°	·47124	·23345	·49527	70°	1·22173	·57358	·46948
28°	·48869	·24192	·49505	71°	1·23918	·58070	·46861
29°	·50615	·25038	·49468	72°	1·25664	·58779	·46776
30°	·52360	·25882	·49431	73°	1·27409	·59482	·46687
31°	·54105	·26724	·49392	74°	1·29154	·60182	·46596
32°	·55851	·27564	·49352	75°	1·30900	·60876	·46506
33°	·57596	·28402	·49311	76°	1·32645	·61566	·46414
34°	·59341	·29237	·49270	77°	1·34390	·62251	·46252
35°	·61087	·30071	·49226	78°	1·36136	·62932	·46321
36°	·62832	·30902	·49181	79°	1·37881	·63608	·46137
37°	·64577	·31730	·49136	80°	1·39626	·64279	·46036
38°	·66323	·32557	·49089	81°	1·41372	·64945	·45939
39°	·68068	·33381	·49040	82°	1·43117	·65606	·45841
40°	·69813	·34202	·48990	83°	1·44862	·66262	·45740
41°	·71559	·35021	·48941	84°	1·46608	·66913	·45652
42°	·73304	·35837	·48889	85°	1·48353	·67559	·45540
43°	·75049	·36650	·48836	86°	1·50098	·68200	·45436
44°	·76794	·37461	·48782	87°	1·51844	·68835	·45332
45°	·78540	·38268	·48725	88°	1·53589	·69466	·45239
46°	·80285	·39073	·48668	89°	1·55334	·70091	·45112
47°	·82030	·39875	·48610	90°	1·57080	·70711	·45016

CHAPTER IV.

RESISTANCE OF MATERIALS TO STRESS.

22. MODULUS OF ELASTICITY.—If a rod of any material, having one square inch of cross section, is subjected to a force S acting in the direction of its length, it will be either shortened or elongated to an amount which will depend on the strength and direction of the force that is applied to it. If the force tends to elongate the rod, it is termed a "tensile stress," and if the force tends to shorten the rod, it is a "compressive stress."

Suppose we put L for the original length of the rod, l the amount of increase or decrease in the length produced by the force S; also that the *elasticity* of the material remains uninjured during the action of the force, so that when the force is removed the rod returns exactly to its original length; then the change of length, l, is to the original length, L, in the proportion of S to some *constant* number which depends on the nature of the material, and is determined by experiment. For this *constant* the letter E is used, and it is called the "modulus of elasticity." Then the above proportion can be expressed thus :—

$$l : L :: S : E.$$

When the elongation or shortening l, due to the force S, has been ascertained, we have—

$$E = S \frac{L}{l} \quad . \quad . \quad . \quad . \quad . \quad (32)$$

Owing to the imperfect elasticity which exists in all building materials, the *modulus* obtained when the experiments are made with a tensile force, differs from that obtained when a compressive force is applied. It is also found that the *modulus* which results from a body subjected to a transverse stress, generally differs from either of the foregoing. We may have therefore three different values of the "modulus of elasticity" for the same material, namely, the *modulus of tensile elasticity* obtained by employing a *stretching* force; the *modulus of compressive elasticity* obtained by a *compressing* force; and the *modulus of transverse elasticity* which is obtained by subjecting a rod to a *transverse* stress. There is often a great difference in the figures given by different experimenters for the values of the modulus, which sometimes arises from the experiments being performed in different ways; some persons using a tensile, others a compressive, and others a transverse force.

If we put S for the force which produces the elongation or contraction l in a rod of one inch sectional area, and A for the number of square inches of section in another rod, then $S \times A = F$, is the force which will produce the same elongation or contraction l in a rod having A inches of section; therefore

$$S = \frac{F}{A},$$

and since $\dfrac{E}{S} = \dfrac{L}{l}$, from equation (32),

we have, $l = \dfrac{S \cdot L}{E} = \dfrac{F \cdot L}{E \cdot A}$ . . (33)

Also, $F = \dfrac{E \cdot A \cdot l}{L}$ . . . (34)

The equation (33) gives the alteration of length for a given force F; and the equation (34) the force F which will produce the above alteration in the length of the rod. Putting these equations into the form

$$\dfrac{F}{A} : \dfrac{l}{L} :: E : 1 \qquad . \qquad . \qquad . \qquad . \qquad (35)$$

where $\dfrac{F}{A}$ represents the tensile or compressive force per unit of area, and $\dfrac{l}{L}$ the proportional change in the length; it follows that the "modulus of elasticity" may be defined as "the constant relation between the tensile or compressive force per unit of area, and the corresponding proportional change in the length which it produces in the rod."

23. MODULUS OF TENSILE ELASTICITY.—The following values of E for different materials expressed in lbs. avoirdupois, have been found by direct experiment with a tensile force. Those for timbers are taken from a work on "Timber and Timber Trees," by T. Laslett, formerly Inspector of Timber to the Admiralty. If we put $l = L$, in equation (32) we get $E = S$; so that E represents the tensile force per square inch of section that would stretch a rod to double its length, supposing such a thing to be possible without injuring the elasticity of the material.

Kind of Material.	E in Lbs. Av.	E in Tons.
English Elm	1,003,280	448
Canadian Elm	2,474,200	1,105
Dantzic Fir	1,737,570	776
Riga Fir	3,009,680	1,344
English Oak	1,545,600	690
French Oak	2,480,880	1,107
Pitch Pine	3,020,940	1,349
Red Canada Pine	2,355,600	1,051
Cast Iron	12,000,000	5,357
Wrought Iron, bars . . .	28,672,000	12,800
,, plates . . .	26,163,200	11,680
Steel, soft cast	30,240,000	13,500
,, hammered Bessemer . .	32,480,000	14,500
,, rolled cast	31,360,000	14,000

24. MODULUS OF COMPRESSIVE ELASTICITY.—As long as the compressive force applied to a rod is not sufficient to injure the elasticity of the material, the modulus of *compressive* elasticity may be considered as equal to that of *tensile* elasticity in timbers.   In cast-iron it was found by Hodgkinson that the two moduli were very nearly the same, as given above, so long as the force was less than six tons per square inch ; but with a force of six tons the modulus of *compressive* elasticity was found to be one-fourth more than the modulus of *tensile* elasticity.   With wrought-iron and steel the two moduli may be practically considered as equal.

25. MODULUS OF TRANSVERSE ELASTICITY.—When a rod is subjected to a transverse stress at right angles to its length, the resistance to bending is proportional to the *modulus of elasticity* as long as the elasticity of the material remains unimpaired, as will be shown in the chapter on " Deflexion " (**47**). The value of this modulus is, however, different in some kinds of material to that of the modulus of tensile or of compressive elasticity, and can be found experimentally by observing the deflexion caused by a known weight. The modulus of *transverse* elasticity for cast-iron is found to vary from 15,000,000 lbs. to 19,000,000 lbs.; or taking the mean of these two, we put $E = 17,000,000$ lbs $= 7,600$ tons, for the modulus of transverse elasticity in cast-iron beams. That for timber, wrought-iron, and steel may be taken as equal to the modulus of *tensile* elasticity given above (**23**).

26. Coefficient of Strength.—When a rod is subjected to a longitudinal stress, F, we find from equation (33) that the change of length, l, varies with the force F. If the force F is sufficient to produce fracture, and l is the alteration of length at the moment of fracture, then F represents the *ultimate strength* of the material. The quantity S which is the force per unit of section, and is equal to $\dfrac{F}{A}$, is called the *coefficient of ultimate strength*. The values of S are determined by experiment, and vary considerably in the same material for tensile and compressive stress. The following table gives the *ultimate tensile strength* of different materials per *square inch* of section; the coefficient for timbers being taken from Laslett's work mentioned above.

Kind of Material.	Coefficient of Ultimate Tensile Strength. S in Lbs. Av.
Brick	300
Slate	9,600 to 12,800
Cast Iron	11,200 to 44,800
,, Malleable	29,120 to 44,800
Plate-Iron . . .	51,000
Bar-Iron	60,000 to 70,000
Hoop-Iron	64,000
Steel	100,000 to 130,000
English Elm	5,460
Canadian Elm	9,180
Dantzic Fir	3,230
Riga Fir	4,050
Spanish Mahogany . .	3,790
American White Oak . . .	7,020
English Oak	7,570
French Oak	8,100
Pitch Pine . . .	4,670
Canada Spruce	3,930

27. Resistance to Crushing.—The resistance of a rod of any material to a *compressive* force follows a very different law to that of the resistance to a *tensile* force; for whereas the latter is independent of the length of the rod, the former varies greatly with its length, being much greater in short than in long pieces. Since long rods are liable to break by bending when subjected to a compressive force in the direction of their length, it is necessary to make our experiments upon very short pieces, as cubes or double cubes, in order to ascertain the actual resisting power of the material to crushing.

The value of the *coefficient of ultimate strength*, S, per square inch of section, for the principal materials used in construction, when subjected to a crushing force, is given below. The numbers for timbers being taken from Laslett's experiments.

Kind of Material.	Coefficient of Ultimate Crushing Strength. 8 in Lbs. Av.
Common Brick	500 to 800
Fire Brick	1,700
Cornish Granite	6,400
Peterhead Granite . . .	8,300 to 10,900
Yorkshire Stone	7,600
Bolsover Stone	8,300
Portland Stone	3,900
Bath Stone	1,500
Cast Iron	44,800 to 215,040
,, Malleable	107,520 to 159,040
Wrought Iron	36,000 to 40,000
Steel	110,000 to 150,000
English Elm	5,786
Canadian Elm	8,586
Dantzic Fir	6,948
Riga Fir	5,248
English Oak, seasoned . . .	7,475
American White Oak . . .	6,451
Dantzic Oak	7,558
French Oak	7,945
Pitch Pine	6,462
Canadian Red Pine . . .	5,600
Canada Spruce	4,852
Spanish Mahogany . . .	6,415

28. COEFFICIENT OF SAFETY.—None of the materials used in construction must ever be subjected to a stress which is anything near the " ultimate-strength," or the stress must never approach to that which would produce fracture. A certain *limit* has been fixed for the amount of stress to which the materials can be subjected without injury to their elasticity, and this limit should never be passed in practice. This limiting stress is called the *coefficient of safety*, and its ratio to the *coefficient of ultimate strength* (**26, 27**) varies for

different materials, being fixed arbitrarily according to
our experience of the effect of stress in producing
strain in those materials.   The proportion is also
varied according as the stress arises from a moving
load or from a stationary one.   In the latter case, the
ratio of the *coefficient of safety* to that of *ultimate
strength* in steel and wrought-iron is 1 to 3 ; for cast-
iron it is 1 to 4 ; in timber 1 to 6 ; and in stone or
brick 1 to 8.

29. Transverse Stress.—When a long beam or
rod of any elastic ma-

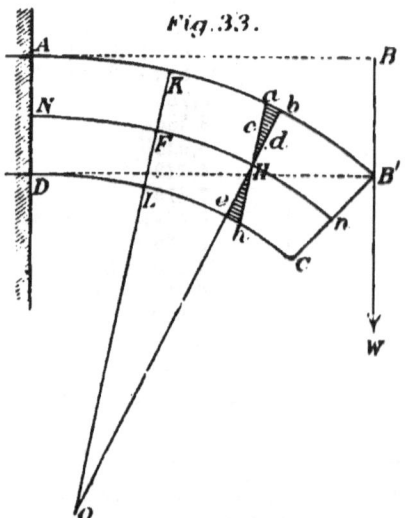

Fig. 33.

terial is fixed or sup-
ported in a horizontal
position, and is sub-
jected to forces acting
vertically, they produce
a transverse strain on
the fibres, and cause
the beam to bend in the
direction of the forces.
Thus, suppose the beam
AB (fig. 33) to be fixed
firmly and horizontally
at one end AD, and to
be loaded at the other
end B with a weight W.   The end B will then be
bent and will take the position B', the upper surface
AB' of the beam becoming convex, while the under-
side DC becomes concave.   It will be seen that
the fibres in the upper part of the beam become
stretched or lengthened in consequence of this bending,
and are therefore subjected to *tensile* stress.   On the

other hand those in the lower part are shortened and are subjected to a *compressive* or *crushing* stress. The lengthening or shortening of the fibres is also greatest at the top or bottom of the beam, on its outer edges, AB and DC, and get less and less as we proceed inwards. Consequently there must be some part or surface near the middle of the beam, as N*n*, where the fibres are neither shortened nor lengthened, the length N*n* being equal to the original length AB of the beam, before the action of the load W. This part of the beam is called its "neutral surface," and is parallel to the surfaces AB, DC. The *neutral surface* may be defined as that surface along which the resultant of the horizontal components of all the diagonal forces is equal to *zero*. The line N*n* is that in which the *neutral surface* cuts the longitudinal section AB'CD of the beam. Let *b*H*e* be the line perpendicular to N*n* in which a transverse section of the beam intersects the longitudinal section ; then a horizontal line through H, where the two planes intersect (perpendicular to the plane of the paper) is called the "neutral axis" of that transverse section. Assuming that the elasticity of the beam is uninjured by the stress to which it is subjected, so that when the load W is removed the beam will return exactly to its original horizontal position ; and also that the resistances to extension and compression are equal ; we may assume that the *neutral axis* will pass through the *centre of gravity* (**16**) of the transverse section *b*H*e*.

Let O be the *centre of curvature* of a small part, HF, of the line N*n*, and assume that HF is very nearly an arc of a circle, so that OF is equal to OH. The forces

E

of resistance to extension and compression will act parallel to FH, and perpendicular to a transverse section at H; the sum of the forces of extension acting above H being equal to the sum of the forces of compression acting in the *opposite* direction below H; or the resultant of these forces must be zero, as above stated.

Produce OeH to b and OLF to K, on the line AB′; and draw hHa parallel to LK. Take any point c on Ha and draw cd parallel to ab; then cd represents the extension of one of the fibres above H. Let f represent the force producing this extension cd in the fibre at c, and call a the area of section of the fibre. Then (**22**) the extension cd is to the original length FH of the fibre as f divided by a to E, the modulus of elasticity; or by equation (34) we have—

$$ f = \text{E} \times a \times \frac{cd}{\text{FH}}. $$

Also since the triangles HOF and dHc are similar, we have—

$$ cd : \text{FH} = \text{H}d : \text{OH} ; $$

so that

$$ f = \text{E} \times a \times \frac{\text{H}d}{\text{OH}}. $$

The *moment* (**8**) of f taken about the point H, is $f \times \text{H}d$; and

$$ f \times \text{H}d = \text{E} \times a \times \frac{(\text{H}d)^2}{\text{OH}}. $$

A similar result is obtained if we take the fibres below H, only they will be subjected to compression. If then we put M for the *moment of resistance* of all the fibres in the section, and put x, x_1, x_2, &c. for Hd, &c., and ρ for OH, we obtain the value of M, namely,

$$M = \frac{E}{\rho} \left(a\, x^2 + a\, x_1^2 + a\, x_2^2 + \ldots \right)$$

Now the sum of all the quantities, $ax^2 + ax_1^2 + $ &c. is called the " moment of inertia " of the transverse section taken about the *neutral axis*, for which we write the letter " I," its value depending on the geometrical form of the section and being determined in different cases by help of the integral calculus. We can therefore put M, the *moment of resistance* of the section, into the form

$$M = \frac{E \times I}{\rho} \qquad . \qquad . \qquad . \quad (36)$$

Also, since by geometry,

$$ab : Ha = FH : OH$$

we have—

$$\frac{1}{OH} = \frac{1}{\rho} = \frac{ab}{FH} \times \frac{1}{Ha};$$

therefore equation (36) becomes—

$$M = E \times \frac{ab}{FH} \times \frac{I}{Ha}.$$

If then we put S for the force per square inch producing the elongation (or compression) ab, and Ha = x the distance of the most extended (or compressed) fibres from the *neutral axis ;* we have by equation (32)

$$E \times \frac{ab}{FH} = S.$$

Therefore by substituting S for $E \frac{ab}{FH}$ in the last equation, we have—

$$M = S \times \frac{I}{z} \quad . \quad . \quad . \quad (37)$$

When the section of the beam is rectangular, putting b for the breadth and d for the depth, z being equal to $\frac{1}{2} d$, we find by integration that $I = \frac{1}{12} b \cdot d^3$; therefore for a rectangular beam

$$M = S \frac{bd^2}{6} \quad . \quad . \quad . \quad (38)$$

30. RECTANGULAR BEAM.—The *moment of resistance*, M, in a beam of rectangular transverse section, as given by equation (38), can also be determined by independent reasoning. For, referring to fig. 33, we see that if Ka and Lh are the lengths of the upper and lower fibres before the stress W is applied; then in consequence of the strain produced by W, Ka is stretched to Kb, and Lh is shortened to Le; and the lines, such as cd, drawn across the two triangles Heh and Hab parallel to ab and eh, represent the alterations of length in the intermediate fibres. The sum of the horizontal stresses exerted by the fibres in either the upper or the lower half of the section be, is the area of the half-section multiplied by the mean unit stress of the fibres. Putting S for the unit-stress in the fibres at b and e we have $\frac{1}{2}$ S for the mean unit-stress of all the fibres. The total stress in each half of the section is therefore $\frac{1}{2}$ S $\times \frac{1}{2} b . d$; and since the horizontal elastic forces in the various fibres are proportional to the horizontal lines in the triangles H*eh* and H*ab*, therefore the centres of the forces of tension and of compression must be at the centres of gravity of those triangles, the distance apart of which (**17**) is two-thirds of the depth *be*, or *d*. Consequently we find as in equation (38),

$$M = \tfrac{2}{3} d \times \frac{S}{2} \times \frac{b \cdot d}{2} = S \frac{b \cdot d^2}{6}.$$

Putting *l* for the length AB of the beam, *x* for the length H*n*, we have by equation (10), M = W × H*n* for the moment of stress at the section through H; so that when the forces are in equilibrium, we have—

$$M = W \times x = S \frac{b \cdot d^2}{6}.$$

When *x* = *l*, we have for the moment of stress at AD,

$$M = W \times l = S \frac{b \cdot d^2}{6}$$

from which we get

$$W = \frac{S}{6} \frac{b \cdot d^2}{l} \qquad . \qquad . \qquad . \quad (39)$$

If S is the *ultimate* resistance per square inch of section, then W is the weight that will produce fracture of the fibres, all the dimensions being expressed in inches. If S is in lbs., then W is in lbs. also.

If the load W is uniformly distributed over the length of the beam AB, instead of being all sustained at B, then (**14**) the moment of strain is only half what it would be if the whole weight were at W; or it requires twice as much weight to produce fracture.

Therefore we have for a distributed load, the breaking weight—

$$W = 2 \times \frac{S}{6} \cdot \frac{b \cdot d^2}{l} = \frac{S}{3} \cdot \frac{b \cdot d^2}{l} \qquad . \quad (40)$$

31. Beam Supported at each End.—When a beam, as ACB (fig. 34), is supported at both ends and loaded at any intermediate point C with a weight W, the stresses are the reverse of what they are when the beam is fixed at one end and loaded at the other (**29**). In this case the fibres in

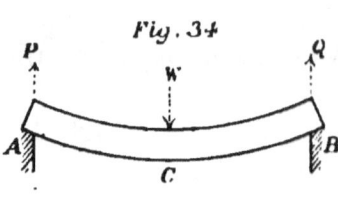

Fig. 34

the upper part are in a state of compression and those in the lower part in extension. If, however, we suppose the point C to be firmly fixed, and the load W to be divided into two parts P and Q, and these parts to act upwards at A and B, the stresses on the beam will be the same as before. The moment of stress at C in this case is given by equation (17), namely,

$$M = W \times \frac{AC \times BC}{AB}.$$

If AC = x, AB = l; then BC = $l - x$, and we have for a rectangular beam,

$$W \frac{(l - x) \cdot x}{l} = S \frac{bd^2}{6} \qquad . \quad (41)$$

When C is the centre of the beam, $l - x = x = \frac{l}{2}$;

then $$W = 4 \frac{S}{6} \cdot \frac{bd^2}{l} = \frac{2}{3} S \frac{bd^2}{l} \qquad . \quad (42)$$

Comparing this with equation (39) it will be seen that the stress for a given weight is one-fourth as great when the load is at the middle of a beam supported at each end, that it is when hung at the end of a beam fixed at one end.

When the load W is uniformly distributed over the whole length of the beam, then by equation (22) the stress at any point C is,

$$M = \frac{W}{2} \times \frac{AC \times BC}{AB} = \frac{W}{2} \times \frac{(l - x) x}{l}$$

$$= S \frac{b \cdot d^2}{6} \qquad . \qquad . \qquad . \quad (43)$$

If C is the centre of the beam, then $l - x = x = \dfrac{l}{2}$,

$$W = 8 \frac{S}{6} \cdot \frac{b \cdot d^2}{l} = \frac{4}{3} S \frac{b \cdot d^2}{l} \qquad . \qquad . \quad (44)$$

32. SAFE LOAD ON A BEAM.—In order to obtain the foregoing equations we assumed that the elasticity of the beam was unimpaired by the stress of the load, and it was only upon this assumption that the neutral-axis could be considered as passing through the centre of gravity of the section. Consequently they ought only to be applied to finding the load which can be laid upon the beams with perfect safety, or the *coefficient of safety* (**28**) must be used for S in place of the *coefficient of ultimate strength*. Now as it is also assumed that the resistances to compression and extension are equal, whereas the coefficients of ultimate tensile and compressive strength differ considerably, as shown by the Tables (**26** and **27**), we shall obtain the most correct results if we take the *average* of the ultimate crushing and tensile strengths, that is, their sum divided by 2, and again divide this average by 6, in the case of beams of timber ; and this will give us the *average coefficient of safety.* The following table gives the value of S in the foregoing formulæ for a safe load on the beam, and is the actual stress per square inch of section that may safely be applied, either in tension or compression, at the top or bottom of the beam.

Kind of Wood.	Safe Load in Lbs. (S).
English Elm	937
Canadian Elm	1481
Dantzic Fir	848
Riga Fir	775
Spanish Mahogany	850
American White Oak	1123
English Oak	1254
French Oak	1337
Pitch Pine	928
Canada Spruce	732

33. EXAMPLES OF APPLICATION.—Let it be required to find the safe-load at the middle of a beam of English oak, supported at each end, whose breadth is 6 inches, depth 12 inches, and length between supports 10 feet or 120 inches, and S = 1254.   Then by equation (42) we have—

$$W = \tfrac{2}{3} S \frac{b \cdot d^2}{l} = \tfrac{2}{3} \times 1254 \frac{6 \times 144}{120} = 6019 \text{ lbs.}$$

Let a beam of Riga fir having the above scantling be 15 feet, or 180 inches in length; S being 775.   To find the safe load under different circumstances of its application.

In this case we have—

$$S \frac{b \cdot d^2}{l} = 775 \frac{6 \times 144}{180} = 3720.$$

First case: let the beam be fixed at one end and loaded at the other; then by equation (39),

$$W = \tfrac{1}{6} S \frac{b \cdot d^2}{l} = \frac{3720}{6} = 620 \text{ lbs.}$$

Second case: let the load W be uniformly distributed over the beam in the last case; then by equation (40),

$$W = \tfrac{1}{3} S \frac{b \cdot d^2}{l} = \frac{3720}{3} = 1240 \text{ lbs.}$$

Third case: let the beam be supported at each end and loaded at a point 6 feet from one end; then by equation (41),

$$W \frac{6 \times 9}{15 \times 15} = \tfrac{1}{6} S \frac{b \cdot d^2}{l} = 620;$$

therefore,

$$W = \frac{25}{6} \times 620 = 2583 \text{ lbs.}$$

Fourth case: let the load in the last case be at the centre of the beam; then by equation (42),

$$W = \tfrac{2}{3} S \frac{b \cdot d^2}{l} = \frac{2 \times 3720}{3} = 2480 \text{ lbs.}$$

Fifth case: let the beam be loaded uniformly over its entire length; then by equation (44),

$$W = \tfrac{4}{3} S \frac{b \cdot d^2}{l} = \frac{4 \times 3720}{3} = 4960 \text{ lbs.}$$

As another example, let us take a beam of English

elm, in which $S = 937$ for safe-load; and let b and d be each 10 inches, and $l = 18$ feet or 216 inches; then we have

$$S \frac{b \cdot d^2}{l} = \frac{937 \times 10^3}{216} = 4338.$$

For the first case: $W = \frac{4338}{6} = 723$ lbs.

In the second case: $W = \frac{4338}{3} = 1446$ lbs.

In the fourth case: $W = \frac{2 \times 4338}{3} = 2892$ lbs.

In the fifth case: $W = \frac{4 \times 4338}{3} = 5784$ lbs.

If the length l is expressed in *feet* instead of *inches* in the above formulæ, the values of S given in the table (**32**) must be divided by 12.

34. BEAM FIXED FIRMLY AT ENDS.—When a beam, AB (fig. 35) is held firmly down at each end, by being built into a wall, or loaded with weights at A and B, it assumes when loaded at any point C, the form of a curve of contrary flexure.* The parts between A and a, B and b, are convex to the direction in which the load W acts, while those between C and a, and C and b, are concave to the direction of W. Consequently the upper parts from A to a and from B to b are in a state of extension, and the lower parts in a state of

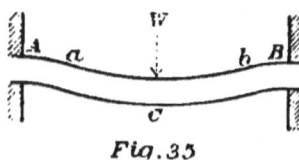

Fig. 35

* Tarn's *Practical Geometry*, p. 108.

compression ; while from a to C and b to C their condition is reversed. At the points a and b where the
flexure changes from concavity to convexity, or the
points of *contrary-flexure*, there will be neither
extension nor compression in the fibres, and the stress
will increase from zero at a and b to a maximum at
A, C and B. The *theoretical* investigation of this
problem is rather complicated, and the result obtained
is that the resistance of the beam in this case to the
stresses produced by a given load is double that of a
beam which is only *supported* at its two ends, as in the
previous cases (**31**). *Experimental* research is not
however in accord with this result, as it is found that
the resistance cannot practically be considered as more
than half as much again as in the other case, or that
the strength of a beam fixed at each end is to that
which is only supported in the proportion of 3 to 2.

35. BEAM SUPPORTED AT THREE POINTS.—This
problem has been treated differently by the various
authorities upon the subject, who have arrived at some-

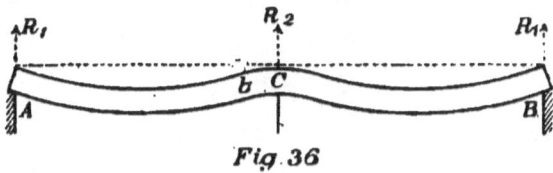

Fig. 36

what varying conclusions. Suppose that ACB (fig. 36)
represents a beam supported at its ends A and B, and
also by a prop at the middle C, and loaded between
these points. Suppose the intermediate prop at C to
be removed, then the reactions at A and B are determined in the manner previously given (**31**), and there is

now no reaction at the point C. The equilibrium will not be affected if the middle support is placed so as just to touch the beam, and the pressure of the support may be gradually increased without disturbing the equilibrium, but the variation of the pressure alters the reactions at A and B. One of the data of the problem is that all three points of support shall be in one straight line. The reactions therefore of three supports to a beam having a distributed load are in general indeterminate.

An attempt has been made to solve this problem theoretically on the supposition that the material is *perfectly elastic;* the result obtained in the case of beams of uniform depth and section, and having a distributed load W over their whole length, is as follows :—For a beam supported at each end and at one point C in the middle, if R_1 is the reaction at A and at B, R_2 the reaction at C; then

$$R_1 = \tfrac{3}{16} W, \quad R_2 = \tfrac{5}{8} W \qquad . \qquad . \quad (45)$$

If w is the load per unit of length, and a is the point of contrary flexure, then

$$Aa = \frac{2 R_1}{w}.$$

If there are *four* supports at points dividing the length into *three* equal parts, R_1 the reaction at each end, R_2 the reaction at each of the two intermediate supports, it is found that

$$R_1 = \tfrac{4}{30} W, \quad R_2 = \tfrac{11}{30} W \qquad . \qquad . \quad (46)$$

If however we look at this question from a practical

and experimental point of view, and suppose the beam (fig. 36) to be severed at C, we then have (14)—

$$R_1 = \tfrac{1}{2}\frac{W}{2} = \frac{W}{4} = \tfrac{4}{16}\,W, \quad R_2 = 2\,R_1 = \tfrac{1}{2}\,W$$
$$= \tfrac{8}{16}\,W \quad . \quad . \quad . \quad . \quad (47)$$

And comparing this with the equations (45) it would appear that the pressure on the intermediate support is greater in a *continuous* beam than in one that is *not continuous:* whereas the contrary ought to be, and is practically, the case, owing to the resistance which the material of the beam offers to its being bent from a straight line. Most of the modern writers who have treated on this subject have adopted these *theoretical* results in cases of beams with distributed loads supported at three or more points. Some however of these writers * admit that there is no material error in considering each point of support as carrying the load on a part of the beam from centre to centre of two adjoining bays; which is the result given by the equation (47). This latter distibution of load appears to be the simplest as well as the most practically correct; and we shall therefore adopt it in all cases where such beams come under our investigations.

In the case of one intermediate support at the centre of the beam, we shall take $\tfrac{1}{4}$ W as the reaction, R_1, at each end, and $\tfrac{1}{2}$ W as that, R_2, at the middle, as in equation (47). If the beam is divided into three equal parts, and has two intermediate supports, then we take

$$R_1 = \tfrac{1}{6}\,W, \quad R_2 = \tfrac{1}{3}\,W \quad . \quad . \quad . \quad (48).$$

* Col. Wray's *Application of Theory to Practice*, p. 201 ; Fenwick's *Mechanics of Construction*, p. 120.

36. BEAMS NOT RECTANGULAR.—From the reasoning used in obtaining the resistance of a rectangular beam **(29, 30)**, it will be seen that much of the material is wasted and does not offer its full resistance to the stresses. For the moment of stress is greatest at the outer edge and diminishes to nothing at the centre of gravity, consequently the inner fibres have less and less work to do according as they are situated nearer and nearer to the centre, although their actual power of resistance is the same as in the outermost fibres. *Theoretically* then the beam of section shown by fig. 37

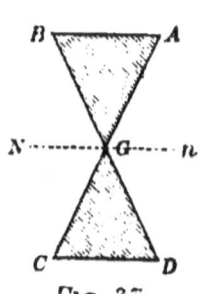

Fig. 37

would be as strong as one of rectangular section, while half the weight and material of the beam would be saved. But *practically* the nature of timber prevents us from adopting this section, as the parts cut out would be useless and the beam itself would break across at the centre G. When however we come to use such a material as iron for our beam, a great saving can be effected by having the top and bottom much wider than the middle part. The form commonly used is that given by fig. 38 and fig. 39, where there are rectangular " flanges " at top and bottom, either equal or unequal as the case may be, which are connected by a thin piece of rectangular metal placed vertically between them called the " web." The angles at which these pieces meet are strengthened by triangular pieces as shown by the dotted lines ; so that the section is a modified form of fig. 37.

37. IRON BEAMS.—The rectangular form of section

is only adapted to beams made of timber, and the form
of section used for beams of iron depends on the kind
of metal used, whether cast-iron, wrought iron or steel.
The three sections given (**20**) in the chapter on " Centre
of Gravity," figs. 29, 30 and 31, are those most commonly in use for this material.   Hav-
ing first determined the position of
the centre of gravity, G, of the
section, we may suppose the *neutral
axis* Nn (fig. 38) to pass through G.
It has been shown by equation (37)
that if I is the *moment of inertia* of
the sectional area taken about the
neutral axis; z the distance of the
most extended or compressed fibres from the neutral
axis, then the *moment of resistance* in a beam subjected
to a load acting at right-angles to its length, is

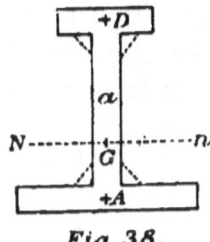

Fig. 38.

$$M = S \frac{I}{z}.$$

To find the moment of inertia of the section (fig. 38)
we must find that of its three parts, namely, the *web* or
vertical part, and the two horizontal *flanges;* and
their *sum* will be the moment of inertia of the whole
section.

Let I_1 be the moment of inertia of the top flange
about Nn; I_2 that of the web; I_3 that of the bottom
flange.   Now it is shown in treatises on Mechanics
that " the moment of inertia of an area about an axis
not passing through its centre of gravity, is equal to its
moment of inertia about a parallel line through its own
centre of gravity *plus* the area of its section multiplied

by the square of the distance of its centre of gravity from the given line." Thus, if D is the centre of gravity of the top flange, b and t_1 its breadth and thickness; B and t_3 the breadth and thickness of the bottom flange whose centre of gravity is A; d and t_2 the depth and thickness of the web whose centre of gravity is at a; then by the above rule, since $\frac{1}{12} b \cdot d^3$ is the moment of inertia (**29**) of a rectangle about an axis through its own centre of gravity, we have—

$$I = \frac{b \cdot t_1^3}{12} + b \cdot t_1 \times DG^2 \qquad . \qquad . \quad (49)$$

$$I_2 = \frac{t_2 \cdot d^3}{12} + t_2 \cdot d \times Ga^2 \qquad . \qquad . \quad (50)$$

$$I_3 = \frac{B \cdot t_3^3}{12} + B \cdot t_3 \times GA^2 \qquad . \qquad . \quad (51)$$

Then we have, $I = I_1 + I_2 + I_3$ . . (52)

The distance z of the fibres in the top flange from G, is

$$z = GD + \frac{t_1}{2} \qquad . \qquad . \qquad . \quad (53)$$

and the distance z of the fibres in the bottom flange from G, is

$$z = GA + \frac{t_3}{2} \qquad . \qquad . \quad (54)$$

It is only in *cast-iron* beams that the top and bottom flanges are made of different size; in *wrought-iron* and *steel* they are generally equal.

38. Cast-iron Beams.—In beams of cast-iron the upper flange is made smaller than the lower one, when the beam is supported at each end ; because the upper fibres are in compression, and the lower in tension, and the *ultimate* resistances of this material to compression and extension are in the ratio of 6 to 1. As, however, we are supposed to apply a load which only strains the material within the *elastic-limit*, or the limit of *safety* (**28**), we may consider that these resistances bear a much less ratio to one another; in fact they ought theoretically to be considered as nearly equal.

For an example we take the case of a cast-iron beam 20 feet long between the supports, and 13 inches deep, the upper flange being 1 inch thick and 5 inches wide, the lower one 2 inches thick and 10 inches wide, and the web 1 inch thick and 10 inches deep. Let the point D be the centre of gravity of the top flange (fig. 38), the point A that of the lower flange, and the point *a* that of the web; then by the method previously given (**20**) we can find the position of G, the centre of gravity of the whole section.

$$GD = \frac{114}{14}, GA = \frac{47}{14}, Ga = \frac{37}{14}.$$

Then by equation (49)—

$$I_1 = \frac{5 \times 1^3}{12} + 5 \times 1 \times \left(\frac{114}{14}\right)^2 = 332 ;$$

By equation (50)—

$$I_2 = \frac{1 \times 10^3}{12} + 1 \times 10 \times \left(\frac{37}{14}\right)^2 = 153 ;$$

By equation (51)—

$$I_3 = \frac{10 \times 2^3}{12} + 10 \times 2 \times \left(\frac{47}{14}\right)^2 = 233 ;$$

F

By equation (52)—

$$I = I_1 + I_2 + I_3 = 718.$$

For the distance, z, from the centre of gravity G of the fibres under greatest compression we have, by equation (53)—

$$z = \frac{114}{14} + \frac{7}{14} = \frac{121}{14} = 8\cdot64 \; ;$$

For the distance, z, of the fibres under greatest extension we have, by equation (54)—

$$z = \frac{47}{14} + \frac{14}{14} = \frac{61}{14} = 4\cdot36.$$

If we suppose that 8 tons per square inch is the safe resistance to compression (S), and put $z = 8\cdot64$, then by equation (37)—

$$M = S \frac{I}{z} = 8 \frac{718}{8\cdot64},$$

And by equation (18)—

$$M = \tfrac{1}{4} W \cdot l = W \times 60 = 8 \times \frac{718}{8\cdot64} \; ;$$

therefore, $W = 6\cdot7$ tons.

If we suppose that 2 tons per square inch is the safe resistance to extension (S), and put $z = 4\cdot36$, then by equation (37)—

$$M = S \frac{I}{z} = 2 \times \frac{718}{4\cdot36},$$

And by equation (18)—

$$M = \tfrac{1}{4} W \cdot l = W \times 60 = 2 \frac{718}{4\cdot36} \; ;$$

therefore, $W = 5\tfrac{1}{2}$ tons.

Hence it appears that we may consider 6 tons as the safe-load to be borne by this beam at the centre.

In the kind of beam known as the Hodgkinson girder, which has the area of the bottom flange six times that of the top flange, or in the ratio of the *ultimate* resistance to compression to the *ultimate* resistance to extension; the rule for the *breaking-weight* at the centre is,

$$W = 2 \cdot 2 \frac{A \cdot d}{l} \qquad \text{in tons} \quad . \quad (55)$$

where A is the area in *square inches* of the bottom flange, d the total depth in *inches* of the beam at the middle, l its length in *feet*. Applying this formula to the above example, although in this case the sectional area of the lower flange is only four times that of the upper one, we find

$$W = 2 \cdot 2 \frac{20 \times 13}{20} = 28 \cdot 6 \text{ tons}$$

for the *breaking-weight* at the centre; so that the safe-load found above, namely, 6 tons, is nearly one-fifth of the breaking-weight as derived from Hodgkinson's formula.

If the beam has no top flange, as shown by fig. 30, we have only to put $I_1 = 0$ in equation (52) and we get the value of I as before.

39. WROUGHT-IRON BEAMS.—Beams of this material are either rolled in one piece, and of the section shown either by fig. 38 or by fig. 39; or they are made of separate plates riveted together by means of *angle-irons*, and formed into a hollow box as shown by fig. 40.

Since the *ultimate* resistances of this material to compression and extension are in the ratio of 2 to 3, we ought theoretically to make the flange which is in tension two-thirds of the area of the flange which is under compression. It seems probable, however, that when the material is subjected to stresses which do not exceed the limit of safety, these resistances are nearly equal, and may be put at 5 tons per square inch of section.

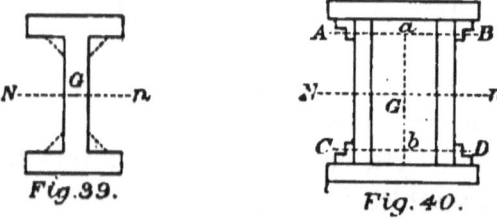

Fig.39.   Fig. 40.

When the two flanges are unequal, we must proceed as in the case of beams of cast-iron (**38**), to find the moments of inertia of the several parts about the axis through the centre of gravity of the section, and adding all these moments together we get the moment of inertia of the whole section.

When, however, as is usually the case, both flanges are alike, the moment of inertia of the section is the difference of that of the box taken as solid, and that of the hollow part of the box. If D is the external depth, *d* the internal depth (fig. 40) between the two flanges, *t* the total thickness of the plates forming the web, *b* the breadth of the flanges; then we have for the moment of inertia of the section,

$$I = \frac{b \cdot D^3 - (b - t)\, d^3}{12}, \text{ and } z = \tfrac{1}{2} D.$$

By equation (37)—

$$M = S\,\frac{I}{z} = 5\,\frac{b \,.\, D^3 - (b - t)\,d^3}{6\,D} \quad . \quad . \quad (56)$$

When the load W is at the centre of the beam, we have by equation (21)—

$$M = \tfrac{1}{4}\,W \,.\, l = 5\,\frac{b \,.\, D^3 - (b - t)\,d^3}{6\,D}.$$

Therefore,

$$W = 5 \times \tfrac{2}{3} \times \frac{b \,.\, D^3 - (b - t)\,d^3}{D \,.\, l} \quad . \quad (57)$$

This is the *safe-load* in tons, when all the dimensions are in inches; the maximum stress on the fibres at top or bottom of the beam being 5 tons per square inch of section.

The following Table gives the *safe-load* for various beams of wrought-iron having equal top and bottom flanges, the sections being either as fig. 39 or fig. 40; W being the load at the middle, and the maximum strain being 5 tons per square inch.

b.	D.	d.	t.	l = 120″.	l = 150″.	l = 180″.	l = 240″.	l = 300″.
Ins.	Ins.	Ins.	Ins.	W in Tons.	W in Tons.	W in Tons.	W in Tons.	W in Tons.
4	8	$6\frac{3}{4}$	$\frac{5}{8}$	$3\frac{1}{2}$	2·8	$2\frac{1}{3}$	$1\frac{3}{4}$	...
5	10	$8\frac{3}{4}$	$\frac{5}{8}$	$5\frac{3}{4}$	4·6	3·83	2·87	...
6	12	$10\frac{1}{2}$	$\frac{3}{4}$	10	8·0	$6\frac{2}{3}$	5	4
7	15	13	1	$19\frac{1}{3}$	15·5	12·9	9·67	$7\frac{3}{4}$
8	16	$13\frac{1}{2}$	$1\frac{1}{2}$	29·12	23·3	19·4	14·56	$11\frac{2}{3}$

40. ANGLE-IRON.—To determine the resistance of a beam of this section (fig. 41), we first find its centre of gravity G by the method previously explained (**20**), making g and g' the centres of gravity of the two arms

DB and BC. Supposing the neutral axis Nn to pass through G, we determine the moment of inertia, I, of the section by the method previously given (**37**); Ga and GA being the vertical distances of the centre of gravity of the two arms from the neutral axis. Putting BD = d, BC = b, t the thickness of the metal in each arm ; we have by equations (49, 50, 51, 52, 53),

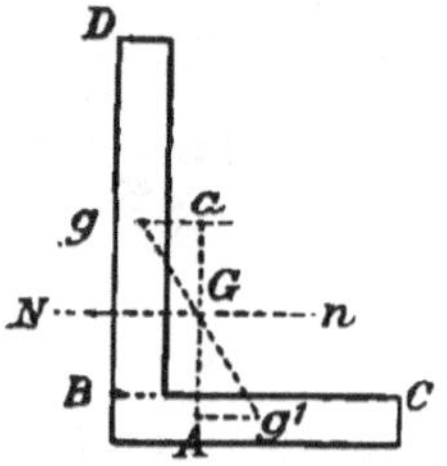

Fig. 41.

$$I_1 = 0,$$

$$I_2 = \frac{t \cdot d^3}{12} + t \cdot d \, (Ga)^2,$$

$$I_3 = \frac{b \cdot t^3}{12} + b \cdot t \, (GA)^2,$$

$$I = I_2 + I_3, \quad z = GA + \frac{t}{2}.$$

Then if W is the load on such a beam at its centre, we have from equations (21 and 37),

$$\tfrac{1}{4} W \cdot l = S \frac{I}{z}.$$

As an example, let $t = \tfrac{3}{4}''$, $d = 4\tfrac{1}{4}''$, $b = 5$ inches. In order to find the position of G, let $\dot{g}$ and g' be the centres of the arms; join gg'. Then we have—

$$\frac{Gg'}{Gg} = \frac{GA}{Ga} = \frac{\tfrac{3}{4} \times 4\tfrac{1}{4}}{5 \times \tfrac{3}{4}} = \frac{17}{20},$$

$$\frac{GA + Ga}{Ga} = \frac{37}{20} = \frac{Aa}{Ga} = \frac{\tfrac{1}{2}(d + t)}{Ga} = \frac{\tfrac{5}{2}}{Ga},$$

Therefore, $$Ga = \frac{20}{37} \times \frac{5}{2} = \frac{50}{37} = 1\tfrac{1}{3}, \text{ very nearly.}$$

$$GA = Aa - Ga = \frac{5}{2} - \frac{4}{3} = \frac{7}{6}$$

$$z = \mathrm{GA} + \frac{\ell}{2} = \frac{7}{6} + \frac{3}{8} = \frac{37}{24} = \frac{3}{2}, \text{ very nearly.}$$

Then, $\quad \mathrm{I}_2 = \frac{\frac{3}{4} \times (4\frac{1}{2})^3}{12} + \frac{3}{4} \times 4\frac{1}{2} \times \left(\frac{4}{3}\right)^2 = 10\frac{1}{2},$

$$\mathrm{I}_3 = \frac{5 \times (\frac{3}{4})^3}{12} + 5 \times \frac{3}{4} \times \left(\frac{7}{6}\right)^2 = 5\frac{1}{4}, \text{ nearly.}$$

Therefore, $\quad \mathrm{I} = \mathrm{I}_2 + \mathrm{I}_3 = 10\frac{1}{2} + 5\frac{1}{4} = 15\frac{3}{4}$

$$z = \frac{3}{2}.$$

For example, let the length be 120 inches; then

$$\frac{1}{4}\,\mathrm{W}\,.\,l = \mathrm{W} \times 30 = \mathrm{S}\,\frac{\mathrm{I}}{z} = 5\,\frac{15\frac{3}{4}}{\frac{3}{2}} = 52\frac{1}{2}.$$

Hence, $\mathrm{W} = \dfrac{52\frac{1}{2}}{30} = 1\frac{3}{4}$ tons, for the *safe-load* at the middle of the beam.

41. GIRDERS WITH ANGLE-IRONS.—Large beams of wrought-iron, whether made in the form of fig. 39 or of fig. 40, are composed of plates held together by means of angle-irons riveted at the meeting of the several pieces; as shown on fig. 40, where A, B, C, D are the angle-irons riveted to the vertical and horizontal plates. In order to calculate the strength of such a beam with accuracy, we must add the resistance of the angle-irons to that of the plates, and therefore must find their moment of inertia about Nn passing through the centre of gravity, G, of the section, by the method given above (**40**). Let Ga and Gb be the distances of the centres of gravity of the angle-irons from Nn, A the area of section of each angle-iron, I' its moment of inertia about an axis through its own centre of gravity. Then if we put I_1 for the total of the moments of inertia of all the angle-irons about Nn, we have—

$$I_1 = 4I' + 2A (Ga^2 + Gb^2) \quad . \quad . \quad (58)$$

If the top and bottom plates are equal, then $Ga = Gb$, and

$$I_1 = 4I' + 4A \times (Ga)^2 \quad . \quad . \quad (59)$$

Then the total moment of inertia, I, of the whole section of the beam is,

$$I = I_1 + I_2 + I_3 + I_4 . \quad . \quad . \quad (60)$$

And when the top and bottom flanges are equal, we have from (**39**)—

$$I = \frac{b \cdot D^3 - (b - t) d^3}{12} + I_1.$$

As an example of application let us take the last of the beams in the Table (**39**), having $D = 16''$, $d = 13\frac{1}{2}''$, $b = 8''$, $t = 1\frac{1}{2}''$, $l = 300$ ins.; and suppose the plates to be held together by angle-irons of the size given in the above example (**40**). Then we have,—

$$I' = 15\frac{3}{4}; \quad A = \frac{37}{4} \times \frac{3}{4} = \frac{111}{16} = 7, \text{ very nearly};$$

$Ga = 5\frac{1}{4}$, very nearly. Then equation (59) becomes—

$$I_1 = 4 \times 15\frac{3}{4} + 4 \times 7 \times (5\frac{1}{4})^2 = 833.$$

Then we have—

$$I = \frac{8 \times 16^3 - 6\frac{1}{2} \times (13\frac{1}{2})^3}{12} + 833$$

$$= 1398 + 833 = 2231.$$

Then as before, when W is the load at the centre,

$$\tfrac{1}{4}\,W \cdot l = S\,\frac{I}{z}, \text{ and } z = \frac{D}{2} = 8$$

$$W = 5 \times \frac{4}{8} \times \frac{2231}{300} = 18\cdot6 \text{ tons};$$

which is 7 tons more than when the resistance of the angle-irons was omitted. A deduction ought however to be made from the area of section of the angle-irons for the rivet-holes. In this example, however, the angle-irons have been taken much larger than would be adopted in practice for a girder of these dimensions.

Suppose the angle-irons for the above girder to be 3″ × 3″ × ½″, then we find by the formulæ given above (**40**)—

$$I_2 = \frac{1}{12} \times \frac{3}{8} + 3 \times \frac{1}{2} \times \frac{9}{16} = \frac{28}{32} = \frac{7}{8} = \frac{42}{48},$$

$$I_3 = \frac{1}{12} \times \frac{1}{2} \times \frac{125}{8} + \frac{5}{2} \times \frac{1}{2} \times \frac{9}{16} = \frac{65}{48},$$

$$I' = I_2 + I_3 = \frac{107}{48} = 2\tfrac{1}{4}, \text{ very nearly,}$$

$$I_4 = 4\,I' + 4 \times \frac{11}{4} \times \left(\frac{23}{4}\right)^2 = 9 + 364 = 372.$$

Therefore, $\quad$ I = 1398 + 372 = 1770,

$$W = 5 \times \frac{4}{8} \times \frac{1770}{300} = 14\tfrac{3}{4} \text{ tons.}$$

This value of W is 3 tons more than was found for the strength of the same girder when the angle-irons were left out of consideration; we may therefore in general add one-fourth to the strength obtained by means of equation (57) for the resistance offered by the angle-irons.

42. STEEL BEAMS.—The mode of calculation which has been used (**39, 41**) for finding the safe-load on beams of wrought-iron is equally applicable to those made of steel, whether rolled in one piece or formed of plates riveted together with angle-irons. The coefficient of safety, S, varies very much according to the kind of metal used. A very hard steel, although having a high coefficient, is unfit for the purposes of the architect; and a comparatively soft or *mild* steel is the best to use, in which the ultimate tensile and compressive strengths are nearly equal and about 40 tons per square inch. Steel has an advantage over ordinary iron in the fact that its *elastic limit* is nearer to the *ultimate strength*, so that the coefficient of safety may be put at one-third of the ultimate strength. The value of S may therefore be taken as about 12 tons per square inch, or more than double the coefficient used for wrought-iron. Then the equation from which to determine the safe-load, W, in tons, in a beam loaded at the centre, is found by combining equations (21) and (37)—

$$\tfrac{1}{4} W \cdot l = 12 \, \frac{I}{z}.$$

If the top and bottom flanges are equal, then $z = \tfrac{1}{2} D$;

$$W = 12 \times \frac{2}{3} \times \frac{b \cdot D^3 - (b - t) \, d^3}{D \cdot l},$$

where b, D, t, and l, are the dimensions as previously given (**39**), all being in inches, and W in tons. If the length is in feet, then—

$$W = \frac{2}{3} \times \frac{b \cdot D^3 - (b - t) \, d^3}{D \cdot l} \qquad . \quad (61)$$

As an example we will take a beam of the dimensions given above (41) for a beam of wrought-iron: where $D = 16''$, $d = 13\frac{1}{2}$ ins., $b = 8''$, $t = 1\frac{1}{2}''$, $l = 25$ ft. If the beam is rolled in one piece, and has no angle-irons, then $I = 1398$, and we have—

$$W = 8 \times \frac{1398}{16 \times 25} = 27 \cdot 96 \text{ tons.}$$

If it is made in the box-form with riveted angle-irons, whose section is $3'' \times 3'' \times \frac{1}{2}''$, then $I = 1770$, and we have—

$$W = 8 \times \frac{1770}{16 \times 25} = 35 \cdot 4 \text{ tons.}$$

43. BEAM OF UNIFORM STRENGTH.—In the beams which we have been considering the depth and breadth have been assumed to be the same throughout the entire length, so that the moment of resistance is the same at all cross sections, whatever their distance from the points of support. But it has been shown by equation (17), that if A and B are the extremities of a beam, and a load is uniformly distributed over its length, then the *moment of stress* at the section at any point, E, is proportional to AE × BE. Consequently, the stress

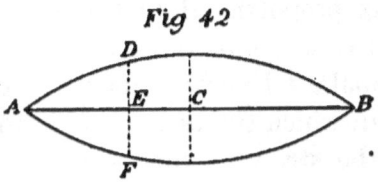

Fig 42

is nothing at A and B, and is a maximum at the centre of the beam. Since the moment of resistance at any section is proportional to the breadth, then

if we vary the breadth DE (fig. 42) in proportion to
AE × BE, keeping the depth uniform, we have a beam
whose moment of resistance at any section is proportional to the moment of stress at that section.   It is a
property of the parabola that the ordinate, or half-
breadth, DE, varies as AE × BE; so that to render
the strength of the beam uniform throughout, or proportional to the stress, we must make the flanges ADB,
AFB (when the beam is of iron) in the form of parabo-
las; or arcs of circles may be used instead without
material error.   This form may be used with advantage
in cast-iron beams, and a great reduction in weight of
metal is thereby obtained; but for wrought-iron it is
not practicable to vary the width of the flanges.

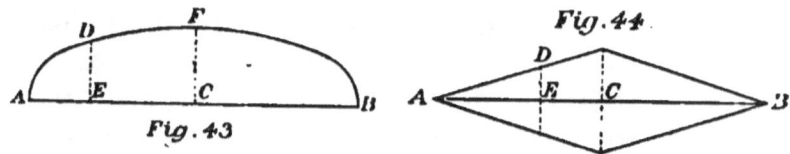

Fig. 44

Fig. 43

When it is necessary to have the width uniform
throughout, we can get a beam of uniform strength by
reducing the depth.   For as the moment of resistance
is proportional to the *square* of the depth, if DE (fig.
43) is the depth at any point E of the beam, and we
make DE² proportional to AE × BE, we have a beam
in which the resistance at any section is proportional to
the stress at that section; and by making the curve
ADB take the form of a semi-ellipse, we obtain a beam
of uniform strength, since in the ellipse the square of
the ordinate DE is proportional to AE × BE.

When the load W is placed at the centre C of the

beam, we find by equation (19) that the stress at any other point E is proportional to its distance from the nearer end of the beam ; so that in this case the breadth ought to be in the same proportion in order that the resistance of the beam should be the same at all points ; the plan of the flanges should be as shown by fig. 44, being in the form of a parallelogram, and widest at the centre C. The depth is here supposed to be uniform throughout.

If the breadth remains the same throughout, we can get a beam of uniform strength, when loaded at the centre, by making the *square* of the depth DE (fig. 43) proportional to AE, the distance of DE from the nearer end of the beam.

The curve ADF which fulfils this condition is the parabola whose vertex is at A; and the curve from B to F will be an equal and similar parabola. This will differ so little from the ellipse, that we can use that curve in either case of load at centre or uniformly distributed.

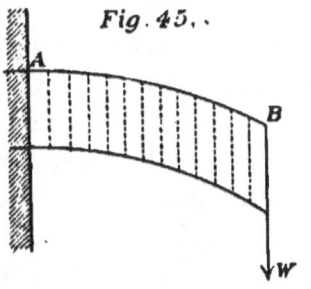

Fig. 45.

44. SHEARING - STRESS. — There is another kind of stress to which beams are subjected by the action of a load, which we will briefly allude to. Suppose the beam AB (fig. 45) to be fixed at A and loaded at B with a weight W ; then a vertical action is set up in the beam which we may suppose to be divided into a number of slices parallel to the direction of the force W. The load W tends to cause the end slice to slide down or to *shear* from the next one, but is prevented from doing so by the lateral resistance of the material. The *shearing-force* acting on the next slice is equal to that which tends to shear it from the next one ; so that a vertical force equal to W is transmitted from one slice to another through the whole length of the beam, and the load W represents the *shearing-stress* at each point or slice along it.

If the load W, or $w \cdot l$, is uniformly distributed over the beam, the *shearing-force* at any point whose distance from B we call x, is represented by $w \cdot x$, which is nothing at B, and increases up to $w \cdot l$, or W, at A.

In a beam AB loaded with W at the middle, C, and supported at each end, a *shearing-force* equal to $\frac{1}{2}$W is produced at every point. If W, or $w \cdot l$, is uniformly distributed, the *shearing-stress* at any point whose distance is x from either end, will be represented by $w (\frac{1}{2} l - x)$, which is $\frac{1}{2} w \cdot l$ at A and B, and is nothing at the centre C.

It is evident from the nature of the force, that shearing-stresses can only exist in pairs ; every shearing-stress on a given plane being necessarily accompanied by a shearing-stress of equal intensity on another plane.

The bolts or rivets used for fastening together the plates of iron girders are especially subjected to shearing-stress ; and also the plates which form the web of the girder.

The *shearing-strength* of wrought-iron, or its power of resisting a *shearing-stress*, is about 18 tons per square inch of section, or about three-fourths of its *tensile* strength. That of mild steel is about 29 tons per inch, or about four-fifths of its *tensile* strength.

CHAPTER V.

45. RESISTANCE TO BENDING.—When a beam is fixed at one end and loaded with a weight W at the outer end, it assumes a curved form, as we have previously seen (**29**); the beam AB (fig. 33) which was horizontal before the application of the load, is now deflected from the straight line by the amount BB′, which is called the " deflexion " due to the weight W. The amount of deflexion under a given load varies considerably in beams of different material; thus, in a beam of stone the deflexion will be almost imperceptible as long as the load is within the *limit of safety* (**28**), or even as the load approaches the *ultimate strength* or breaking-weight, fracture generally taking place suddenly and without any warning. Almost the same may be said of beams made of cast-iron, but with them there is a measurable deflexion before fracture takes place. With beams of timber or wrought-iron, a very considerable amount of deflexion may be produced long before the load approaches the breaking-weight, and without any serious injury to the material. As however it is essential to the stability of a building that there should be as little deviation as possible from the straight line in the beams used in its construction, it becomes a matter of much importance to determine

the load that such beams will bear without being
deflected to any great amount. Consequently, we have
not merely to calculate the safe-load that may be laid
on a beam, but also the amount of deflexion which that
load will produce, so that we may regulate the size of
the beam accordingly.

The resistance which a beam offers to *fracture* by a
load acting transversely, is called its "strength;"
while its resistance to *deflexion*, or to any deviation
from a straight line, is called its "stiffness." These
two properties of a beam are, as we shall proceed to
show, governed by very different laws.

46. RADIUS OF CURVATURE.—When considering
the case of a beam fixed at one end and loaded at the
other (**29**), we saw that the neutral surface N*n* (fig. 33)
assumed a curved form, and that any small portion as
FH might be taken as very nearly an arc of a circle
whose centre was at O, OF or OH being the radius, ρ,
of this arc. The circle of radius ρ which most nearly
approaches to the curvature of N*n* at any point, is
called its *circle of curvature* at that point, and its
radius, ρ, is called the *radius of curvature* of a small
arc. This *radius* varies in length when the whole line
N*n* differs from a circular curve, and can be found for
a point at any distance from N by help of the differential calculus.

From equation (36) we have for M the *moment of
resistance* to flexure at any point of the beam—

$$M = \frac{E \cdot I}{\rho},$$

where E is the *modulus of elasticity*, I the *moment of*

inertia of the section about the *neutral axis*, ρ the *radius of curvature* of the neutral surface at a given point. The above equation can be put into the form—

$$\frac{1}{\rho} = \frac{M}{E \cdot I} \quad . \quad . \quad . \quad (62)$$

47. BEAM SUPPORTED AT EACH END.—Suppose a straight rod or beam whose length between the supports is AB, or *l*, to be supported at A and B (fig. 46)

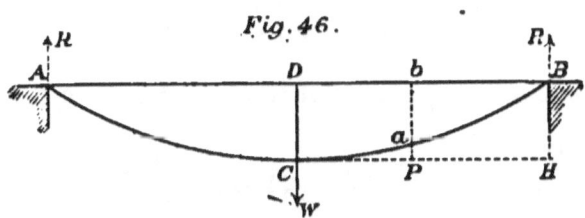

Fig. 46.

and to be loaded with a weight W at the centre C. If ADB is its position before the load is applied, then the effect of the load is to cause it to bend downwards, as ACB, producing the *deflexion* DC at the middle. Let R be the reaction at each of the supports A and B, then $R = \frac{1}{2} W$; and the deflexion will be the same if we suppose the beam to be fixed at C and drawn upwards by a force R at each end. Draw CPH parallel to ADB (P being any point on the line CH) and P*ab* perpendicular thereto. Let CP = *x*, P*a* = *y*; then *x* and *y* are called the *co-ordinates* of the curve ACB, and vary in length with the position of the point P. In order to determine the deflexion at any point, we must find the relation between *x* and *y*, or the *equation to the curve*. Since the curvature is but small, we are

G

able to put for the quantity $\dfrac{1}{\rho}$ the second *differential coefficient* of y with respect to x.* From equation (62) we have—

$$\frac{1}{\rho} = \frac{M}{E \cdot I}, \text{ and } M = R \times Bb = \frac{W}{2}\left(\frac{l}{2} - x\right),$$

Since $CH = \dfrac{l}{2}$, $CP = x$, $Bb = PH = \dfrac{l}{2} - x$; therefore we get—

$$\frac{1}{\rho} = \frac{W}{2\,E \cdot I}\left(\frac{l}{2} - x\right).$$

After two integrations of this equation, we obtain—

$$y = \frac{W}{2\,E \cdot I}\left(\frac{l \cdot x^2}{4} - \frac{x^3}{6}\right) = \frac{W}{24\,E \cdot I}(3\,lx^2 - 2x^3) \quad (63)$$

which is the *equation to the curve* ACB. When the point whose coordinates are x and y is at the middle of the beam, we have $x = CH = \frac{1}{2}\,l$, and $y = HB = CD = \delta =$ the deflexion of the beam at the middle. Substituting these values for x and y in equation (63) we find—

$$\delta = \frac{W}{2\,E \cdot I}\left(\frac{l^3}{16} - \frac{l^3}{48}\right) = \frac{1}{48} \cdot \frac{W}{E} \cdot \frac{l^3}{I} \quad . \quad (64)$$

In the case of a beam of rectangular section of which b is the breadth and d the depth, we have (**29**),

$$I = \tfrac{1}{12}\,b \cdot d^3, \text{ and}$$

$$\delta = \frac{1}{4} \cdot \frac{W}{E} \cdot \frac{l^3}{b \cdot d^3} \quad . \quad . \quad (65)$$

---

* See the author's edition of *Tredgold's Carpentry*, p. 40.

Since the *stiffness* of a beam, or its *resistance to deflexion*, is *inversely* as the amount of deflexion for a given load, it will vary as $\frac{1}{\delta}$, or directly as the breadth and the cube of the depth and inversely as the *cube* of the length. Comparing this with equation (39) which gives the *strength* of a beam, we see that the *strength* varies as the breadth and *square* of the depth and inversely as the length; so that we have then the following proportion :

$$\text{Stiffness : strength} = \frac{d^3}{l^3} : \frac{d^2}{l} = \frac{d}{l^2} = d : l^2.$$

The deflexion ab at any point P on the beam, where $CP = x$, $Pa = y$, is evidently $CD - Pa$, or $\delta - y$; and we have by subtracting equation (63) from equation (64)—

$$ab = \delta - y = \frac{1}{48} \cdot \frac{W}{E \cdot I} (l^3 - 6 l \cdot x^2 + 4x^3) \quad (66)$$

$$= \delta \left(1 - 6 \frac{x^2}{l^2} + 4 \frac{x^3}{l^3} \right).$$

If $x = \frac{1}{4} l$, then $y = \frac{W}{2 E \cdot I} \times \frac{5 l^3}{384} = \frac{5}{16} \delta$

$$ab = \delta - y = \frac{11}{16} \times \frac{W \cdot l^3}{48 E \cdot I} = \frac{11}{16} \delta,$$

which is the deflexion at a point half way between B and D or A and D.

48. BEAM WITH DISTRIBUTED LOAD. — When the load is uniformly distributed over the entire length of the beam, we have from equation (22) the moment of stress M at any point b (fig. 46) :

$$M = \frac{W}{2} \cdot \frac{Ab \times Bb}{AB} = \frac{W}{2l}\left(\frac{l}{2} + x\right)\left(\frac{l}{2} - x\right)$$

$$= \frac{E \cdot I}{\rho}, \text{ from equation (36).}$$

Therefore, $\quad \dfrac{1}{\rho} = \dfrac{W}{2\,E \cdot I \cdot l}\left(\dfrac{l^2}{4} - x^2\right).$

Then proceeding as before (47), and substituting the second differential coefficient for $\dfrac{1}{\rho}$, we obtain, after two integrations—

$$y = \frac{W}{2\,E \cdot I \cdot l}\left(\frac{l^2 x^2}{8} - \frac{x^4}{12}\right)$$

$$= \frac{1}{48} \cdot \frac{W \cdot l^3}{E \cdot I}\left(3\frac{x^2}{l^2} - 2\frac{x^4}{l^4}\right) \qquad . \qquad . \quad (67)$$

At the centre of the beam, $x = \dfrac{l}{2}$ and $y = \delta$; therefore

$$\delta = \frac{1}{48} \cdot \frac{W \cdot l^3}{E \cdot I}\left(\frac{3}{4} - \frac{2}{16}\right)$$

$$= \frac{5}{8}\left(\frac{1}{48} \cdot \frac{W \cdot l^3}{E \cdot I}\right) \qquad . \qquad . \quad (68)$$

Comparing equation (68) with equation (64) we see that the deflexion at the middle with a distributed load is five-eighths of that produced by the same load at the centre. If then a beam is loaded with a weight W at the centre, and w is the weight of the beam itself, we must add $\frac{5}{8}$ w to the load W in order to obtain the true deflexion, where the section is the same throughout.

49. BEAM FIXED AT ONE END. — When the beam

is fixed at one end and loaded by a weight W at the other end, we have for the moment of resistance, M, at a point whose distance is x from the fixed end—

$$M = W (l - x);$$

and by equation (62)—

$$\frac{1}{\rho} = \frac{M}{E \cdot 1} = \frac{W}{E \cdot 1} (l - x).$$

Proceeding as before (47) we have after two integrations, for the deflexion at the point x,—

$$y = \frac{1}{6} \frac{W \cdot l^3}{E \cdot 1} \left(3 \frac{x^2}{l^2} - \frac{x}{l^3} \right) \quad . \quad . \quad (69)$$

For the deflexion, δ, at the end of the beam, when $x = l$, we find—

$$\delta = \frac{W \cdot l^3}{3 E \cdot I} \quad . \quad . \quad . \quad (70)$$

Let δ_1 be the deflexion when the load is uniformly distributed; then it is found by the same method as in the case of the beam supported at each end (48), that—

$$\delta_1 = \frac{W \cdot l^3}{8 E \cdot I} = \frac{3}{8} \frac{W \cdot l^3}{3 E \cdot I} = \frac{3}{8} \delta \quad . \quad (71)$$

therefore the deflexion caused by the distributed load is three-eighths of that caused by the same load placed at the outer end, or is the same as if $\frac{3}{8}$ W instead of W were placed on the end of the beam. Therefore, to get the true deflexion caused by a load W at one end, we must add three-eighths of the weight of the beam to the weight W.

50. TREDGOLD'S RULE. — In considering the deflexion of beams of timber, Tredgold lays down the rule that in practice the deflexion at the middle of a beam supported at both ends, should not exceed one-fortieth of an inch for every foot of length, or one inch in a length of 40 feet, when the greatest load that the beam is required to sustain is laid upon it. It is therefore advisable to have an expression for the load W which will produce a deflexion of this amount in a beam of timber. The deflexion of such a beam is given by equation (65), the load being at the middle, namely—

$$\delta = \frac{1}{4}\ \frac{W}{E}\cdot\frac{l^3}{b\cdot d^3};$$

where all the dimensions are in inches. If we put L for the length in feet, then $L = \frac{1}{12}\,l$, or $l = 12\,L$; if then we put $\delta = \frac{1}{40}\,L$ in inches, we obtain the value of W which will produce the deflexion δ; namely—

$$W = \frac{L}{40}\times\frac{4\,E\cdot b\cdot d^3}{12^3\,L^3} = \frac{E}{17280}\times\frac{b\cdot d^3}{L^3}\ .\quad(72)$$

where the load W must include $\frac{5}{8}$ths of the weight of the beam. If the load W is given, and also the length and depth, we can find the breadth b from the equation—

$$b = \frac{17280}{E}\times\frac{W\cdot L^3}{d^3}\qquad.\quad.\quad(73)$$

If the depth d is required, when W, L and b are given, we have—

$$d = \sqrt[3]{\left(\frac{17280}{E}\times\frac{W\cdot L^3}{b}\right)}\ .\quad(74)$$

51. Practical Examples.—As an application of the foregoing rules, we will take the case of the beam of Riga fir, whose breadth was 6 inches, depth 12 inches, length 180 inches, the safe-load W at the centre was found (**33**) to be 2,480 lbs. The modulus of elasticity is given in the Table (**23**) as 3,009,680 lbs., and if we put the weight of the material at 30 lbs. per cubic foot, we have w = 225 lbs. for the weight of the beam ; so that (**48**) we must put the actual weight at the centre as 2,480 + $\frac{5}{8}$ × 225 = 2,621 lbs. Then by equation (65) we have for the deflexion at the middle—

$$\delta = \frac{1}{4} \times \frac{W}{E} \times \frac{l^3}{b \cdot d^3}$$

$$= \frac{1}{4} \times \frac{2621}{3009680} \times \frac{180^3}{6 \times 12^3} = \cdot 1225 \text{ in.}$$

If we apply Tredgold's rule (**50**) to this example, we have the limit of deflexion equal to $\frac{18}{48}$ or $\cdot 375$ inch, which is just three times the amount found above; so that the load might in this case be trebled. By means of equation (72) we can determine exactly the load that will produce the deflection $\cdot 375$ inches ;

$$W = \frac{3009680}{17280} \times \frac{6 \times 12^3}{15^2} = 8026 \text{ lbs.}$$

Next, let us apply the equations to the example of English Elm (**33**) in which b and d are each 10 inches, and the length 18 feet. The safe-load at the centre was found to be 2,892 lbs. In this case we have E = 1,003,280 lbs. ; and if we put the weight of a cubic foot at 36 lbs. we have w = 450 lbs., and $\frac{5}{8} w$ =

281 lbs., or the load at the middle is ·3173 lbs. Then by equation (65)

$$\delta = \frac{1}{4} \times \frac{3173}{1003280} \times \frac{12^3 \times 18^3}{10 \times 10^3} = \cdot 8 \text{ in.}$$

But by Tredgold's rule (50) the maximum deflexion should be $\frac{18}{40}$ or ·45 inch, so that the load should be reduced in the proportion of 45 to 80, which gives $W + \frac{5}{8} w = 1,785$ lbs., or $W = 1,504$ lbs. instead of 2,892 lbs.

We will apply equation (72) to the following example:—a beam of Dantzic fir whose weight per cubic foot is 30 lbs., and where $E = 1,737,570$ lb., has $b = 10$ inches, $d = 15$ inches, $L = 20$ feet; required to find the load at the centre, that will produce a deflexion of $\frac{20}{40}$ or half an inch. Here $w = 625$ lbs. is the weight of the beam, and $\frac{5}{8} w = 391$ lbs. From equation (72)

$$W = \frac{1737570}{17280} \times \frac{10 \times 15^3}{20^2} = 8484 \text{ lbs.}$$

Deducting $\frac{5}{8} w$ from this value of W, we have for the required load at the centre, $8,484 - 391 = 8,093$ lbs. And if the load is uniformly distributed, the weight would be $\frac{8}{5} \times 8,093 = 12,949$ lbs.

Apply the equation (73) to a similar beam whose depth is 12 inches; to find the *breadth* when the load, including the weight of the beam, is 5,000 lbs. By equation (73)

$$b = \frac{17280}{1737570} \times \frac{5000 \times 20^2}{12^3} = 11\frac{1}{2} \text{ ins.};$$

and $w = 575$ lbs., $\frac{5}{8} w = 358$ lbs. So that the load

at the centre producing a deflection of $\frac{1}{2}$ inch, is
5,000 − 358 = 4,642 lbs. If the load is uniformly
distributed, it will be $\frac{8}{5} \times 4,642 = 7,427$ lbs.

To find by equation (74) the *depth* of the beam of
Dantzic fir whose breadth is 10 inches, and length 20
feet; W being 8,484 lbs.

Then by equation (74) we have—

$$d = \sqrt[3]{\left(\frac{17280}{1737570} \times \frac{8484 \times 20^3}{10} \right)} = 12 \sqrt[3]{1 \cdot 05}$$
$$= 15 \text{ ins.}$$

52. Scantlings of Floor Timbers.—The following
Table gives the *scantlings* (breadth and depth) of floor
joists of *yellow fir* (where E = 1,737,570 lbs.) which
shall carry a load of 120 lbs. per foot of length
uniformly distributed; the deflexion not to exceed
one-fortieth of the length of bearing, according to
Tredgold's rule (50). The joists are supposed to be
placed 12 inches apart from middle to middle. Single-
joisted floors should never exceed a bearing of 18 feet.
The above may be considered as the ordinary load
which house floors have to support, but in the case of
warehouses, workshops, or rooms where very heavy
loads are to be sustained, the strength must be proportionally greater. Thus, if the joists are required
to carry 240 lbs. per foot, we must either *double*
the *breadth* of the joist, or else multiply the depth
by the cube root of 2, or 1·26. If the load is
360 lbs. we must either treble the breadth, or else
multiply the depth by the cube root of 3, or 1·44;
and so on.

If the joists are placed further apart we must either

diminish the load or increase their scantling proportionally; if placed at 18 inches from middle to middle, they would have to bear half as much again, so that the *breadth* must be increased in that proportion if the depth is unaltered; or else the *depth* increased by multiplying it by the *cube-root* of 1·5, or 1·145.

Bearing.	Breadth 2 Ins.	Breadth 2½ Ins.	Breadth 3 Ins.	Breadth 3½ Ins.	Breadth 4 Ins.
	Depth.	Depth.	Depth.	Depth.	Depth.
5 ft.	4 ins.	4 ins.	4 ins.	4 ins.	4 ins.
6	4¼	4	4	4	4
7	5	4¼	4¼	4	4
8	5½	5	4½	4¼	4¼
9	6½	6	5	5	5
10	7	6½	6	6	5½
11	8	7	6½	6½	6
12	8½	8	7	7	6½
13	9	8½	8	7½	7
14	10	9¼	9	8	8
15	10½	9¾	9¼	9	8½
16	11	10½	9¾	9½	9
17	11½	10¾	10¼	10	9¼
18	12	11¼	10½	10¼	9½

When the span exceeds 18 feet it is usual to form a *framed-floor*, consisting of large beams of wood or iron called "Girders," which are placed across from wall to wall at distances of about 10 feet apart; on these rest some smaller beams called "binders," placed about 6 feet apart; and on these again rest the "bridging" joists on which the floor is laid; these latter being about 12 inches apart. The scantling of the *bridging joists* is determined by the above Table; that of the *Binders* is found in the same way by supposing the load on each of them to be six times as great per linear foot as on a bridging-joist. The following Table gives scantling for *binders* of *fir*, on the supposition that the load per foot is 720 lbs.

Length of Bearing.	6″ Deep.	7″ Deep.	8″ Deep.	9″ Deep.	10″ Deep.	11″ Deep.
	Breadth.	Breadth.	Breadth.	Breadth.	Breadth.	Breadth.
6 ft.	4½ ins.	4 ins.	3 ins.	2 ins.	2 ins.	2 ins.
7	7	5	4	3	2	2
8	10¼	7	5	4	2½	2
9	...	9	6½	5	3½	2½
10	...	12	8¾	6	4½	3½
11	...	...	11½	8	6	4½
12	...	...	...	10½	8	6

When the load per foot is greater than 720 lbs., the scantling must be increased in the proportions mentioned above for common joists; if *doubled*, then either by *doubling* the *breadth*, or multiplying the depth by the cube root of 2, or 1·26; and so on as before.

The *girders* support the ends of the *binders* at points about 6 feet apart, consequently a girder 10 to 12 feet long will support the end of one *binder* at the middle, and will therefore carry half the weight of the floor, the other half resting on the two walls; so that the scantling is that of a beam loaded at the centre by half that part of the weight of the floor which lies between two girders; and by equation (73) we obtain b when L and d are given—

$$b = 6 \frac{L^3}{d^3}.$$

When the length of the girder is from 15 to 18 feet, we have a binder supported at two points, and one-third of the weight of floor resting at each point; then by combining equation (26) with equation (73) we get—

$$b = \frac{16}{3} \cdot \frac{L^3}{d^3}.$$

For girders whose length is from 20 to 24 feet the binders rest upon three points, at each of which one-fourth of the weight is sustained; then by combining equation (27) with equation (73) we have—

$$b = 6 \frac{L^3}{d^3}.$$

When the span exceeds 24 feet, iron girders should be used in place of timber where practicable, the section of which can be determined for a given deflexion by the equation (64). The following Table gives the scantling of fir girders when the load is as given above, namely 120 lbs. on every square foot of flooring. If the load is 240 lbs. per square foot, then either the breadth of the girder must be doubled or its depth increased by one-fourth.

Span.	12″ Deep.	14″ Deep.	15″ Deep.	16″ Deep.	18″ Deep.
	Breadth.	Breadth.	Breadth.	Breadth.	Breadth.
10 ft.	4 ins.	3 ins.	3 ins.	3 ins.	3 ins.
11	5	$3\frac{1}{2}$	3	3	3
12	6	4	$3\frac{1}{2}$	3	3
13	7	5	4	4	3
14	9	6	5	$4\frac{1}{2}$	$3\frac{1}{2}$
15	11	7	6	5	4
16	13	$9\frac{1}{2}$	7	$5\frac{1}{2}$	$4\frac{1}{2}$
17	15	10	8	6	5
18	16	11	9	7	$5\frac{1}{2}$
19	18	13	$10\frac{1}{2}$	8	6
20	21	15	12	9	7
21	...	17	14	$10\frac{1}{2}$	8
22	...	19	16	12	9
23	...	21	18	14	10
24	...	...	20	16	11

53. FLANGED BEAMS. — Beams with " flanges " connected by a vertical " web " are made of cast-iron, wrought-iron, or steel; and of section shown by figs. 38, 39, 40. By equation (64) we get the deflexion (δ) of such a beam at the middle,

$$\delta = \frac{1}{48} \times \frac{W}{E} \times \frac{l^3}{I}$$

where all the dimensions are expressed in inches, W and E in lbs. or tons.

Let us apply this formula to find the deflexion of the cast-iron beam (**38**) whose depth is 13 inches, top flange 5" × 1", bottom flange 10" × 2", web 10" × 1", the length of bearing being 20 feet, the weight will be just 1 ton. Let W at the centre be 6 tons including $\frac{5}{8}$ of the beam's own weight, or the actual load on the centre will be $5\frac{3}{8}$ tons. Taking E = 7600 tons, (**25**) and the value of I being 718, we have

$$\delta = \frac{1}{48} \times \frac{6}{7600} \times \frac{(240)^3}{718} = \cdot 31 \text{ in., or nearly } \tfrac{1}{3}\text{rd in.}$$

By Tredgold's rule (**50**) the deflexion should not exceed $\frac{2.0}{4.0}$, or $\frac{1}{2}$ inch in this case.

Take the example of the wrought-iron box-girder (**41**), whose safe-load was found to be $14\frac{3}{4}$ tons; the weight of this beam is 2 tons, five-eighths of which is $1\frac{1}{4}$ tons; therefore W = 16 tons. Taking I = 1770, l = 25 feet, E = 10,714 tons, we have for the deflexion in the middle,

$$\delta = \frac{1}{48} \times \frac{16}{10714} \times \frac{25^3 \times 12^3}{1770} = \cdot 4743 \text{ in.}$$

By Tredgold's rule (**50**) the deflexion should not exceed

$\frac{2.5}{4.0}$ or ·625 inch, so that the deflexion in this case is well within the rule.

As an example of a *rolled* iron beam, we will take the first beam in the table (**39**), where the length is 10 feet or 120 inches, and the load at the centre $3\frac{1}{2}$ tons. Here the weight of the beam is $\frac{1}{16}$ ton, five-eighths of which is $\frac{1}{16}$ ton, so that W = $3\frac{9}{16}$ ton, the value of I is found to be 84.

Then the deflexion at the middle is,

$$\delta = \frac{1}{48} \times \frac{57}{16 \times 10714} \times \frac{120^3}{84} = \cdot142 \text{ in.}$$

By Tredgold's rule (**50**) the maximum deflexion is $\frac{2.0}{4.0}$ or ·25 of an inch.

Let us apply the formula to the case of the steel box-girder (**42**), which has the same dimensions as the wrought-iron girder given above; the load at the middle being 35·4 tons, and five-eighths of its weight being 1·2 tons, we have W = 36·6 tons; taking E = 13,500 tons, l = 300 inches, we find the deflexion at the middle,

$$\delta = \frac{1}{48} \times \frac{36·6}{13500} \times \frac{300^3}{1770} = \cdot86 \text{ in.}$$

By Tredgold's rule (**50**) the maximum deflexion is ·625 inch, so that here the deflexion is considerably in excess, and the load ought to be reduced in the proportion of 860 to 625, or nearly as 3 to 2.

54. BEAM OF UNIFORM STRENGTH.—It has been shown in the last chapter that a beam may be made of uniform strength throughout, or have its resistance the same at every section, by making the plan of the

flanges in the form of a parabola (**43**), being widest at
the centre, and the depth being the same throughout.
In such a beam the contraction or shortening of the
fibres in the upper flange, and their extension in the
lower one, will be uniform throughout, and consequently
the curve which the neutral surface takes when the
beam is loaded will be an arc of a circle.

Let ABCD (fig. 47) be such a beam, loaded in the
middle, and let O be the
centre of curvature of the
neutral surface, AKBH the
circle of curvature of the
top flange. Then if AB
($= l$) is a horizontal line,
EK is the deflexion, δ, at
the middle, $\rho = OA = OK$
is the radius of curvature,
AC ($= d$) the depth of the
beam. Draw AF perpen-
dicular to CFD, then CF
is half the difference in
length between the most
compressed and most ex-

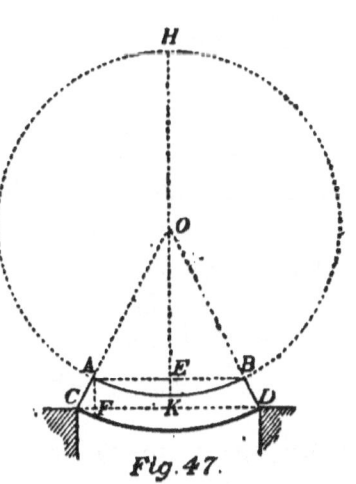

Fig. 47.

tended fibres. Let $CF = \frac{1}{2}\lambda$. Then HK and AB are
two chords of the same circle intersecting in E; and
by Euc. Bk. 3. 35, the rectangle under the segments
HE and EK equals the rectangle under the segments
AE and EB. And since AE = EB, we have,

$$HE \times EK = AE^2,$$
$$\text{or, } (2\rho - \delta) \times \delta = \frac{1}{4}l^2 = 2\rho \cdot \delta - \delta^2$$

But since δ is very small as compared with ρ, we may put $\delta^2 = 0$. Then we have for the deflexion at the middle,

$$\delta = \frac{l^2}{8\rho}, \text{ very nearly.}$$

Also, since the triangle ACF is similar to triangle OAE, therefore OA : AE = AC : CF;

or, $\qquad \frac{\rho}{\frac{1}{2}l} = \frac{d}{\frac{1}{2}\lambda}$, $\qquad$ therefore $\rho = \frac{d \cdot l}{\lambda}$.

Hence we have $\qquad \delta = \frac{l^2 \cdot \lambda}{8d \cdot l} = \frac{l \cdot \lambda}{8d}$. $\qquad$. (75)

The value of λ depends on the modulus of elasticity, E, of the material. If we put c for the compressive stress per square inch in the extreme upper fibres, and t for the tensile stress per square inch in the extreme lower fibres, we have from equation (32), putting $c + t$ for S, λ for l, l for L,

$$\lambda = \frac{c + t}{E} \times l . \qquad . \qquad . \quad (76)$$

Substituting this value of λ in equation (75) we obtain :

$$\delta = \frac{c + t}{E} \times \frac{l^2}{8d} \qquad . \qquad . \quad (77)$$

The resistance of the web is not taken into account in this formula. If l is expressed in feet, then d must also be in feet, or if d is in inches, l must also be in inches.

Take for example the cast-iron beam (**38**) of which the length was 20 feet or 240 inches, the depth 13 inches, $c = 8$ tons, $t = 2$ tons, E = 7,600 tons. Then, if we suppose the flanges to diminish from the centre towards the ends in the form of a parabola, we have by equation (77) for the deflexion in the middle,

$$\delta = \frac{10}{7000} \times \frac{400 \times 144}{8 \times 13} = \cdot 73 \text{ in.};$$

which is nearly half as much again as the deflexion allowed by Tredgold's rule (**50**), and more than double of that found previously (**53**) for a beam having flanges of equal width throughout.

CHAPTER VI.

55. LONG PILLARS.—By this term we mean pillars whose length exceeds 30 times the diameter or least width. When such a pillar, having a uniform section throughout, is placed in a vertical position with the lower end resting on a hard or impenetrable surface, and a load W is placed on the upper end, it will assume a slightly curved form as ACB (fig. 48), being deflected from the straight line ADB by the amount CD which we call δ; and we may assume, without material error, that for a small value of δ the curve is very nearly an arc of a circle. We can therefore find the amount of deflexion in the same way as in the case of the " Beam of uniform strength " (**54**), taking λ to represent the difference of the lengths of the extreme fibres on the concave and convex sides. Let d be the least diameter of the pillar, l its length, f the longitudinal unit of stress in the extreme fibres of either side in a horizontal section across its middle. Then the forces acting on the pillar are, W at A, the longitudinal tensile stress on the convex side at C, and the longitudinal compressive stress on the concave side at C.

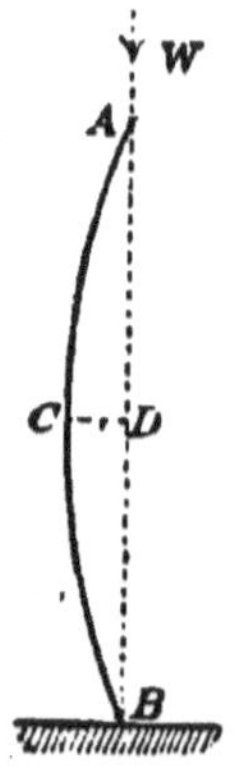

Fig. 48

Taking moments about the point C, we have for the moment of resistance M,

$$M = W . \delta;$$

and by equation (75)—

$$\delta = \frac{\lambda . l}{8\,d};$$

Therefore,

$$W = \frac{8\,d}{\lambda . l} \times M;$$

and by equation (76)—

$$\lambda = \frac{2f}{E} \times l;$$

Therefore, $W = \dfrac{4d . E}{f . l^2} \times M = 4\,\dfrac{d}{f} \times \dfrac{E}{l^2} \times M$ (78)

For a *square* section, we have by equation (38)—

$$M = f\,\frac{d . d^2}{6}, = \frac{d'}{6} \times \frac{f}{d};$$

In which case, from equation (78)—

$$W = \frac{2}{3}\,\frac{d'}{l^2} \times E \qquad . \qquad . \quad (79)$$

And for a circular section whose radius is $r = \dfrac{d}{2}$, we have from equation (37)—

$$M = f . \frac{I}{z}$$

where I is the *moment of inertia* of the section about its centre, and $z = r$.

For a circle, $I = \frac{1}{4}\,\pi\,r^4$, consequently

$$M = \frac{\pi\,r^4}{4} \times \frac{f}{\dfrac{d}{2}} = \frac{\pi}{2} \times \frac{d'}{16} \times \frac{f}{d}.$$

Therefore, $W = \dfrac{\pi}{8} \times \dfrac{d^4}{l^2} \times E$.   .   (80)

Hence it appears by comparison of equations (79) and (80), that the strength of a long solid pillar whose section is *square*, is to that of one whose section is the *inscribed circle* of the *square*, in the proportion of 2 × 8 to 3 × π, or as 17 to 10.

For long hollow pillars of circular section, of which *d* is the internal diameter,

$$W = \dfrac{\pi}{8} \left(\dfrac{d^4 - d_1^4}{l^2} \right) \times E \quad . \quad (81)$$

Experiment shows that this *theoretical* result only applies in the case of cast-iron to pillars whose length exceeds 50 diameters ; and in the case of wrought-iron where it exceeds 80 diameters. For pillars of steel or timber the above formula gives correct results for those whose length is not less than 30 diameters.

The foregoing investigation has however but little interest for the architect, as he seldom, if ever, uses pillars in which the length exceeds 30 diameters.

56. HODGKINSON'S EXPERIMENTAL FORMULÆ. — Taking the theoretical result above obtained in equations (80) and (81) as the basis on which to work, Mr. Eaton Hodgkinson* proceeded to determine by numerous experiments the laws which govern the resistance of pillars. In the case of long round pillars of cast-iron the strength was found to vary in a rather less degree than the *fourth power* of the diameter, and inversely in a less degree than the *square* of the length.

Comparing the strength of two pillars 10 feet long,

* *Phil. Trans.*, 1840 and 1857.

their diameters being $1\frac{1}{2}$ inches and $2\frac{1}{2}$ inches respectively, it was found that the ratio of their breaking-weights was as 6 : 1. Since the ratio of their diameters is as 1·6667 : 1, if we put n for the power of the diameter by which the strength varies when the length is the same, we have—

$$1^n : 1\cdot6667^n = 1 : 6,$$

therefore, $\qquad n = \dfrac{\log. 6}{\log. 1\cdot6667} = 3\cdot5.$

Putting x for the inverse power of the length by which the strength varies when the diameter is the same, it is found that two pillars which are each $2\frac{1}{2}$ inches in diameter, and whose lengths are 10 feet and $7\frac{1}{2}$ feet respectively, have their breaking-weights in the ratio of 1 to 1·598. And as the diameters are in the ratio of 1 to 1·3333, we have to determine the value of x from the equation—

$$1^x : 1\cdot3333^x = 1 : 1\cdot598,$$

or, $\qquad x = \dfrac{\log. 1\cdot598}{\log. 1\cdot3333} = 1\cdot63.$

The formula for the breaking-weight of a round pillar of cast-iron, whose length is at least thirty times its diameter, is

$$W = S \frac{d^{3\cdot5}}{l^{1\cdot63}}.$$

As the result of numerous experiments it was found that the average value of S was 42, when W is in tons, d in inches, l in feet, so that the formula for the breaking-weight in tons of a round solid pillar of cast-iron is

$$W = 42 \, \frac{d^{3\cdot5}}{l^{1\cdot6\overline{8}}} \qquad . \qquad . \qquad (82)$$

a.	$a^{1\cdot33}$.	a^2.	$a^{3\cdot5}$.	$a^{3\cdot35}$.	a^4.
3	6·0	9·0	46·8	49·4	81
3·25	6·8	10·6	60·5	65·6	111·6
3·5	7·7	12·3	80·2	85·4	150·1
3·75	8·6	14·1	102·1	109·1	197·8
4	9·6	16·0	128·0	137·2	256·0
4·25	10·6	18·1	158·3	170·1	326·3
4·5	11·6	20·3	193·3	213·3	410·1
4·75	12·7	22·6	233·6	252·5	509·1
5	13·8	25·0	279·0	303·6	625
5·25	14·9	27·6	331·6	360·2	759·7
5·5	16·1	30·3	390·2	424·9	915
5·75	17·3	33·1	455·9	497·5	1093
6	18·6	36·0	529·0	578·8	1296
6·25	19·8	39·1	610·4	668·9	1526
6·5	21·1	42·3	700·2	768·9	1785
6·75	22·5	45·6	799·0	879·1	2076
7	23·9	49·0	907·5	1000	2401
7·25	25·3	52·6	1026·1	1133	2763
7·5	26·7	56·3	1155·3	1278	3164
7·75	28·2	60·1	1295·8	1436	3607
8	29·7	64·0	1448·2	1607	4096
8·25	31·2	68·1	1612·8	1792	4632
8·5	32·7	72·3	1790·5	1993	5220
8·75	34·3	76·6	1981·7	2209	5862
9	35·9	81·0	2237·9	2441	6561
9·25	37·6	85·6	2407	2690	7321
9·5	39·2	90·3	2643	2957	8150
9·75	40·9	95·1	2894	3243	9037
10	42·7	100·0	3162	3548	10000
10·5	46·2	110·3	3751	4219	12161
11	49·8	121	4414	4977	14641
11·5	53·6	132	5158	5828	17497
12	57·4	144	5986	6778	20736
12·5	61·4	156	6905	7835	24422
13	65·4	169	7921	9005	28561
13·5	69·6	182	9040	10296	33225
14	73·8	196	10267	11716	38416
14·5	78·2	210	11610	13270	44216
15	82·6	225	13071	14967	50625

where the length l is expressed in feet, and the diameter d in inches.

For hollow pillars, in which d_i is the internal diameter,

$$W = 42 \frac{d^{3\cdot5} - d_1^{3\cdot5}}{l^{1\cdot63}} \qquad . \qquad . \quad (83)$$

or, the strength of the hollow pillar is the difference between that of a pillar whose diameter is d and one whose diameter is d_1.

As these powers of d and l can only be found by means of logarithms, the preceding table of the values of $d^{3\cdot5}$ and $l^{1\cdot63}$ will help to facilitate calculation.

As an example of the application of equation (82) let the diameter of a round column be 6 ins. and its length 15 feet. Then looking in the table in the column under $a^{3\cdot5}$ we find $6^{3\cdot5} = 529$; and looking in the column under $a^{1\cdot6}$ we find $15^{1\cdot63} = 82\cdot6$. Therefore for breaking weight by equation (82)

$$W = 42 \frac{529}{82\cdot6} = 269 \text{ tons.}$$

For an application of equation (83) let the outer diameter be 6 ins., the inner diameter $4\frac{1}{2}$ ins., and the length 15 feet. Then the table gives $4\cdot5^{3\cdot5} = 193\cdot3$, and the breaking weight is

$$W = 42 \frac{529 - 193\cdot3}{82\cdot6} = 171 \text{ tons.}$$

The strength of the two pillars is nearly in the ratio of 3 to 2; but the quantity of metal is in the proportion of $4\frac{1}{2}$ to 2, from which it will be seen that for a given amount of metal the hollow pillar is much stronger than the solid one.

When wrought-iron is used for pillars, they are generally made square; and Hodgkinson's experiments gave the strength as varying according to the power 3·55 of the diameter, and inversely as the *square* of the length; or the breaking weight in tons is for long square pillars,

$$W = 134 \frac{d^{3\cdot55}}{l^2} \qquad . \qquad . \qquad . \qquad (84)$$

Taking the first example given above, where $d = 6$ inches, and $l = 15$ feet, we find in the table under the column $a^{3\cdot55}$, that $6^{3\cdot55} = 578\cdot8$, and $l^2 = 225$. Therefore the breaking weight by equation (84) is—

$$W = 134 \frac{578\cdot8}{225} = 344\cdot7 \text{ tons.}$$

Experiments on long pillars of steel and timber show that the strength is according to the 4th power of the diameter, and inversely as the square of the length, following the theoretical law obtained above (**55**) and shown by equations (79) and (80).

For long square steel pillars the breaking weight in tons is given by the formula,

$$W = 100 \frac{d^4}{l^2} \qquad . \qquad . \qquad . \qquad (85)$$

For example, let the diameter be 6 inches and the length 15 feet; then by the table we find $6^4 = 1296$, and $15^2 = 225$; therefore the breaking weight is

$$W = 100 \frac{1296}{225} = 576 \text{ tons,}$$

which is more than half as much again as that of the wrought-iron pillar of the same dimensions.

For long square pillars of timber the breaking weight in tons is given by the formula

$$W = S \frac{d^4}{l^3} \qquad . \qquad . \qquad . \quad (86)$$

and for Dantzic oak, $S = 11$; for red deal, $S = 8$.

A pillar 6 ins. square and 15 ft. long, will have for its breaking-weight, $11 \times \frac{1296}{225} = 63\frac{1}{3}$ tons, if of Dantzic oak; and $8 \times \frac{1296}{225} = 46$ tons, if of red deal.

The *safe-load* for pillars of cast-iron may be taken at one-sixth of the *breaking-weight* where there are no vibrations or sudden shocks to be sustained; but where these have to be encountered the safe-load should be from one-eighth to one-tenth of the breaking weight. In the case of wrought-iron pillars the safe-load may be taken at one-fourth the breaking-weight where there is no vibration, and at one-sixth where the pillar is subjected to vibration. The safe-load for pillars of steel may be taken at from one-third to one-fifth of the breaking-weight, according as they are free from vibration or subjected thereto. In pillars of timber the safe-load should not exceed one-eighth or one-tenth of the breaking-weight.

57. SHORTER PILLARS.—By this term we mean pillars whose length is less than 30 times their diameter. Mr. Hodgkinson * found that the formula given above for long pillars had to be modified in the case of those whose length was less than 30 diameters, because the resistance of the material to crushing now

* *Phil. Trans.*, 1840 and 1857.

comes into play, so that it will be partly crushed before
it can break merely by bending. Since the resistance
to crushing is much greater in cast than in wrought
iron, we find that short pillars of cast-iron have a great
advantage over those of wrought-iron.

Suppose we put c for the force which would crush the
pillar without flexure, d for the utmost pressure the pillar
as flexible would bear to break it without being weakened
by crushing, b for the breaking-weight as calculated by
the formula for long pillars, y the true breaking-weight.
Then supposing a portion of the pillar equal to what
would just be crushed by pressure d to be taken away
(if such a thing were possible) we have $c-d =$ the
crushing strength of the remaining part, and $y-d =$
the weight actually laid upon it.   Therefore $\dfrac{y-d}{c-d}$ is
the part of this remaining portion of the pillar which
has to resist crushing, and

$$1 - \frac{y-d}{c-d} = \frac{c-y}{c-d}$$

is the part to sustain flexure.

But the strength of the pillar if rectangular may be
supposed to be reduced by reducing either its breadth,
or the calculated strength of the whole, to the degree
indicated by the fraction last obtained.   In the circular
pillar this mode is not strictly applicable, but we
obtain a near approximation to the breaking-weight y
by reducing the calculated value of b in that proportion.
Consequently,

$$c - d : c - y = b : y,$$

$$\text{or, } y = \frac{b \cdot c}{b + c - d}.$$

It is found that about one-fourth of the crushing-weight is the utmost a flexible pillar can be broken with without crushing the material ; so that we put $d = \frac{c}{4}$, and obtain the equation

$$y = \frac{b \cdot c}{b + \frac{3}{4} c}.$$

The rule then which we get for calculating the breaking-weight of short pillars, may be expressed as follows :—First calculate the breaking-weight by the equations (82, 83, 84, 85, 86), as if for long pillars, and call w the weight thus obtained. Let c be the crushing strength of very short pieces of the material per square inch, a the area of the transverse section of the pillar in inches ; then we find the true breaking-weight W from the equation

$$W = \frac{w \cdot c \cdot a}{w + \frac{3}{4} c \cdot a} \quad . \quad . \quad . \quad (87)$$

W, w and c must all be expressed in the same denomination, either tons or lbs. The values of c in lbs. are given in the table (27), or if w and W are in tons, we may put $c = 50$ for cast-iron, $c = 16$ for wrought-iron, $c = 60$ for steel, $c = 3$ for Dantzic fir, $c = 3\frac{1}{2}$ for English oak.

We will apply this rule to the first example given above (56), where the diameter of a solid cast-iron pillar is 6 inches, and suppose the length to be 12 feet ; then by equation (82) we find,

$$w = 42 \, \frac{529}{57 \cdot 4} = 387 \text{ tons;}$$

And by equation (87),

$$W = \frac{387 \times 50 \times \pi \cdot 3^2}{387 + \frac{3}{4} \times 50 \times \pi \cdot 3^2} = 392 \text{ tons,}$$

which is nearly half as much again as the strength of the same column 15 feet long, as calculated above.

To find the strength of a wrought-iron pillar 6 inches square and 12 feet long, we have by equation (84)—

$$w = 134 \frac{578 \cdot 8}{144} = 539 \text{ tons;}$$

and by equation (87)—

$$W = \frac{539 \times 16 \times 6^2}{539 + \frac{3}{4} \, 16 \times 6^2} = 320 \text{ tons,}$$

which is considerably *less* than the strength of the pillar 15 feet long, and also than that of the cast-iron pillar 12 feet long.   This difference is due to the smaller resistance to crushing that wrought-iron offers as compared with cast-iron.

Apply the rule to find the strength of a fir pillar 6″ square and 12 feet long.   Here we have by equation (86)—

$$w = 8 \frac{1296}{144} = 72 \text{ tons;}$$

and by equation (87)—

$$W = \frac{72 \times 3 \times 36}{72 + \frac{3}{4} \times 3 \times 36} = 51 \text{ tons,}$$

which is greater by 5 tons than that obtained for the same pillar 15 feet long.

58. TABLE OF STRENGTH OF PILLARS.—The following table gives the breaking-weight in tons of solid round cast-iron pillars of various lengths and diameters, from 10 diameters in length up to 40 diameters, calculated from the equations (82, 87). The crushing strength of cast-iron per square inch of section is taken at 50 tons in the shorter pillars.

Diam.	LENGTH IN FEET.									
	6 Ft.	7 Ft.	8 Ft.	9 Ft.	10 Ft.	12 Ft	15 Ft.	18 Ft.	20 Ft.	24 Ft.
Inches.	Tons.	Tons.	Tons.	Tons.	Tons.	Tons.	Tons.	Tons.	Tons.	Tons.
3	98½	84	65	55	46	...	...	...	...	...
3½	161	135	115	94	79	59	...	...	...	...
4	239	203	175	152	126	94	...	...	...	...
4½	337	290	250	219	192	141	98	...	...	...
5	455	394	343	302	267	213	142	105	...	...
6	752	660	584	520	465	392	269	200	168	...
7	1131	1011	907	817	737	607	461	343	289	...
8	...	1444	1270	1190	1083	906	707	570	461	342
9	...	...	1813	1664	1525	1300	1056	831	712	538
10	...	...	...	2193	2015	1735	1387	1130	1000	747

59. GORDON'S FORMULÆ.—The formulæ given by Hodgkinson being rather inconvenient for calculation, a simpler form has been adopted which has the advantage of being applicable to all pillars whose length exceeds 10 diameters. Suppose f to represent the breaking-weight per unit of the section, or the "unit-strength" of the pillar, and let r be the ratio of the length to its diameter, or least breadth; a and b are constants depending on the material and form of section of the pillar, a being a close approximation to the crushing strength per unit of section, or the crushing "unit-strength" of the material. Then Gordon's formula for pillars whose ends are flat and carefully bedded is,

$$f = \frac{a}{1 + b \cdot r^2} \qquad \cdot \quad \cdot \quad (88)$$

For round columns of cast iron, whether solid or hollow, $a = 36$ tons, $b = \frac{1}{400}$. For square cast-iron pillars and stanchions, $b = \frac{1}{500}$. The following table gives the value of f for round and square pillars of cast-iron ; then to find the actual breaking-weight of the pillar, multiply the value of f by the area of section in inches. For solid wrought-iron square pillars, $a = 16$ tons, $b = \frac{1}{3000}$.

r	10	12	15	20	25	30	35	40	
f	Tons. 29	Tons. 27	Tons. 23	Tons. 18	Tons. 14	Tons. 11	Tons. 9	Tons. 7	Round cast-iron.
f	30	28	25	20	16	13	10	9	Square cast-iron.
f	15½	15¼	15	14	13	12	11	10	Wrought-iron, square.

For pillars whose section is angle, tee, channel or cross section, in wrought-iron, $a = 19$ tons, $b = \frac{1}{900}$.

For example, take the case of a round cast-iron pillar 4 inches diameter and 10 feet long. Here we have $r = 30$; A, the area of section is $\pi \times 2^2 = 12\cdot5664$, and $f = 11$; the breaking-weight is $f \times A = 138\frac{1}{4}$ tons. By Hodgkinson's formula, equation (82), the breaking-weight is only 126 tons. Let the diameter be 6 inches and the length 10 feet ; then $r = 20, f = 18$, $A = \pi \times 3^2 = 28\cdot2744$; the breaking-weight is therefore $f \times A = 509$ tons; and by Hodgkinson's formula it is only 465 tons. Take 6″ for the diameter of pillar 15 feet long ; then $r = 30, f = 11$; and the breaking weight is $f \times A = 311$ tons ; the breaking-weight of the same pillar by Hodgkinson's formula is only 269 tons.

Take the case of a wrought-iron pillar 6 inches square and 15 feet long; here $r = 30$, $f = 12$, A $= 36$; therefore the breaking-weight is $f \times A = 432$ tons; the Hodgkinson formula, equation (84), gives the breaking-weight of such a pillar as only 344·7 tons. The same column 12 feet long, has $r = 20$, $f = 14$, and $f \times A = 504$ tons for the breaking-weight, which is half as much again as that found for the same pillar by Hodgkinson's formula. It appears then that Gordon's formula gives a much higher result both for cast and for wrought-iron pillars than is obtained by Hodgkinson's formulæ; and we are disposed to think that Hodgkinson's method of calculating the strength of pillars is the more accurate one.

60. Short Pillars.— Pillars whose length is less than 10 diameters are liable to have their material crushed before any bending can take place, and therefore their breaking-weight is found by multiplying A, the area of section in inches, by the quantity S for that particular material as given by the table (**27**), or the formula for *very short pillars* is

$$W = S \times A \qquad . \qquad . \qquad . \qquad . \qquad (89)$$

Thus, a very short square wrought-iron pillar whose section is 6^2 or 36 inches, will have a breaking-weight of $16 \times 36 = 576$ tons. And a cast-iron round pillar 6 inches diameter, will have a breaking-weight of $50 \times \pi \times 3^2 = 1414$ tons. A short pillar of fir 6 inches square has a breaking-weight of $3 \times 6^2 = 108$ tons; and one of oak has a breaking-weight of $3\frac{1}{2} \times 6^2 = 126$ tons.

CHAPTER VII

61. Roofs.—The covering of a building, whether of slate, tile, lead, zinc, or other material, has to be carried upon beams of wood or iron laid across from wall to wall, and generally placed at a considerable angle with the horizontal, called the "pitch" of the roof, so as to allow the rain-water to run off freely. The angle of *pitch* depends upon the kind of covering used, and for slates or tiles should not be less than 27°, which is the pitch adopted by Tredgold; but where lead or zinc is used the pitch may be as low as 4°. The weight of the covering per square foot also varies with the material employed, being about $1\frac{1}{2}$ lb. for zinc, exclusive of the boarding on which it is laid; for lead it is about 7 lbs., for slates from 7 to 11 lbs., and for tiles from 18 to 24 lbs. In addition however to the constant load of the covering there is the occasional load of snow to be sustained, and above all that of the wind pressure, which is greater for roofs of high than of low pitch. Tredgold puts down 40 lbs. per foot for the wind pressure and other occasional forces, including snow, for roofs of 27° pitch, and the scantlings of timbers given in his "Carpentry" are calculated for an allowance of 66 lbs. on the square foot, including the weight of timbers, boarding, &c.

Some authors however prefer to take the pressure of the wind separately from the dead load, as the wind can only act on one side of a roof at a time, and consequently subjects it to a racking force tending to strain the timbers unequally. We shall however in the following pages adopt Tredgold's method and take 66 lbs. as the load per foot which the timbers of a roof must be prepared to sustain with safety, when the pitch is about 27°, or the value of the *cotangent* of the angle of pitch is 2.

Roofs may be divided into two distinct classes, namely, *untrussed* roofs and *trussed* roofs. In the former we have simple *rafters* placed either from wall to wall, as in the " lean-to " roof (**62**), or the " span roof" (**63**) where the rafters are in pairs meeting in the middle and forming a *ridge*. These rafters are covered with battens or boards nailed upon them, on which the material of the covering is fixed. Such roofs are only adapted to small spans, or where the walls have a considerable thickness. Trussed roofs on the other hand may be used for spans of almost unlimited amount, and are formed of a number of beams fitted together in such a manner as to be a mutual help to one another, and to prevent any outward thrust being caused by the weight of the covering. In these roofs the " trusses " are placed about 10 or 12 feet apart, and the load of the roof presses on them by means of horizontal beams called *purlins*.

62. Lean-to Roof.—The simplest form of untrussed roof is that in which the rafters are supported by two walls, one at a higher level than the other, as AB

I

(fig. 49), the lower end B being laid upon a piece of timber called a "wall-plate," which lies horizontally

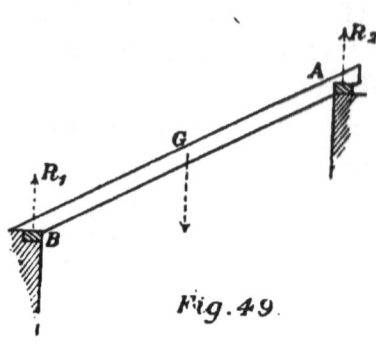

Fig. 49.

along the wall and distributes the load uniformly over its entire length. The upper end A is notched out and bedded upon a similar piece of timber. If W is the load upon the rafter, then half W is supported by the plate at A and at B. Putting R_1 for the reaction of the plate at B, R_2 for that at A, we have—

$$R_1 = \tfrac{1}{2} W = R_2 \qquad . \qquad . \qquad . \qquad . \quad (90)$$

Tredgold says* of a rafter placed in this manner, "cut so as to rest upon two level plates at A and B, the beam would have no tendency whatever to slide, notwithstanding its inclined position, and consequently it would have no horizontal thrust." "By cutting the rafters of a shed-roof, so that they may rest level upon the plates, the roof will have no tendency to push out the lower wall."

Let G be the centre of the rafter AB, w the load per foot length upon the rafter whose length is l; then we have $W = w \cdot l$, and the transverse stress P at G is, $P = w \cdot l \cdot \cos \theta$.   .   . (91)

where θ is the angle of pitch; and the load being uniformly distributed, the deflexion will be that produced by a load at the centre (48) equal to $\tfrac{5}{8} \times$

* *Tredgold's Carpentry,* art. 42, 7th edition, by E. W. Tarn.

$w \,.\, l \,.\, \cos.\, \theta$. Putting $\theta = 27°$, we have, $w \,.\, l \,.\, \cos.\, \theta =$ ·89 $w \,.\, l$; and the deflexion is ·55 $w \,.\, l$. Supposing the rafters to be placed one foot apart from middle to middle, and w to be 66 lbs., we have for the transverse stress, 36·3 l in lbs. Then from equation (73), if we put $\frac{17280}{E} = \frac{1}{100}$, we have by Tredgold's rule (**50**), for the scantling of the rafter when the deflexion is $\frac{1}{40}$th of the length—

$$b = \frac{36·3}{100} \times \frac{l^3}{d^3} = ·363 \frac{l^3}{d^3} \quad . \quad . \quad (92)$$

where b and d are its breadth and depth in inches, l its length in feet. The following table of scantlings of rafters for a pitch of 27° has been calculated from this rule.

Length of Rafter.	Breadth 2″.	Breadth 2½″.	Breadth 3″.
	Depth.	Depth.	Depth.
5 ft.	3 ins.	3 ins.	3 ins.
6	3½	3½	3
7	4	3¾	3½
8	4¾	4¼	4
9	5¼	4¾	4½
10	6	5¼	5
11	6¼	6	5½
12	6¾	6½	6
13	7½	7	6½
14	8	7½	7
15	8½	8	7½

If instead of notching the rafter AB on the plate at A, we cause it simply to *lean against* the face of the wall, we have a similar case to that shown by fig. 7, where a beam AB is supported at B and rests against a vertical wall at A, which supports it by

means of the horizontal reaction T at A.   The load W
acting vertically at C meets the direction of the reaction
T and of the reaction R of the wall B, in the point D;
since, as has been already demonstrated (4), the directions of three forces in equilibrium must meet in one
point.   If the line DE (fig. 7) represents W, then BE
will represent T, which will be the horizontal thrust of
the rafter at B; and we have (putting θ for the angle
of pitch)—

$$T : W = BE : DE$$
$$= BE : 2\ CE,$$

or, $T = W \dfrac{BE}{2\ CE} = \frac{1}{2}\ W\ .\ \text{cotan.}\ \theta$.   (93)

We can obtain the same result by taking moments
(8) of T at A and W at C about B (fig. 7); and in
equilibrium we have—

$$T \times DE = W \times BE;\ \text{or since}\ DE = 2\ CE,$$

$T = W \dfrac{BE}{2\ CE} = \frac{1}{2}\ W\ .\ \text{cotan.}\ \theta$, as in equation (93).

When $\theta = 30°$, we find $T = \cdot 866\ W$; for $\theta = 45°$,
$T = \frac{1}{2}\ W$; for $\theta = 60°$, $T = \cdot 29\ W$.

If we put as before, $W = w\ .\ l$, and $s = $ span of
roof, we have $s = l\ .\ \cos.\ \theta$, or, $l = \dfrac{s}{\cos.\ \theta}$; and equa-

tion (93) becomes—

$$T = \frac{wl}{2} \cdot \frac{\cos.\ \theta}{\sin.\ \theta} = \frac{w}{2} \cdot \frac{s}{\sin.\ \theta}$$.   .   (94)

Then since *sin. θ diminishes* with the angle, θ, we see that the thrust T *increases* with *decrease* of θ. When $\theta = 30°$, T $= w \cdot s$; when $\theta = 20°$, T $= 1\cdot46 \, w \cdot s$; for $\theta = 10°$, T $= 2\cdot88 \, w \cdot s$; and when $\theta = 5°$, T $= 5\cdot75 \, w \cdot s$.

63. SPAN ROOF.—This is the common form of roof, where the rafters are in pairs, and meet together in a "ridge" as AB, AC (fig. 50), the lower ends resting

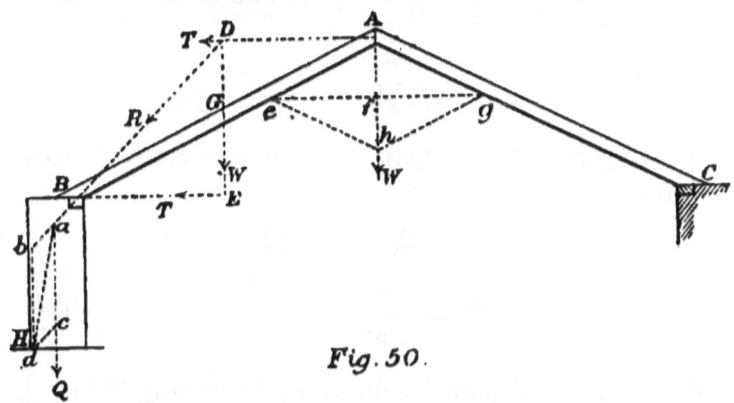

Fig. 50.

on walls at B and C. If we suppose the load, W, to be uniformly distributed over each rafter, w the load per square foot, and l the length of the rafter in feet; then W $= w \cdot l$ is the load on each, and $\frac{1}{2} w \cdot l$ the vertical pressure at B and C, $w \cdot l$ or W the vertical pressure at A. Let the vertical line Ah represent W at A; draw he parallel to AC, and hg parallel to AB; and draw the horizontal line efg cutting Ah in f. Then by the parallelogram of forces, Ae represents the compression P down the rafter AB, ef the horizontal thrust, T, which it produces at B. Then we have (θ being the angle of *pitch*)—

$$A f = \tfrac{1}{2} A h = A e \times \sin. \, \theta,$$

or, $$\tfrac{1}{2} W = P \cdot \sin \theta ;$$

therefore, $$P = \frac{W}{2 \sin \theta} \qquad . \qquad . \quad (95)$$

Also since $$ef = Ae \times \cos \theta,$$

therefore, $$T = P \times \cos \theta$$

$$= \frac{W}{2} \cdot \frac{\cos \theta}{\sin \theta} = \frac{W}{2} \times \cotan \theta \quad . \quad (96)$$

$$= \frac{w \cdot l}{2} \cdot \frac{\cos \theta}{\sin \theta},$$

or, putting s for the "span" of the roof from B to C, we have—

$$\tfrac{1}{2} s = l \cdot \cos \theta, \quad s = 2 l \cdot \cos \theta ;$$

therefore, $$T = \frac{w \cdot s}{4 \sin \theta} \qquad . \qquad . \quad (97)$$

The same result can be obtained by taking T for the mutual pressure of the rafters at A, W for the load on the rafter acting vertically at G; the directions of these two forces will meet in the point D, consequently the direction of the reaction at B must meet them in D. Then if DE represents W, EB will be the horizontal thrust T at B, BD the reaction R at B; or, by taking moments of T and W about B, we have—

$$T \times DE = W \times BE ;$$

or, $$T \times 2 EG = W \times BE ;$$

$$T = \frac{W}{2} \times \frac{BE}{EG} = \frac{W}{2} \cotan \theta$$

as obtained before, in equation (96); also we have—

$$R = \sqrt{W^2 + T^2} \qquad . \quad . \quad (98)$$

Produce DB to b, take the point a on the vertical line through the middle of the wall BH, and ab to represent on any convenient scale the value of R from equation (98) : and the vertical line ac the value of Q the weight of 1 foot length of the wall BH, on the same scale. Complete the parallelogram $abcd$, and the diagonal ad will represent the resultant of R and Q both in direction and magnitude (2). Produce ad to cut the base of the wall; then if the point where it cuts the base lies within the thickness of the wall it will be in a condition of stability under the action of the forces; but if it lies outside the point H the wall will be overturned.

We can determine the conditions of stability of the wall, or the thickness necessary to be given to it to insure its stability, by taking moments of the forces T, W, and Q, about the point H.

Let h be the height and t the thickness of the wall. Suppose W and Q to act vertically down the centre of the wall. Then we have for the moments in equilibrium about H—

$$T \times h = Q \times \frac{t}{2} + W \times \frac{t}{2}.$$

Let p be the weight of a cubic foot, then $Q = p \cdot h \cdot t$, and the equation becomes

$$T \times h = \frac{p}{2} h \cdot t^2 + \frac{W}{2} t \quad . \quad . \quad (99)$$

from which equation we get the value of t, which will just suffice for equilibrium, when the height h is given.

For example, suppose the pitch of a roof to be 27°, then, since cotan. $\theta = 2$, we have $T = W$ from equation (96). Suppose $l = 10$ feet, and $W = 660$ lbs. the pressure on one foot length of wall, $p = 120$ lbs., $h = 20$ feet. Then equation (99) becomes

$$660 \times 20 = 60 \times 20 \times t^2 + 330\,t,$$

or,
$$t^2 + \frac{11}{4}t - 11 = 0;$$

from which we find, $t = 2\frac{1}{8}$ feet, which is the least thickness the wall must have for *equilibrium*. The "span" of the roof in this case is very nearly 18 feet. For *stability* we put 2W for W in this equation, and we then find $t = 3\frac{1}{2}$ feet.

The value of T as given by equation (96) may also be found in another way, namely, by taking moments of the forces about the vertex A. These forces are as follows: the vertical reaction upwards at B which is equal to W; the load W acting at G; and the thrust T acting at B. Then we have for the moments of these forces in equilibrium about A—

$$W \times l \cdot \cos.\ \theta - W \cdot \frac{l}{2} \cdot \cos.\ \theta = T \times l \cdot \sin.\ \theta,$$

or,
$$T = \frac{W}{2} \cdot \text{cotan. } \theta, \text{ as before.}$$

64. Collar - Roof.—It is a common practice in making roofs of the form given in fig. 50, to place a horizontal beam across from rafter to rafter as DE (fig. 51). This is called a "collar" when placed between A and B, and a "tie-beam" if placed at the feet, B and C, of the rafters. Its usual place as a collar is about half way up the rafters, and it is then halved and spiked to the rafters. This collar may act either as a " tie " to prevent the *spreading* of the rafters at the feet, or else as a " strut " to prevent the rafters from bending in the middle when the scantling is in-

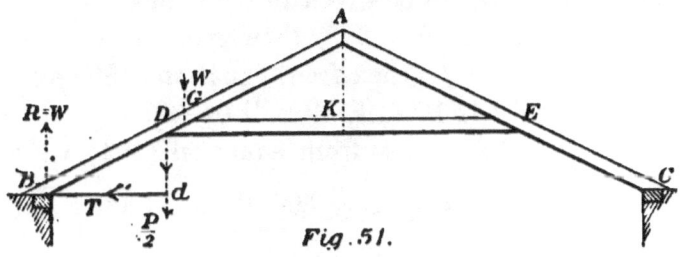

Fig .51.

sufficient to enable them to bear the load without " sagging." We shall first assume that the collar acts as a *tie*; and suppose $AB = l$, $AD = x$, $BD = l-x$.

We can find F the tensile stress in the collar, in the same way as we found T in the former case (**63**), namely, by taking moments about the vertex A of the reaction R ($= W$) at B, of the weight W at G the centre of rafter, and of the horizontal stress F in the collar DE; then the equation we get, for equilibrium, is,

$$W \times \tfrac{1}{2} \text{span} = W \times \tfrac{1}{4} \text{span} + F \times AK$$

or, $\quad W \times l \cdot \cos \theta = W \times \dfrac{l}{2} \cos \theta + F \times x \cdot \sin \theta.$

From this we obtain

$$F = \frac{W}{2} \cdot \frac{l}{x} \cdot \cot \theta \qquad . \quad (100)$$

Putting S for the safe tensile stress on a beam per

square inch of section we have for the safe stress in a beam whose transverse section is $b \cdot d$,

$$F = S \times b \cdot d \quad . \quad . \quad (101)$$

Equating these two values of F, we obtain the scantling necessary to be given to the collar.

For example, let $\theta = 27°$, then $cotan. \theta = 2$; and suppose that $x = \frac{1}{2} l$; then from equation (100) we get $F = 2 W$. If l is 10 feet, W will be 660 lbs., therefore $F = 1,320$ lbs. Also from equation (101) we get $b \cdot d = \dfrac{F}{S}$, and putting $S = 500$ lb. we have $b \cdot d = 2·63$ square inches for the transverse section of the collar, but this must be doubled at least in order to allow for the reduction by halving the ends on the rafters.

It is usual to make the collar of the same scantling as the rafters themselves, which in this case will be, by the foregoing table (**62**), either $2'' \times 6''$, $2\frac{1}{2}'' \times 5\frac{3}{4}''$, or $3'' \times 5''$.

When $x = l$, or the collar becomes a tie-beam, we have from equation (100), $F = \frac{1}{2} W \cdot cotan. \theta$, which corresponds with the value of T given in equation (93) representing the horizontal thrust at the foot of the rafter in the case where there is no tie.

In order to find the horizontal thrust T of the feet of the rafters on the wall at B and C in the case of a collar-roof, we shall assume that all the horizontal thrust of the upper part between D and A is counteracted by the tension F in the collar; so that we have only to consider the thrust of the part DB. Now the load P on DB is $w (l - x)$, half of which acts vertically at D and may be represented by the vertical line Dd, and causing the horizontal thrust T represented by Bd. Then, since $\dfrac{\mathrm{B}d}{\mathrm{D}d}$ = cotan. θ, we have,

$$\mathrm{T} : \tfrac{1}{2} \mathrm{P} = \mathrm{B}d : \mathrm{D}d,$$

or, $$\mathrm{T} = \frac{w}{2} (l - x) \cdot \text{cotan. } \theta \quad . \quad (102)$$

And when $x = 0$, this gives $\mathrm{T} = \dfrac{w}{2} l \cdot \text{cotan. } \theta$, which was the horizontal thrust of a pair of rafters as found by equation (93). When $x = l$, $\mathrm{T} = 0$, or there is no thrust on the walls; the tie-beam completely counteracting the thrust of the rafter feet. When $\theta = 27°$, and $x = \tfrac{1}{2}l$, we find $\mathrm{T} = \tfrac{1}{4} \mathrm{W}$.

We can now determine the thickness t to be given to a wall of given height h, when subjected to the thrust T. Let Q be the weight of the wall 1 foot long, p its weight per cubic foot; then $\mathrm{Q} = p \cdot h \cdot t$. Suppose the load W of one foot length of roofing to act vertically down the middle of the wall. Then by taking moments about its outer edge, we have—

$$\mathrm{T} \times h = \frac{p}{2} h \cdot t^2 + \mathrm{W} \times \frac{t}{2} \cdot \quad . \quad (103)$$

which gives a quadratic equation from which t can be found. For example, let $\mathrm{W} = 660$ lbs., $p = 120$ lbs., $h = 20$ feet, and the angle $\theta = 27°$; then $\mathrm{T} = \tfrac{1}{2} \mathrm{W} = 330$ lbs., and equation (103) becomes,

$$330 \times 20 = 60 \times 20 \, t^2 + 330 \, t;$$

or, $$t^2 + \frac{11}{40} t - \frac{11}{2} = 0;$$

therefore, $t = 2\frac{1}{4}$ feet for *equilibrium;* and taking 2T for T in the equation, we find $t = 3\frac{1}{8}$ feet for *stability.*

The value of T as given in equation (102) can also be determined by taking moments about D, of P acting vertically half way between D and B, and of T acting horizontally at B. Then,

$$T \times \mathrm{D}d = \mathrm{P} \times \frac{\mathrm{B}d}{2} ;$$

or, $T (l - x) \sin. \theta = \frac{\mathrm{P}}{2} (l - x) \cos. \theta ;$

therefore, $T = \frac{w}{2} (l - x)$ cotan. θ, as in equation (102).

When the feet of the rafters are kept from spreading by means of a tie-beam as BC (fig. 52), the rafters have a tendency to bend in the middle; and to prevent

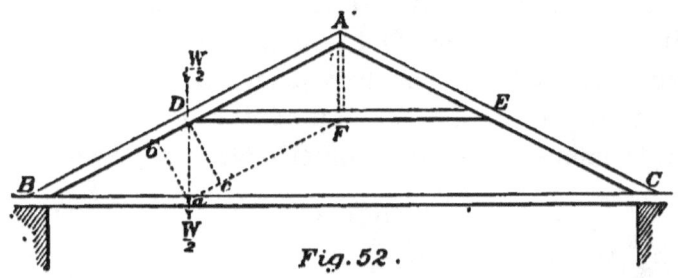

Fig. 52.

this from happening a collar is sometimes introduced at the middle, as DE. In this case the collar acts as a *strut* and is in compression, having to support the transverse stress arising from the weight of the covering acting at the middle point D of the rafter. If there were no collar or strut at DE, we should have the load W producing a deflexion at D, and as W acts vertically it can be resolved into two forces, namely, a transverse stress at right angles to the rafter, and a compressive one down the rafter. Since $\frac{1}{4}$ W is supported at A and at B, we may consider $\frac{1}{2}$ W to act at D, and represent it by the vertical line Da. Draw ac parallel to BD, Dc and ab perpendicular to BD; then Dc is the transverse stress at D, when there is no collar, Db the compression down DB. Now suppose the collar DE to be placed in position, then the pressure at D is supported by the two beams DE and DB; we must therefore resolve $\dfrac{W}{2}$ in the direction of the beams as was done at A in fig. 50, by drawing a F parallel to BD, and Ba parallel to DF; then DF represents the compression V in the collar when it acts as a strut.

Then, $V : \frac{1}{2} W = DF : Da = Ba : Da$;

and since $\dfrac{Ba}{Da} =$ cotan. θ, we have for the compression V in DE,

$$V = \frac{W}{2} \text{ cotan. } \theta \quad . \quad . \quad (104)$$

If we suppose the collar to have a square section, d being its breadth and depth in inches, l its length in feet; then by equation (86) we have

$$d^4 = \frac{V}{S} \times l^2 \quad . \quad . \quad . \quad (105)$$

where V is expressed in tons, and the value of S is 1 ton for fir, and $1\frac{1}{2}$ ton for oak. Putting $V = \frac{1}{3}$ ton, $S = 1$, $l = 9$ feet, we find by equation (105), $d = 2·28$

inches; so that a collar $3'' \times 3''$ will suffice for the purpose.

The following method for finding the horizontal thrust on the tie-beam when there is a collar acting as a strut, as in fig. 52, is employed by some writers on this subject. Suppose the collar to be placed anywhere between A and B, and that $AD = x$; then $w . x$ is the load on AD, $w (l - x)$ that on BD; and the vertical pressure at A from the load on AD and AE is $w . x$. If P is the pressure down AD, we have from equation (95)

$$P = \frac{w . x}{2 \sin. \theta}.$$

The vertical pressure at D by the load on BD is $\frac{1}{2} w \times (l - x)$, and by the load on AD it is $\frac{1}{2} w . x$; therefore the whole load at D is the sum of these two quantities, or $\frac{1}{2} w . l$. Let Q be the pressure down DB from the load at D, then by equation (95)—

$$Q = \frac{w . l}{2 \sin. \theta}.$$

The whole compression down the rafter is $P + Q$, and the horizontal thrust, T, at B and C is therefore,

$$T = (P + Q) \cos. \theta = \frac{1}{2} w (l + x) \cotan. \theta \quad . \quad (106)$$

which is the tensile stress produced upon the tie-beam. The value of T varies with that of x, or the distance of the collar from the ridge; being least when $x = 0$, when there is *no* collar, in which case $T = \frac{1}{2} W . \cotan. \theta$, as in equation (96). It appears therefore from equation (106) that the effect of the collar-strut is to *increase* the tension in the tie-beam. For when $x =$

$\frac{1}{2}$ l, θ being 27°, we have T = $\frac{3}{4}$ W ; but when there is
no strut we get T = W. In this latter case, however,
the bending of the rafters will tend to increase the
thrust on the tie-beam. So that we have the alterna-
tive of either increasing the scantling of the rafters, or
of introducing a collar-strut to prevent them from
bending.

A nother method of determining the horizontal thrust
at the rafter feet when the " collar " acts as a " strut,"
is as follows :—Take the moments of the forces about
D, namely, R = w . l acting vertically upwards at B,
the weight on BD, w $(l - x)$, acting vertically downwards at a point half-way between B and D, and the
thrust T acting horizontally at B. The moments of
these forces in equilibrium give the equation,

$$ w \cdot l \, (l - x) \, \cos. \; \theta \; - \; w \; (l - x) \times \tfrac{1}{2} \, (l - x) \cos. \; \theta $$
$$ = T \times (l - x) \sin. \; \theta $$

whence we get

$$ T = \frac{w}{2} (l + x) \cot. \; \theta, $$

which is the same as given by equation (106).

65. Hammer-beam Roof.—This is a form of roof
frequently found over buildings of the late Gothic
period, the largest in this country being that which
covers Westminster Hall, and having a span of 68 feet.
The hammer-beam roof is generally a modification of
the " collar " roof (**64**) the upper part usually having
a collar, as DH (fig. 53) placed across horizontally
between the rafters and at about one-third or one-half
of the distance from the vertex to the springing. The

feet of the rafters are framed into a horizontal piece called the "hammer-beam" which rests at one end on the wall and is supported at its outer end by a strut. Thus, in fig. 53, BF is the hammer-beam,

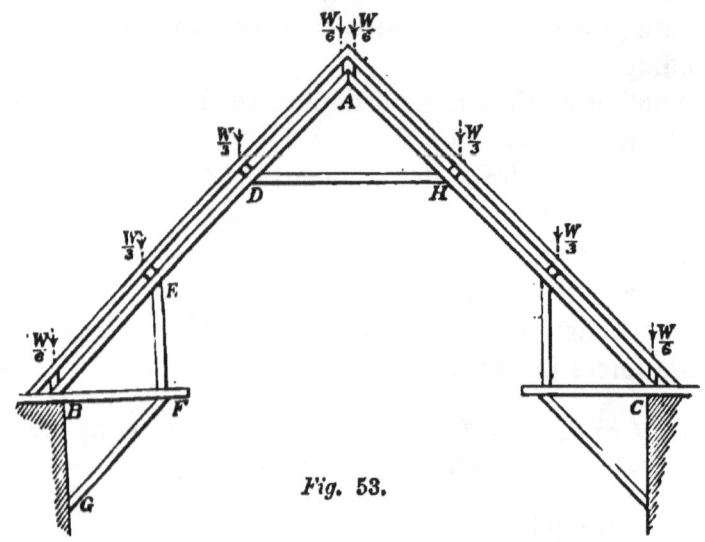

Fig. 53.

and GF the inclined strut supporting its outer end F and resting against the wall at G. The foot of the rafter AB is framed into the hammer-beam at B, and is further stiffened by a vertical strut EF resting on the outer end of the hammer-beam.

In order to investigate the action of the load in a roof of this kind, we will suppose that the weight of the roof covering is supported by purlins at A, D, E and B; and that $\frac{W}{6}$ is supported by each rafter at A, $\frac{W}{6}$ at B, $\frac{W}{3}$ at D and E. We will assume as before

(**64**) that the collar takes all the horizontal thrust arising from the pressure down AD; also that the strut EF takes the load $\frac{1}{3}$ W at E ; so that the only horizontal thrust, T, on the hammer-beam at B will arise from the action of the load $\frac{1}{3}$ W at D. Then if θ is the angle of pitch, we have, as in fig. 51,

$$T = \frac{W}{3} \times \text{cotan.} \, \theta \quad . \quad . \quad (107)$$

Since the beams BF and FG form a *bracket* (as shown by fig. 8), where the load $\frac{1}{3}$ W acts on the outer end F by means of the strut EF, there must be a tensile stress F, produced in the beam BF (**4**) acting from B towards F. Taking moments about G of the load $\frac{W}{3}$ at F and of the tension F in BF, we have in equilibrium,

$$F \times BG = \frac{W}{3} \times FB,$$

or,
$$F = \frac{W}{3} \times \frac{BF}{BG} . \quad . \quad (108)$$

In order that there may be no outward thrust at B, we must have F in equation (108) equal to T in equation (107), or it is requisite that—

$$\frac{BF}{BG} = \text{cotan.} \, \theta,$$

or that the strut FG shall be parallel to BE. If BG is *increased* then F becomes less than T, and if BG is *diminished*, F becomes greater than T, so that the greater the inclination of FG to the wall BG the greater

K

will be the inward stress upon BF. By means of
the strut FG the horizontal thrust is taken lower
down the wall, and therefore the tendency to overturn
it about its base is diminished. If BG represents the
load $\frac{1}{3}$ W acting at F, then BF will represent the
horizontal thrust F, on the wall at G, or we have—

$$F : \tfrac{1}{3} W = BF : BG,$$

or, $$F = \frac{W}{3} \times \frac{BF}{BG} = \frac{W}{3} \text{ cotan. BFG.}$$

Take, for example, a roof having a pitch of 45°, and
let AB = 18 ft., the span being 25$\frac{1}{2}$ ft. Taking the
load on the purlins at 660 lbs. per foot of rafter, the
principals being supposed 10 ft. apart, we have W =
11,800 lbs., $\frac{1}{3}$ W = 3960 lbs., tan. θ = cotan. θ = 1, BF =
BG when FG is parallel to BE, so that there is no thrust at
B. The horizontal thrust F at G will be $\frac{1}{3}$ W or 3960 lbs.

Suppose the wall to be 20 feet high, BG being equal
to BF, or to one-sixth of the span, or 4$\frac{1}{4}$ ft.; so that
the horizontal thrust, F or T, will act at a distance of
15$\frac{3}{4}$ feet from the base. Let t be the thickness of the
wall, which it is required to find; and suppose that the
load W of the roof acts vertically down the middle of
the wall. Let Q be the weight of 10 feet length of
wall, w its weight per cubic foot, say 120 lbs.; then if h
is the given height of the wall, we have Q = $w \cdot h \cdot t \cdot 10$
= 120 × 20 × 10t = 24,000 t. Taking moments
about the outer edge of the wall, of the forces W, Q,
and T, we have in equilibrium—

$$(Q + W) \frac{t}{2} = T \times 15\tfrac{3}{4} = \frac{W}{3} \cdot \frac{63}{4},$$

or, $\qquad 24000\,\dfrac{t^2}{2} + 11880\,\dfrac{t}{2} = 3960 \times \dfrac{63}{4}$,

or, $\qquad t^2 + \cdot5\cdot t - 5\cdot2 = 0$,

from which we find $t = 2\cdot04$ft. as the minimum thickness of the wall that will keep it from being overturned. In order that the wall may have sufficient *stability* its thickness should be 3 feet.

The tension in the hammer-beam in this case is 3,960lbs., which is also the compression down EF. The compression down the strut FG is $\dfrac{3960}{\sin.\ \theta} = \dfrac{3960}{\cdot707} = 5600$ lbs. The tension in the collar DH is $\frac{1}{6}$ W or 1980 lbs. and the compression down AD is $\dfrac{1980}{\cdot707} = 2800$ lbs.; that down DEB will be 8400 lbs.; the load $\frac{1}{6}$ W at E being taken by the strut EF. The transverse stress at right angles to the rafter at D is $\dfrac{W}{3}$ cos. $\theta = 2800$ lbs.; then by Tredgold's rule (50) we can determine the scantling of the rafter from the equation (72), namely, $2800 = 100\,\dfrac{b\,.\,d^2}{12^2}$.

Putting the breadth $b = 5$ inches, we find $d = 9\frac{1}{4}$ inches, in order that the rafter may not deflect more than three-tenths of an inch at D.

66. Trussed Roofs.—This term is applied to those roofs in which a number of beams are connected together in such a manner as to form what is called a "truss." These pieces of framing are placed on the walls at intervals of 10 or 12 feet, and support the covering of the roof by means of longitudinal

"purlins," as explained before (65). The whole
weight of the roof is therefore borne by the trusses at
the points where the purlins rest upon them. For
roofs of moderate span it will be sufficient to make the
truss of a pair of "principal rafters," as AB and AC

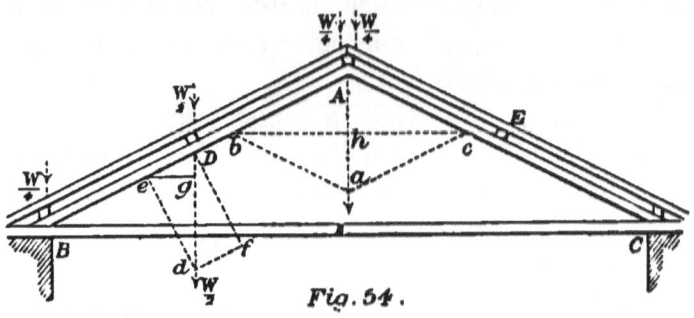

Fig. 54.

(fig. 54), which are prevented from thrusting out the
walls by having their feet B and C framed into a tie-
beam BC, to which they are secured by iron straps and
bolts. The purlins are fixed on the "principals" at
A, B, C, D and E; and if W is the load on each
side, we have ½ W sustained at each of the points A, D
and E, and ¼ W at B and at C. To find the stress on
the beams, take the vertical line Aa to represent ½ W,
draw ab parallel to AE, ac parallel to AD, and bc
horizontal; then the compression P down the rafter
AD is represented by the line Ab; and we have—

$$P \cdot \sin \theta = \frac{W}{4},$$ where θ is the angle of "pitch;"

or, $$P = \frac{W}{4 \sin \theta}.$$

Let the load ½ W at D be represented by the vertical
line Dd, draw df parallel to the rafter, Df at right

angles to the rafter, *ed* parallel to D*f*. Then the line
D*e* represents the compression Q down the rafter from
the load at D, and the line D*f* or *ed* the transverse
stress on the rafter at D. Then we have—

$$Q = \frac{W}{2} \text{ sin. } \theta,$$

and the total compression down DB is represented by
A*b* + D*e*, or by P + Q; and we have compression in
rafter DB

$$= P + Q = \frac{W}{4 \text{ sin. } \theta} + \frac{W}{2} \text{ sin. } \theta \qquad . \quad (109)$$

The tension, T, in the tie-beam is represented by
bh + *eg*, or by (A*b* + D*e*) cos. θ; therefore—

$$T = (P + Q) \text{ cos. } \theta$$

$$= \frac{W}{4} \text{ cotan. } \theta + \frac{W}{2} \text{ sin. } \theta \text{. cos. } \theta \qquad . \quad (110)$$

The transverse stress, V, at D is represented by D*f*
or D*d* . cos. θ, or we have—

$$V = \frac{W}{2} \text{. cos. } \theta \ . \qquad . \qquad . \quad (111)$$

For example, taking θ = 27°, let AB = 10 ft., W =
6600 lbs., then cotan. θ = 2, cos. θ = ·89, sin. θ = ·45,
½ W = 3300 lbs.

Then from equation (111) we find V = 2937 lbs.;
and by Tredgold's rule for the deflexion (**50**) we
determine the scantling of the rafter from the equation,

$$2937 = 100 \frac{b \cdot d^3}{L^3},$$

or, $\qquad\qquad\qquad b \cdot d^3 = 2937.$

If we assume the breadth, b, to be 5 inches, we get $8\frac{1}{3}$ ins. for the depth; or if $b = 4$ ins., then $d = 9$ ins. Such a scantling will be found more than sufficient to resist the compression given by equation (109).

The tension, T, in the tie-beam is found by equation (110),

$$T = \frac{W}{2} + \frac{W}{2} \times \cdot 4 = \cdot 7 \times 6600 = 4620 \text{ lbs.}$$

which is about the safe tensile stress on 9 square inches of fir, so that a tie-beam $3'' \times 3''$ would be sufficient for this purpose. In practice, however, the breadth of the tie-beam must equal that of the rafter which is framed into it. It must also be of sufficient depth so as not to bend by its own weight more than $\frac{1}{40}$ inch for each foot of length. Taking the scantling as $4'' \times 4''$, the weight per foot will be about 4 lbs., and the length being 18 feet, the total weight will be 72 lbs., and we must take $\frac{5}{8}$ of this as the load at the centre causing deflection, or 45 lbs.; we have then by Tredgold's rule (50),

$$w = 100 \frac{4 \times 4^3}{18^2} = 79 \text{ lbs.}$$

Therefore it is evident that a scantling of $4'' \times 4''$ is more than sufficient to resist any stress that the tie-beam has to bear. In such a case, where the tie-beam has no load to carry, an iron rod or bar may be used in its place.

When the tie-beam has to carry the weight of a ceiling the load per foot must be taken at 120 lbs., if the trusses are 10 feet apart; in which case we have $w = 120 \times 18 = 2160$ lbs. Then we find the scantling from the equation—

$$\frac{5}{8} \times 2160 = 100 \frac{b \cdot d^3}{324},$$

which gives $b \cdot d^3 = 4374$; and if we put $b = 5$, then we find $d = 9\frac{1}{2}$ ins.

When there is a floor to be carried by the tie-beam, as well as a ceiling, the beam must be made strong enough to bear a distributed load of 1200 lbs. per foot length; or $b \cdot d^3$ will be ten times as great as when there is only a ceiling.

67. KING-POST ROOF.—When the tie-beam of a trussed roof has a considerable load to carry, it is advisable to support it in the middle by means of a piece of timber called a "king-post," as AF (fig. 55).

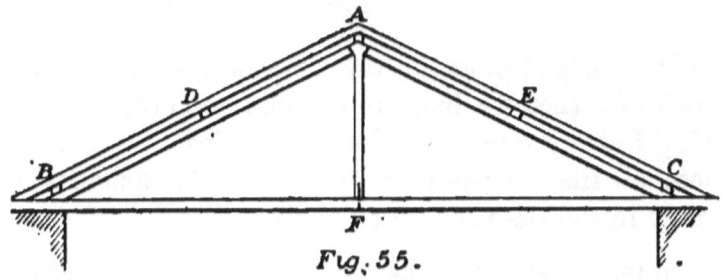

Fig. 55.

The heads of the principal rafters are framed into the head of the post at A, so as to hold up the post, and the tie-beam is held up in the middle by means of a strap passed round it and bolted to the post. By this means half the load on the tie-beam is carried by the rafters, which causes an increase in the compression down the rafters, and a corresponding increase in the tension in the tie-beam. This beam will then be divided into two equal parts, each of which has half the load to carry, and is half the length of the whole beam; consequently the moment of stress is reduced to one-fourth, and the strength of the tie-beam to bear a transverse strain need only be one-fourth of what it was when there was no king-post. A light iron rod may be used for a king-post instead of wood, when it has only a tensile stress to bear. If w is the load on the whole length of the tie-beam, then we can determine the scantling necessary according to Tredgold's rule (**50**) by the equation,

$$\frac{5}{8} \times \frac{w}{2} = 100 \cdot \frac{b \cdot d^3}{\left(\frac{L}{2}\right)^2},$$

or,

$$b \cdot d^3 = \frac{w}{1280} \times L^2 \quad . \quad . \quad (112)$$

If $w = 2160$ lbs. as in the last example (**66**), where $L = 18$ ft., then we find from equation (112), $b \cdot d^3 = 547$. Putting $b = 4$, we find $d = 5\frac{1}{2}$ ins.; so that the effect of the king-post is to reduce the quantity of timber in the tie-beam by one-half.

The tensile stress in the king-post will be $\frac{w}{2}$, and the sectional area can be found from the equation—

$$\frac{w}{2} = S \cdot b \cdot d, \text{ or } b \cdot d = \frac{w}{2 \cdot S} \quad . \quad (113)$$

where S is the coefficient of safety. Taking $S = 500$ lbs. per square inch for fir, we have—

$$b \cdot d = \frac{2160}{1000} = 2 \cdot 16 \text{ inch}$$

in the foregoing example.

As however the rafters are framed into the head of the king-post, its breadth must be the same as that of the rafters. If an iron rod is used for a king-post, we can put S = 4 tons, or about 10 times as much as when fir is used; so that a rod of iron ½ inch in diameter will serve the purpose.

The compression R down each rafter from the stress on the king-post is

$$R = \frac{w}{4 \sin. \theta},$$

which must be added to P + Q in equation (109) to get the total compression down the rafter DB, namely,

$$P + Q + R = \frac{W}{4 \sin. \theta} + \frac{W}{2} \sin. \theta + \frac{w}{4 \sin. \theta}$$

$$= \frac{W + w}{4 \sin. \theta} + \frac{W}{2} \sin. \theta \qquad . \quad (114)$$

The tensile stress in the tie-beam is now increased by R. cos. $\theta = \frac{w}{4}$ cotan. θ; so that the total tension, T, in the tie-beam is,

$$T = \frac{W + w}{4} \text{cotan.} \theta + \frac{W}{2} \sin. \theta . \cos. \theta . \quad (115)$$

which in the foregoing example is 5700 lbs., or the safe tensile stress on 12 square inches of fir; so that the scantling given above is amply sufficient.

The final step to be taken in completing the king-post truss is to insert a strut between the foot of the king-post and the middle points D and E (fig. 56) of the rafters, by which means the greater part of the load at those points is taken off the rafters, and we are enabled to reduce their scantling in the same way as we reduced that of the tie-beam by inserting the king-post. There will then be no transverse stress on the rafter to be taken into consideration, but only the compression down it, which will be increased by the

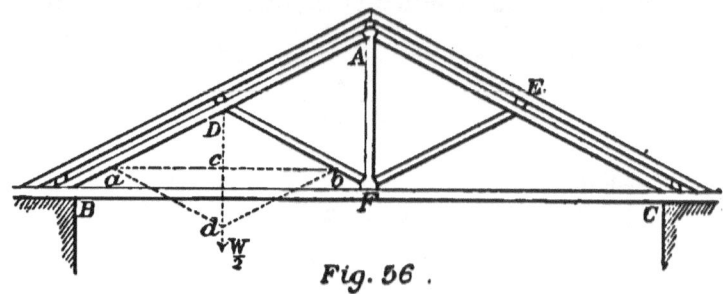

Fig. 56 .

strut conveying the load at D and E to the king-post, and thence to the head of the rafters at A.

When there was no strut at D, as in fig. 54, the load $\frac{W}{2}$ at D was resolved parallel and perpendicular to the rafter; but the insertion of the strut alters the directions of the forces, and we must treat the beams DB and DF as if they were a pair of rafters, and resolve $\frac{W}{2}$ down each, as we did in fig. 54. Taking Dd to represent $\frac{1}{2}$ W, and drawing da parallel to DF, db parallel to BD, we have Da representing the compression down DB, Db that down the strut. Drawing the horizontal diagonal ab, we have $\frac{1}{2}$ ab or ac representing the horizontal thrust on the foot of the rafter and consequently the additional tension in the tie-beam. The line Dc will represent the additional stress in the king-post arising from the pressure down one strut, or Dd will be the additional stress in the king-post arising from the two struts; so that an addition of $\frac{1}{2}$ W must be made to the stress in the king-post. The total stress in the king-post is therefore $\dfrac{W}{2} + \dfrac{w}{2}.$

Taking Q for the compression down the strut, we have—

$$Q \cdot \sin. \theta = \tfrac{1}{2}\frac{W}{2} = \frac{W}{4},$$

or, $$Q = \frac{W}{4 \cdot \sin. \theta} \qquad . \qquad . \quad (116)$$

This is also the corresponding compression down the rafter DB due to the load at D. If $W = 6600$ lbs., $\theta = 27°$, sin. $\theta = \cdot454$, we find $Q = 3633$ lbs. Applying the equation (105), and putting $Q = 1\frac{2}{3}$ tons, we have for fir, $d^4 = \dfrac{1\frac{2}{3}}{1} \times 5^2 = 42$, when the length of the strut is 5 ft., or, $d = 2\cdot546$ inches.

The additional tension in the king-post caused by the two struts, is 2 Q sin. $\theta = \dfrac{W}{2}$. Putting F for the total tensile stress in the king-post, we have—

$$F = \frac{W + w}{2} \qquad . \qquad . \quad (117)$$

In the foregoing example $W = 6600$ lbs., $w = 2160$ lbs., therefore $F = 4380$ lbs. To find the minimum scantling of the king-post we must put, as in equation (113)—

$$F = S \cdot b \cdot d.$$

Therefore, $\qquad b \cdot d = \dfrac{W + w}{2\,S}$.   . (118)

Taking $S = 500$ for fir, we get $b \cdot d = 8\cdot8$ square inches, and as the breadth must be 4 inches, the scantling will be $4'' \times 2\frac{1}{4}''$.

Putting R for the compression down each rafter caused by the tensile stress in the king-post, we have—

$$R \cdot \sin \theta = \dfrac{F}{2}\,;$$

or, $\qquad\qquad R = \dfrac{W + w}{4 \sin \theta}.$

The total compression down the rafter A D is therefore—

$$P + R = \dfrac{W}{4 \sin \theta} + \dfrac{W + w}{4 \sin \theta}$$

$$= \dfrac{2\,W + w}{4 \sin \theta} \qquad . \qquad . \quad (119)$$

In the above example we find $P + R = 8458$ lbs.

To find the compression down DB, we have to add Q as found by equation (116) to $P + R$, so that the pressure down D B is—

$$P + Q + R = \dfrac{3\,W + w}{4 \sin \theta} \qquad . \quad (120)$$

In the above example we find $P + Q + R = 12,100$ lbs. The tension, T, in the tie-beam is—

$$T = (P + Q + R)\cos \theta$$

$$= \dfrac{3\,W + w}{4} \times \operatorname{cotan} \theta \ . \qquad . \quad (121)$$

In the foregoing example we have, since cotan. $\theta = 2$ nearly, $T = 10{,}781$ lbs. To find the scantling of the tie-beam to resist this tensile stress, we must put

$$T = S \cdot b \cdot d,$$

as in equation (113); and in the above example this will give—

$$b \cdot d = \frac{10781}{500} = 21\tfrac{1}{2} \text{ square ins.}$$

so that a scantling of $5'' \times 4\tfrac{1}{2}''$ will be sufficient.

To find whether the scantling $5'' \times 4\tfrac{1}{2}''$ is sufficient for the rafter, we must consider the part BD as a pillar 5 feet long with a pressure of 12,100 lbs. or 5·4 tons; and that it will bear one-fourth more than a pillar 4 ins. square. The breaking weight of a fir pillar 4 ins. square and 5 feet long can be found by equations (86, 87) to be equal to 39 tons; and the strength of one $5'' \times 4''$ will be nearly 49 tons, which is about 9 times the actual pressure; so that this scantling may be considered as sufficient.

We have now seen how by the addition of each piece of the truss the stress on the other parts is changed, and also the kind of stress to which each part is subjected. There are other stresses which have to be borne besides those we have been considering, such as the compression of the head of the king-post by the heads of the principal rafters being framed into it, and the compression on the foot of the king-post by the heels of the struts. There is also the compression on the tie-beam by the feet of the rafters.

Summing up all the results obtained in the foregoing

example, where $\theta = 27°$, and the rafter is 10 ft. long, the span being 18 feet, and the tie-beam carrying the weight of a ceiling only, we have—

Compression down rafter $AD = P + R = 8458$ lbs.
      do.   do. $DB = P + Q + R = 12,100$ lbs.
      do.   do.   strut $DF = Q = 3633$ lbs.
Tension in the tie-beam $BC = T = 10,781$ lbs.
      do.   king-post $AF = F = 4380$ lbs.

From these we find the minimum scantling of the beams to be as follows :—

Tie-beam	$5'' \times 4\frac{1}{2}''$
Principal rafters . . .	$4'' \times 5''$
do. struts . . .	$3'' \times 2\frac{1}{4}''$
do. king-post . . .	$4'' \times 2\frac{1}{4}''$

The scantling of the purlin at D, which is placed square with the rafter, can be determined by Tredgold's rule (**50**) the load $\dfrac{W}{2}$ being uniformly distributed along it, and its bearing being 10 feet. The actual transverse stress is therefore $\dfrac{W}{2} \times \cos. \theta$, and for deflexion under a distributed load we take five-eighths of this for the load at the middle. Consequently the equation from which to determine its scantling is—

$$\frac{5}{8} \times \frac{W}{2} \times \cos. \theta = 100 \frac{b \cdot d^3}{10^2}$$

from which we get $b \cdot d^3 = 1856$; and if we put $b = 4$ inches we find $d = 8$ inches, or the scantling of the purlin is $4'' \times 8''$.

For the purlin at A, we have—

$$\frac{5}{8} \times \frac{W}{2} = 100 \frac{b \cdot d^3}{10^2}$$

or, $\qquad b \cdot d^3 = 2003, \cdot$

and the same scantling, $4'' \times 8''$, will do in this case.

68. STRESS-DIAGRAM.—It has been previously shown (**5**) & (**7**) that when a *jointed polygonal frame* is in equilibrium from forces acting at the joints, the relative magnitude of the stresses they produce in the directions of the several bars can be found by drawing a "stress-diagram" whose sides are respectively parallel to the directions of the impressed forces, and from the vertices of the new polygon drawing lines parallel to the sides of the frame. This method is now generally

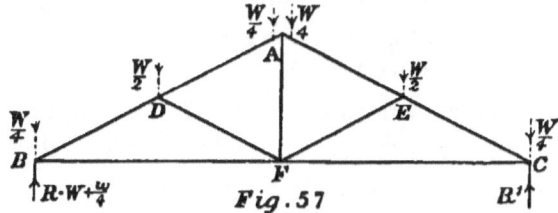

Fig. 57

employed for determining the stress in every bar of a truss, and we shall proceed to show its application to the king-post roof which we have just been considering. First draw an outline of the truss, as fig. 57, and mark the loads supposed to act at each of the joints. Thus at A we have $\frac{1}{4}$ W supported by the end of each rafter, so that $\frac{1}{2}$ W is the total pressure at A; at D and E we have $\frac{1}{2}$ W; and at B and C we have $\frac{1}{4}$ W. All these forces act vertically *downwards*. At B and C there is also the reaction R acting vertically *upwards*, and if x is the load on the tie-beam, we have $R = W + \dfrac{w}{4}$.

In the example given above we have $W = 6600$, $w = 2160$, $\frac{1}{2}W = 3300$, $\frac{1}{4}W = 1650$, $\frac{1}{4}w = 540$, $R = 7140$.

To form the stress-diagram, draw a vertical line $b\,X\,g$ (fig. 58) to represent on any convenient scale the value of $2\,W$, and bisect this line in X. Take Xd, Xe, gf and ba each to represent $\frac{1}{4}$ W; gk and bl each to represent R. Draw horizontal lines through k and l, as ln and km. Then draw an parallel to the rafter BD; fm parallel to the rafter CE, meeting ln and km in the points n and m. Draw dp parallel to the rafter AD, meeting np parallel to the strut DF. Draw eq parallel to the rafter AE, meeting mq parallel to the strut EF. Then, if the figure is correctly drawn, the line joining p and q, will be vertical.

To determine whether the stress in any member of the frame is compressive or tensile, we have only to consider that each of the points A, B, C, &c. (fig. 57) is a centre, to or from which the forces act; and by transferring the forces in the *stress-diagram* (fig. 58) to the members in the *frame-diagram* (fig. 57) with which they are parallel, the forces which act *towards* the central point are *compressive* and those which act *from* it are tensile.

For the forces in equilibrium at the point B (fig. 57) we have in fig. 58 ab representing $\frac{1}{4}$ W, bl the reaction R, ln the tension in the tie-beam BF, and na the compression down the rafter DB, forming a closed polygon; so that ab and bl being known we can find

ln and *na* by measurement.   Starting from *a* towards
b, and from *b* towards *l*, from *t* towards *n* and from *n*
back to *a* again, we see that *ln* acts *away* from A and
is therefore a pulling force, or BF is in tension; while
na acts *towards* B, or is a pushing force, and BD is in
compression.   In like manner for the other side of the
truss we have *fy, gk, km, mf,* representing the polygon
of forces at C.

Proceeding to the point D (fig. 57) we have the forces
at D represented in fig. 58 by *da* for $\dfrac{W}{2}$ at D, *an* the

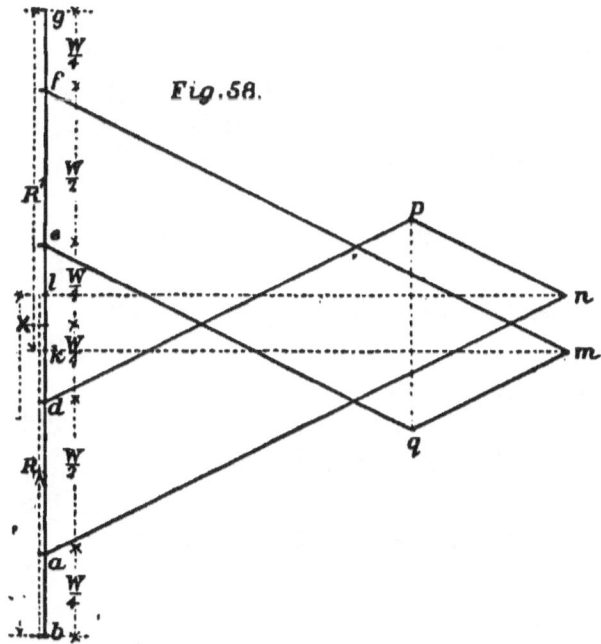

Fig. 58.

compression in BD acting *towards* D as before found;
np the compression in the strut FD acting *towards*
D; *pd* the compression in AD acting *towards* D, and

L

forming a closed polygon. From this we obtain by measurement the compression *pn* in the strut DF, and the compression *pd* in the rafter AD. And for the other side of the truss the lines *ef, fm, mq, qe,* represent the polygon of forces at E; *mq,* being the compression in EF, and *qe,* that in AE.

Proceeding to the point A (fig. 57) we have the forces represented in fig. 58 by X*d, dp, pq, qe, e*X, where X*d* and *e*X are each ¼ W, acting at A; and we now determine the tension in the king-post which is represented by the line *pq* and acts *from* A, all the other forces acting *towards* A.

In this manner we obtain for a roof whose pitch is 27°, the following results from measuring the lines in the stress-diagram, where the rafters are 10 ft. long, W being taken at 6600 lbs., and *w* = 2160 lbs.

Compression in rafter AD, represented by
line *pd* . . . . . . 8,534 lbs.
Compression in rafter DB, represented by
line *na* . . . . . . 12,200 lbs.
Compression in strut DF, represented by
line *np* . . . . . . 3,667 lbs.
Tension in the tie-beam BF, represented by
line *ln* . . . . . . 10,980 lbs.
Tension in the king-post AF, represented by
line *pq* . . . . . . 4,380 lbs.

If we compare these figures with those calculated in the last article (**67**) we see that the stresses obtained by this method agree pretty closely with those obtained by the analytical method.

The *strong* lines in the stress-diagram indicate forces producing *compression*, and the dotted lines those producing tension.

This method has the merit of simplicity and of requiring no algebra or trigonometry in its application, and is therefore largely used by draughtsmen in finding the stresses in the parts of complicated trusses.

69. QUEEN-POST ROOF.—The king-post roof (**67**) is only suitable for spans not exceeding 30 feet, and for longer spans the tie-beam is held up in *two* places by means of " queen-posts," as DH and FK (fig. 59).

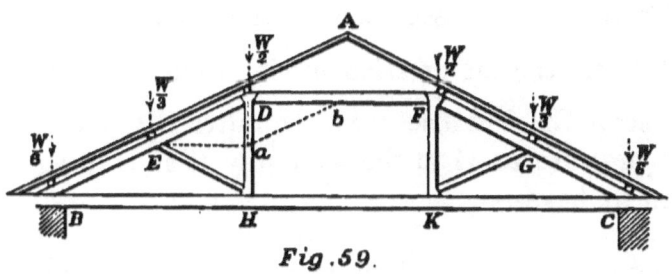

Fig .59.

In this truss the heads of the rafters do not meet, but are framed into the heads of the queen-posts, which are kept apart by means of the " straining-piece " DF. The common rafters which carry the roof-covering meet in a " ridge-piece " at A, and rest upon purlins at D, E, B, F, G and C. If W is the load on each side, then $\frac{1}{2}$ W is supported at D and at F, $\frac{1}{3}$ W at E and at G, and $\frac{1}{6}$ W at B and at C. If w is the weight on the tie-beam, as of a ceiling, then $\frac{1}{3}$ w is supported at H and at K, and $\frac{1}{6}$ W at B and C.

Let the vertical line Da represent $\frac{1}{2}$ W, and draw ab parallel to DE, and Ea horizontal ; then ab represents the compression down DE, Db that in DF. Then since $\dfrac{Db}{Da} = \text{cotan. } \theta$, we have for the compression in DF caused by the load $\frac{1}{2}$ W at D—

$$\frac{W}{2} \times \text{cotan. } \theta \, ;$$

which is equal to W when $\theta = 27°$. And the compression P down the rafter is—

$$P = \frac{W}{2 \sin. \theta} \qquad . \qquad . \quad (122)$$

To find the stresses caused at E by the load $\frac{1}{3}$ W; put Q for the compression arising from $\dfrac{W}{3}$ at E down the strut EH and also down the rafter EB, and resolve the pressure at E in the same way as at D in fig. 56 ; then we have—

$$Q = \frac{W}{6 \sin. \theta} \qquad . \qquad . \quad (123)$$

The tension which the pressure of the strut produces in the queen-post is—

$$Q \sin. \theta = \frac{W}{6},$$

so that the total tension, F, in the queen-post is—

$$F = \frac{W + 2w}{6} \qquad . \qquad . \quad (124)$$

The tension in the queen-post produces a compression in the rafters and straining-piece; the compression along the straining-piece is $F \times \text{cotan. } \theta$, so that the total pressure (which we will call V) in the straining-piece is—

$$V = \left(\frac{W}{2} + F\right) \text{cotan.}\ \theta$$

$$= \frac{4\ W + 2\ w}{6} \times \cot.\ \theta \quad . \quad (125)$$

Putting R for the pressure down the rafter produced by the tension F in the queen-post, we have—

$$R = \frac{F}{\sin.\ \theta} = \frac{W + 2\ w}{6 \sin.\ \theta}.$$

The total pressure down DE is therefore—

$$P + R = \frac{W + 2\ F}{2 \sin.\ \theta} = \frac{4\ W + 2\ w}{6 \sin.\ \theta} \quad . \quad (126)$$

The total pressure down EB is --

$$P + Q + R = \frac{W}{6 \sin.\ \theta} + \frac{4\ W + 2\ w}{6 \sin.\ \theta}$$

$$= \frac{5\ W + 2\ w}{6 \sin.\ \theta} \quad . \quad . \quad (127)$$

The tension, T, in the tie-beam is—

$$T = (P + Q + R)\ \cos.\ \theta$$

$$= \frac{5\ W + 2\ w}{6} \times \text{cotan.}\ \theta \quad . \quad (128)$$

As an example of the application of these equations, let the length of the common rafter be 15 feet, the angle $\theta = 27°$, cos. $\theta = ·891$, cot. $\theta = 2$, sin. $\theta = ·454$, the load $W = 9900$ lbs. Then the span is 27 feet, and the load on the tie-beam from a ceiling is, $w = 27 \times 120 = 3240$ lbs.

Then the following results are obtained from the foregoing equations :—

Compression, V, in straining-piece, DF,
    by equation (125) .   .   .   .   15,360 lbs.
Compression, P + Q, down rafter, DE,
    by equation (126) .   .   .   .   16,916 lbs.
Compression, P + Q + R, down rafter,
    EB, by equation (127) .   .   .   20,551 lbs.
Compression, Q, in strut, EH, by
    equation (123)   .   .   .   .   3,634 lbs.
Tension, T, in tie-beam, BH, by equation
    (128)   .   .   .   .   .   .   18,311 lbs.
Tension, F, in queen-post, DH, by equa-
    tion (124)   .   .   .   .   .   2,730 lbs.

To find the strength of the rafter, we consider the part BE as a pillar, and calculate by equations (86, 87) its ultimate strength, when 4 inches square, to be $33\frac{1}{2}$ tons; or if taken as 4″ × 8″, the ultimate strength will be about 67 tons, or 7 times the compression given by equation (127). That of the straining-piece can be found by first taking it as a pillar 4 inches square and 9 feet long, of which, by equations (86, 87) the ultimate strength is 20 tons, or nearly 3 times the pressure found by equation (125); it will therefore be advisable to make the scantling of this beam not less than 4″ × 8″. The size of the strut can be determined by the equation (105), $d^4 = Q . l^2$, l being 5 feet; which gives $d = 2\cdot515$, or we can put $2\frac{1}{2}$″ × 3″ for its scantling.

The sectional area of the tie-beam can be found by dividing 18,311 by 500, which gives $36\frac{2}{3}$ square inches, so that a scantling of 4″ × $9\frac{1}{4}$″ will suffice in this case.

Dividing 2730 by 500 we get 5½ square inches for the section of the king-post at its narrowest part, or a scantling of 4″ × 2″ will be ample. All the timbers are supposed to be of fir.

70. SCANTLING OF ROOF TIMBERS.—The following tables give the scantling that will generally be found sufficient for fir timbers in roofs of 27° pitch; the tie-beam being supposed to carry the weight of a ceiling only, and the trusses placed 10 feet apart.

King-post Roofs.

Span.	Tie-Beam.	King-post.	Principal Rafter.	Struts.	Purlins.
Feet.	Ins.	Ins.	Ins.	Ins.	Ins.
20	7 × 4	4 × 3	4 × 5	3½ × 2½	8 × 4
24	8 × 4	4 × 3½	5 × 4½	4 × 3	8 × 5
30	8 × 5	5 × 4	6 × 5	4¾ × 3	9 × 5

Queen-post Roofs.

Span.	Tie-Beam.	Queen-post.	Principal Rafter.	Straining Beam.	Struts.	Purlins.
Feet.	Ins.	Ins.	Ins.	Ins.	Ins.	Ins.
32	9 × 4	4 × 2	8 × 4	8 × 4	3½ × 2¼	8 × 4
36	9 × 5	5 × 2	8 × 5	8 × 5	4¼ × 2½	8 × 5
40	10 × 5	5 × 3	9 × 5	9 × 5	4½ × 2½	9 × 5
46	11 × 5½	5½ × 3½	9 × 5½	9 × 5½	4¾ × 3	9½ × 5

71. IRON ROOFS.—By the application of iron in the construction of roof-trusses, we are enabled to adopt various forms of truss that it would be impossible to use with timber. The simplest form of iron truss is that shown by ABDC (fig. 60), which consists of principal rafters, AB, AC, and sloping tie-rods BD, CD, held up by a king-bolt AD. The roof covering is supported by purlins at A, B and C; and if W is the load on each side, we shall have ½ W supported at B

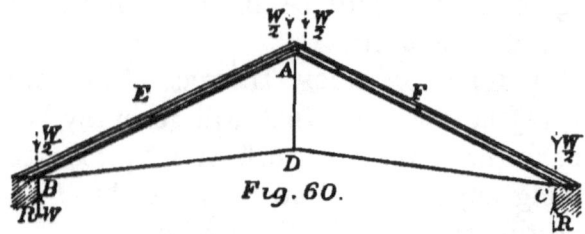

Fig. 60.

and at C, and also on each side of the vertex A, or W is the total load at A.   The reaction R at B and C is equal to W.

The stress-diagram for this truss is very simple; draw bd (fig. 61) vertical and representing on any scale the load $2W$; bisect bd at X, and take ab, aX, Xc, and cd, each to represent ½ W. Then bX represents the reaction R at B. From a draw al parallel to AB, meeting Xl parallel to BD.

Also from c draw cm, parallel to AC, meeting Xm parallel to CD.   Join lm, which line ought to be vertical if the figure is correctly drawn.

For the forces in equilibrium at B, we have the polygon *ab*, *bX*, *Xl*, *la*, where *Xl*, acting *from* B, represents the tension in BD; *la*, acting *towards* B, the compression in AB (**68**). So also *Xm* is the tension in CD, *mc* the compression in AC. For the forces in equilibrium at D, we have the triangle *Xm*, *ml*, *lX*, all acting *from* D and therefore being tensile, from which we get *ml*, the tension in the king-bolt AD. The strong lines indicate the forces producing compression, and the dotted lines those producing tension (**68**). Suppose for example that $\frac{1}{2}$ W = 1 ton, then we find by measuring *al*, that the compression in the rafter is 3 tons, the tension in each tie-rod BD or CD is $2\frac{3}{4}$ tons, the tension in the king-bolt $\frac{3}{5}$ ton.

If there was a purlin half way down each rafter, at E and F, the same pressure would be produced at A and B, but we should also have to consider the transverse stress on the rafter at E from a force $\frac{1}{2}$W × cos. *θ*. The compression down the lower half of the rafter will also be greater than down the upper half.

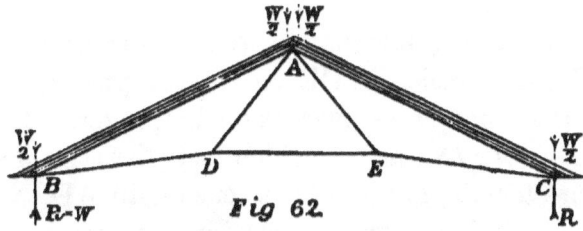

Fig 62.

Another simple form of roof is shown by fig. 62, the king-bolt being removed and *two* inclined braces taking its place in holding up the tie-rod, which is here made in three parts. Proceeding as before we draw the stress-diagram (fig. 63) by taking $bd = 2W$, and bisecting bd at X, then taking ab, aX, Xc, cd, each to represent $\frac{1}{2}$ W; then bX is the reaction R at B, and equals W. Draw al parallel to BA, meeting Xl parallel to BD; also cm parallel to AC, meeting Xm parallel to EC. Draw lo and mo parallel to AD and AE, and meeting the horizontal line from X in the point o.

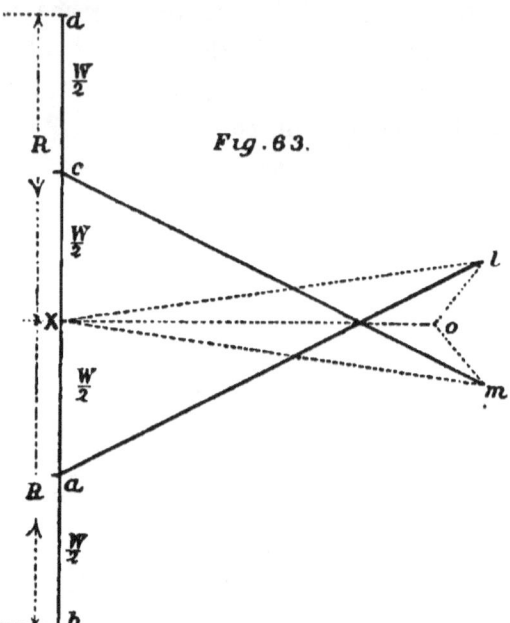

Fig. 63.

The polygon of forces at B is formed by ab, bX, Xl, la, whence we get Xl acting *from* B for the tension in BD, la acting *towards* B, the compression in AB. Also Xm the tension in CE, mc the compression in AC.

For the point D we have the triangle formed by the lines Xo, ol, lX, all of which act *from* D, and are therefore tensile, giving ol the tension in AD, Xo that in DE. Similarly we get mo for the tension in AE. If $\frac{1}{2}W = 1$ ton, then we find by measurement that al, the compression down the rafter, is 3·6 tons, the tension Xl in BD is 3·25 tons, the tension Xo in DE is 2·75 tons, the tension lo in AD or AE is 0·7 ton. The strong lines in the stress-diagram show compressions, and the dotted lines tensions.

Comparing the stress-diagrams (fig. 61 and fig. 63), we see that by introducing two inclined braces in place of the king-bolt, the stresses in the rafters and tie-rods are increased; but this arrangement has the advantage of holding up the tie-rod in two places instead of one, and thereby preventing it from bending under its own weight; so that a smaller tie-rod will do in the truss (fig. 62) than in fig. 60.

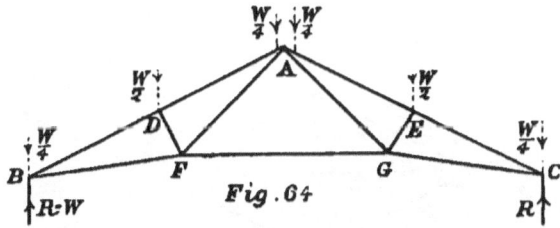

Fig. 64

A common form of iron-roof is shown by fig. 64, which only differs from fig. 62 in having struts at D and E. Supposing W to be the load on each rafter, we have $\frac{1}{4}$ W supported on each side of A, $\frac{1}{4}$ W at B and C, and $\frac{1}{2}$ W at D and E. For the stress-diagram draw the vertical bh (fig. 65) to represent 2W, and bisect it in X. Take $ab = \frac{1}{4}$ W, $ad = \frac{1}{2}$ W, $dX = \frac{1}{4}$ W, $Xe = \frac{1}{4}$ W, $ef = \frac{1}{2}$ W, $fh = \frac{1}{4}$ W. Then bX is the reaction R at B, and equals W. Draw am parallel to BD, meeting Xm parallel to BF; fn parallel to EC, meeting Xn parallel to CG. Draw dl parallel to DA, meeting ml parallel to DF; ep parallel to AE, meeting np parallel to EG. Draw lo and po parallel to AF and AG, meeting the horizontal line from X in the point o.

For the polygon of forces at B, we have, ab, bX, Xm, ma, in which Xm acts *from* B and is *tensile*, ma acts *towards* B and is *compressive*; from which we determine the compression ma in BD, and the tension Xm in BF. Also in like manner nf, the compression in EC is found, and the tension Xn in GC.

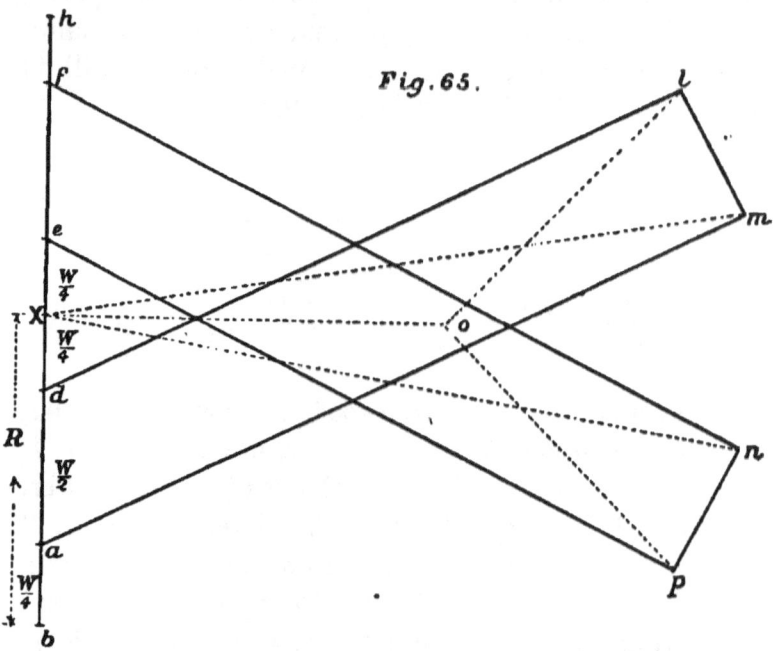

Fig.65.

For the forces in equilibrium at D, we have da, am, ml, ld, all acting *towards* D and being compressive; from which we get ld for the compression in AD, ml the compression in the strut DF. So also pe for the compression in AE, pn for that in EG.

For the forces at A, we have the polygon Xd, dl, lo, op, pe, eX, from which we get the tension lo in AF, and po in AG, both acting *from* A.

For equilibrium at F, the polygon of forces is formed by the lines X*o*, *ol*, *lm*, *m*X; which gives X*o* for the tension in FG, acting *from* F.

If we take ½ W = 1 ton, we find *am* the compression down DB to be 5 tons; that down AD 4½ tons: that in each strut 0·9 ton; the tension X*m* or X*n* in the tie-rods BF and CG 4½ tons; that in AF and AG 2·15 tons, and that in the tie-rod FG 2·5 tons. The angle of pitch is 27°.

The rafter is usually made of tee-iron, and if the top flange is 4″ × ½″, and the web 3″ × ½″, and we suppose AB = 10 feet, or DB = 5 feet, then by Gordon's formula (**59**) the breaking-weight as a pillar will be about 50 tons, which is ten times the compression found above. Taking 3 tons per square inch as the safe tensile stress, we see that the oblique ties BF and GC require a sectional area of 1½ inches, and the horizontal tie-rod an area of ·83 inch.

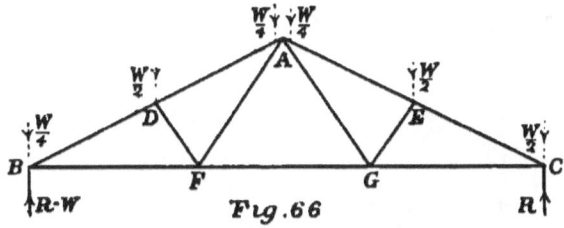

Fig.66

The truss shown by fig. 66 is similar to fig. 64, only the tie-rod is horizontal throughout, which very much simplifies the stress-diagram. Take *bh* (fig. 67) as before, to represent 2 W, *ab* = ¼ W, *ad* = ½ W, *d*X = ¼W, and so on. Draw *am* parallel to BD, meeting the horizontal line X*m*; so also *fm* parallel to CE. Draw *dl* parallel to DA, meeting *ml* parallel to DF ; so also *ep* parallel to AE, meeting *mp* parallel to EG. Draw *lo* and *po* parallel to AF and AG. Then for the forces in equilibrium at B, we have the polygon formed by the lines *ab*, *b*X, X*m*, *ma*, from which we get *ma* acting *towards* B for the compression in DB, X*m* acting *from* B for the tension in BF and GC.

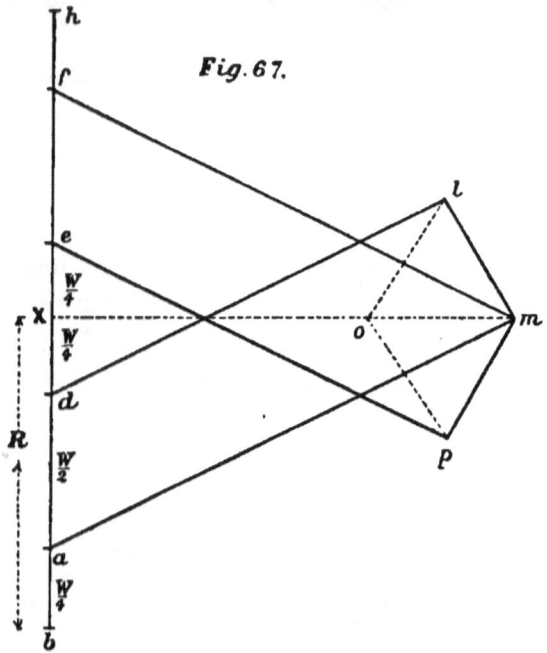

Fig. 67.

The forces at D are *da*, *am*, *ml*, *ld*, all acting *towards* D and being compressive; from which we get *ld* the compression in AD, and *lm* that in the strut DF ; so also *mp* is the compression in the strut EG, and *pe* that in the rafter AE.

Proceeding to the point A, we have the forces in equilibrium represented by the polygon X*d*, *dl*, *lo*, *op*, *pe, eX*; in which we get *lo* and *op* acting *from* A for the tensions in the braces AF and AG.

At the joint F we have the polygon of forces, X*o*, *ol*, *lm*, *m*X, which gives X*o* acting *from* F for the tension in the tie-rod FG.

Taking, as before, $\frac{1}{2}$ W = 1 ton, we find *am* the compression in the rafter DB is 3·3 tons; that in the rafter AD is 2·8 tons; the compression in each strut 0·9 ton; the tension in the tie-rods BF and GC is 3 tons, and that in the tie-rod FG 2·1 ton. The tension in the braces AF and AG is 0·9 ton.

Comparing this form of roof with that shown by fig. 64, we see by the diagrams figs. 65 and 67 that the stresses are greatly reduced by having horizontal tie-rods instead of inclined ones.

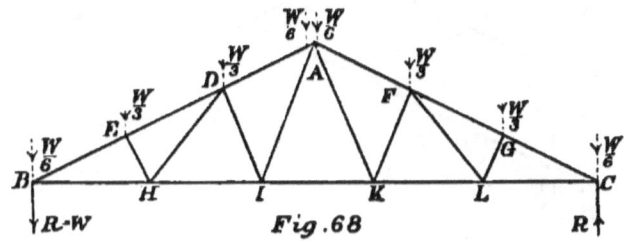

Fig. 68

72. Roofs of Large Span.—As the span of the roof is increased, it becomes necessary to have a larger number of struts and braces than are shown in the foregoing examples. In ABC (fig. 68) we have a truss similar to that of fig. 66, but for a longer span, and having seven purlins instead of five. Here we suppose the purlins to rest on the principal rafter at A, B, C, D, E, F, and G, and the rafter to be stiffened by struts and braces at D, E, F and G. We may consider that $\frac{1}{3}$ W is the vertical force at each of these last named points, and $\frac{1}{6}$ W on each rafter at A, or $\frac{1}{3}$ W the total pressure at A, and $\frac{1}{6}$ W at B and at C. The reaction, R, at B and C is equal to W, which represents the weight on *one* side of the roof.

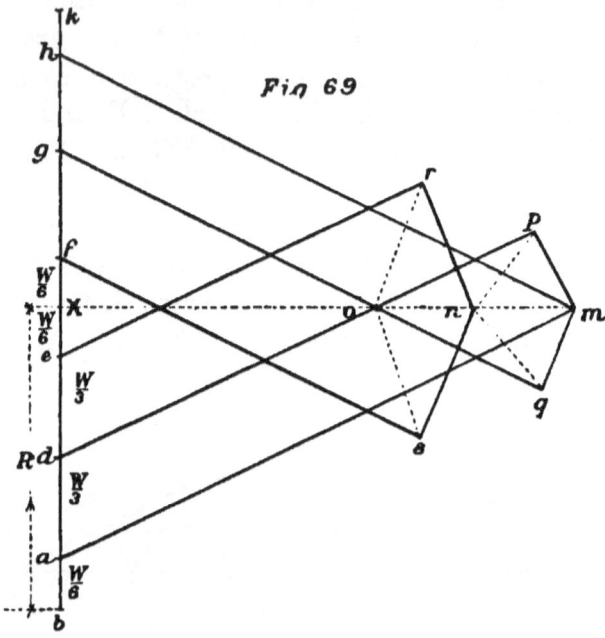

In order to find the stress-diagram, we must assume as before that the rafters and tie-beam are *jointed* at the vertex of each triangle, or that BE, ED, BH, HI, &c. are distinct bars connected by hinges or joints. Draw the vertical bXk (fig. 69) representing on scale the load 2 W, and let X be its middle point. Then Xb and Xk represent the reactions, R, at B and C. Take ab to represent $\frac{1}{6}$ W, ad and de to represent $\frac{1}{3}$ W each, Xe and Xf one-sixth of W; fy and gh each $\frac{1}{3}$ W, hk

½ W. Draw a horizontal line through X, and draw *am* parallel to BE, meeting this line in *m*, and draw *hm* parallel to CG. Draw *dp* parallel to ED, meeting *mp* parallel to EH; also *gq* parallel to FG, meeting *mq* parallel to GL. Draw *pn* and *qn* parallel to DH and FL, meeting the horizontal line from X in the point *n*. Draw *er* and *fs* parallel to AD and AF, meeting *nr* and *ns* parallel to DI and FK. Draw *ro* and *so* parallel to AI and AK, meeting X*m* in the point *o*.

For the forces in equilibrium at B, we have the polygon *ab*, *b*X, X*m*, *ma*, in which we find *ma* acts *towards* B and gives the compression in EB, while X*m* acts *from* B and gives the tension in BH. Proceeding to the joint E, we have the forces in equilibrium represented by the polygon *da*, *am*, *mp*, *pd*, all acting *towards* E, from which we get *pd* for the compression in DE, *mp* for that in the strut EH. For the joint at H we have the polygon X*n*, *np*, *pm*, *m*X, which acts *from* H and gives X*n* the tension in HI, *np* acts *from* H and is the tension in DH. At D the polygon of forces is *ed*, *dp*, *pn*, *nr*, *re*, which gives *re* acting *towards* D for the compression in AD, and *nr* that in DI. For equilibrium at I, we have the polygon X*o*, *or*, *rn*, *n*X, from which we get X*o*, acting *from* I, for the tension in IK, and *or* that in the brace AI. For equilibrium at A, we have X*e*, *cr*, *ro*, *os*, *sf* and *f*X; in which *er* and *sf* act *towards* A and represent the compression in AD and AF, while *ro* and *os* act *from* A and represent the tension in AI and AK.

In the same way we complete the diagram for the other side of the truss. The strong lines show where

м

the stresses are compressive, and the dotted lines where they are tensile.

Suppose for example that W = 3 tons, then from fig. 69 we find that *am*, which represents the compression in the rafter BE, is $5\frac{2}{3}$ tons; *dp*, the compression in DE, is $5\frac{1}{3}$ tons; *er*, the compression in AD, is 4 tons. The compression in EH, represented by *mp*, is $\frac{9}{10}$ ton; and the compression in DI, represented by *nr*, is $1\frac{1}{3}$ ton. The tension in BH is $5\frac{1}{4}$ tons, that in HI is $4\frac{1}{8}$ tons, and that in IK is $3\frac{1}{8}$ tons. The tension in the brace DH, which is represented by *pn*, is 1 ton; that in AI, represented by *or*, is $1\frac{1}{3}$ ton.

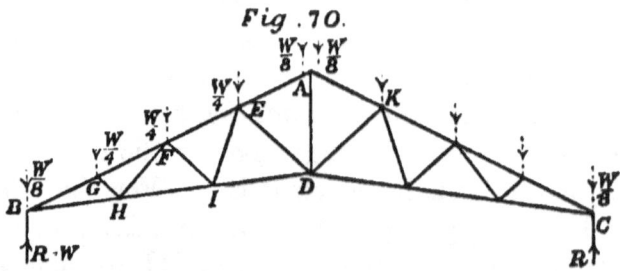

Fig. 70.

The truss shown by ABDC (fig. 70) is for a roof of still larger span than the last, and has purlins resting on nine points. If W is the load on one side, then we have $\frac{1}{4}$ W at E, F and G; $\frac{1}{8}$ W at B and C; and twice $\frac{1}{8}$ W, or $\frac{1}{4}$ W, at A. There is a king-bolt at A holding up the centre D of the tie-rods, which latter are inclined in this example. For the stress-diagram (fig. 71), draw the vertical line *bXk* representing on scale 2 W, having X for its middle point; then *bX* and *kX* represent the reaction, R, or W, at B and C.

Take *ab*, *ki*, X*e* and X*f*, each to represent $\frac{1}{8}$ W; *ac*, *cd*, *de*, *fg*, *gh*, *hi*, each to represent $\frac{1}{4}$ W. Draw *al*

parallel to BG, meeting X*l* parallel to BH; *cr*
parallel to GF, meeting *lr* parallel to GH ; *rm* parallel to
FH, and *mq* parallel to FI, meeting *dq* parallel to FE ;
qn parallel to EI, and *no* parallel to DE, meeting *eo*
parallel to AE. In the same way for the other side of
the truss, draw the lines *is, ht, gu, fp, pv, vu, uz, zt,
ts*, X*s*. Draw the vertical line *op*.

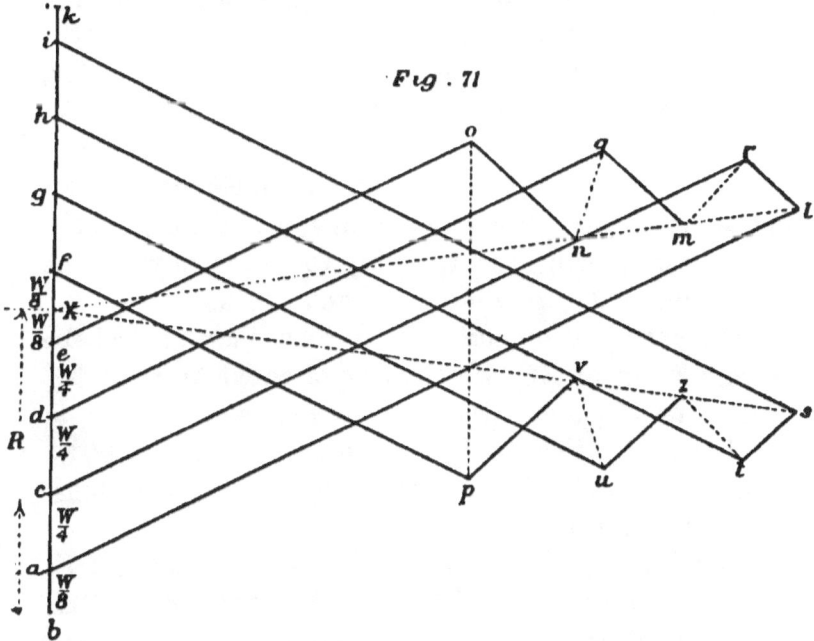

Fig. 71

For the point B (fig. 70) we have the polygon of
forces (fig. 71) *ab, bX,* X*l, la,* which gives *la* acting
towards B for the compression in BG, X*l* acting *from*
B for the tension in BH. At the joint G, we have *ca,
al, lr, rc ;* from which we get *rc* and *lr* acting *towards*
G, for the compression in FG and in GH. Proceeding
next to the joint H, we have the polygon X*m, mr, rl,*

*l*X, which gives X*m* and *mr* acting *from* H, for the
tension in III, and in the brace FH. At F we get
the polygon of forces *dc*, *cr*, *rm*, *mq*, *qd*, which gives
qd and *mq* acting *towards* F, for the compression in EF,
and in the strut FI. At I, the polygon is X*n*, *nq*,
qm, *m*X, giving X*n* and *nq* acting *from* I, for the
tension in DI, and in the brace EI. At the joint E,
we have, *cd*, *dq*, *qn*, *no*, *oc* ; giving *oc* and *no* acting
towards E, for the compression in AE, and in the strut
DE. At the ridge A, we have the polygon of forces,
X*c*, *co*, *op*, *pf*, *f*X, from which we get *op* acting *from*
A and representing the tension in the king-bolt AD.

The stresses in the other half of the truss are represented by the lines drawn from *f*, *g*, *h*, *i*, &c. ; but the
line *fp* is the only one necessary to be drawn when the
two sides of the truss are equal and similar.

For example, suppose W = 4 tons, then we find the
compression in BG, which is represented by *al* in fig.
71, to be $10\frac{3}{4}$ tons ; that in FG, represented by *cr*, to
be $10\frac{1}{8}$ tons ; that in EF, represented by *dq*, to be $8\frac{1}{8}$
tons ; that in AE, represented by *co*, to be 6 tons.
The compression in GH, represented by *lr*, is 1 ton ;
that in FI, represented by *mq*, is 1·4 ton ; that in DE,
represented by *no*, is 1·8 ton. The tension in BH,
represented by X*l*, is 10 tons ; that in III, represented
by X*m*, is $8\frac{1}{2}$ tons ; that in DI, represented by X*n*, is
7 tons. The tension in the brace FH, represented by
mr, is $1\frac{1}{8}$ ton ; that in the brace EI, represented by *nq*,
is $1\frac{1}{4}$ ton ; that in the king-bolt AD, represented by
op; is $4\frac{1}{2}$ tons.

73. Bow-string Truss.—For roofs of railway
stations and other buildings of wide span, the form of

truss called the "Bow-string" is commonly used. In this truss the rafters are made to form a part of a polygon, or nearly an arc of a circle, as AD, DE, EC, &c. (fig. 72), which is the "bow." The tie-rods are

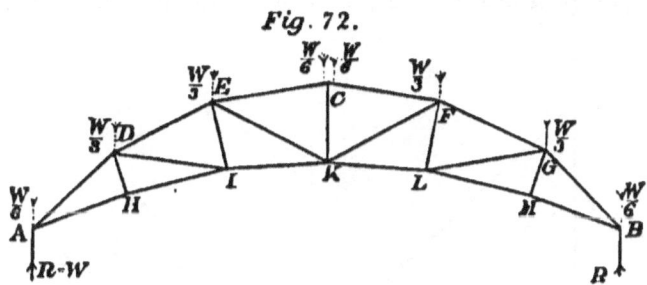

Fig. 72.

arranged either in a straight line, or else in the form of a circular arc or part of a polygon, as AH, HI, IK, &c.; and this is the "string" of the "bow." The truss is formed by struts and braces connecting the "bow" and the "string," as DH, DI, &c.; and the load of the roof-covering is carried by means of purlins at A, D, E, C, &c.

If we suppose 2 W to be the load supported by the whole truss, then ⅓ W is carried at the points D, E, C, F and G; and ⅙ W at A and B. The reaction, R, at each of the points of support A and B, is equal to W.

To find the stress in each part of this truss, take the vertical line bXh (fig. 73) to represent 2 W on any scale, X being the middle point, so that Xb and Xh each represent the reaction, R, or W. Take ab, Xd, Xe, each equal to ⅙ W; ac, cd, ef, fg, each equal to ⅓ W. Draw ai parallel to AD, meeting Xi parallel to AH; also gs parallel to BG, meeting Xs parallel to BM. Draw il parallel to DH, meeting Xl parallel to

HI; also *sr* parallel to GM, meeting X*r* parallel to LM.   Draw *ck* parallel to DE, meeting *lk* parallel to DI; also *fq* parallel to FG, meeting *qr* parallel to GL. Draw *km* parallel to EI, meeting X*m* parallel to IK; also *pq* parallel to FL, meeting X*p* parallel to KL.

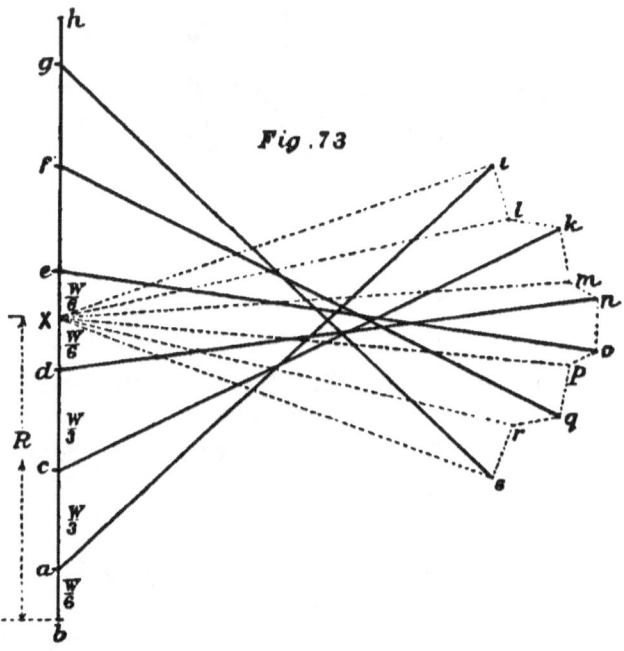

Fig. 73

Draw *dn* parallel to EC, meeting *mn* parallel to EK; also *eo* parallel to CF, meeting *op* parallel to FK; and draw the vertical line *no*.

Beginning at the point A, the polygon of forces in equilibrium at A is, *ab*, *b*X, X*i*, *ia*, in which *ia* acting *towards* A, gives the compression in the rafter AD, and X*i* acting *from* A is the tension in the tie AH.

Proceeding to the joint H, we have the polygon formed by the lines X*l*, *li*, *i*X, all acting *from* H, which gives *li* the tension in the brace DH, and X*l* that in HI.

At D, the polygon of forces is, *ca*, *ai*, *il*, *lk*, *kc*, which gives *kc* acting *towards* D for the compression in DE, and *lk* acting *from* D for the tension in the brace DI.

At I, we have the polygon, X*m*, *mk*, *kl*, *l*X, all acting *from* I, which gives *mk* the tension in EI, and X*m* that in IK.

At E, the polygon of forces is, *dc*, *ck*, *km*, *mn*, *nd*, giving *nd* acting *towards* E for the compression in CE, and *mn* acting *from* E for the tension in EK.

At the joint C, the forces in equilibrium are represented by X*d*, *dn*, *no*, *oe*, *e*X, from which we get *no* acting *from* C and representing the tension in the king-bolt CK.

The strong lines in the diagram (fig. 73) represent compressive forces, and the dotted lines tensile forces; from which it appears that the only parts of this truss that are in compression are the rafters forming the "bow;" all the others being in tension.

For example, suppose W = 3 tons, then it will be found by measuring the diagram (fig. 73) that the compression in the rafter AD, which is represented by the line *ai*, is 5·8 tons; that in DE, represented by *ck*, is 5·5 tons; that in the rafter EC, represented by *dn*, is 5·4 tons. The tension in the tie AH, which is represented by the line X*i*, is 4·5 tons; that in the tie HI, represented by X*l*, is 4·5 tons; and that in IK, represented by X*m*, is 5·1 tons. The tension in DH, represented by *il*, is ·55 ton; that in EI, represented by *km*, is ·55 ton; that in CK, represented by *no*, is

·55 ton. The tension in DI, represented by *kl*, is ·55 ton; that in EK, represented by *mn*, is one-third of a ton.

74. WARREN GIRDER. — The simplest form of trussed girder is that represented by figs. 74 and 76, consisting of horizontal bars at top and bottom, called the "booms," which are connected by cross struts and braces. In fig. 74 (showing half the

Fig. 74

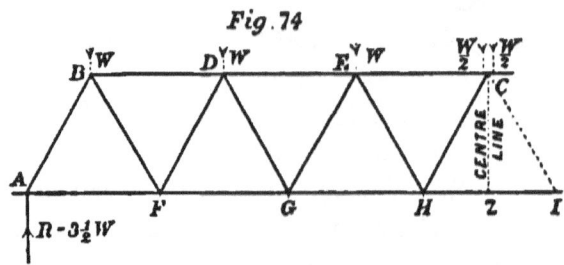

girder) the lower *boom* AFGHI rests upon the supports at each end, and the girder is loaded on the upper *boom* at the points B, D, E, C, &c., with equal weights, W. The reaction, R, at the supports will be equal to the sum of the load at B, D and E, and half the load at C; and in this case R = $3\frac{1}{2}$ W.

To form the stress-diagram for such a girder, take the vertical line *eXa* (fig. 75) to represent 4 W, X*a* being half the total load on the girder, or $3\frac{1}{2}$ W. Make X*e* and X*d* each equal to $\frac{1}{2}$ W; *dc*, *cb*, and *ba*, each equal to W. Draw the horizontal lines X*o*, *dr*, *cq*, *bp*. Draw *al* parallel to AB, meeting the horizontal line X*o* in *l*. Draw *lp* parallel to BF, meeting *bp* in *p*. Draw *pm* parallel to FD, meeting X*o* in *m*. Draw *mq* parallel to DG, meeting *cq* in *q*. Draw *qn* parallel to EG, meeting X*o* in *n*. Draw *nr* parallel to EH, meeting *dr* in *r*. Draw *ro* parallel to CH, meeting X*o* in the point *o*.

Starting from the joint A (fig. 74) we have the polygon of forces in equilibrium (fig. 75) *a*X, X*l*, *la*,

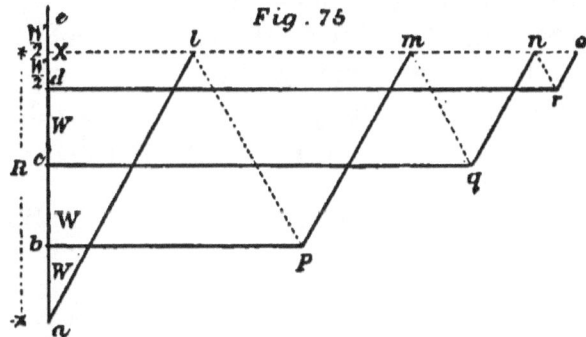

Fig. 75

where X*l*, acting *from* A, represents the tension in AF; and *la* acting *towards* A, the compression in AB.

Proceeding to the joint B, we have the polygon, *ba*, *al*, *lp*, *pb*, which gives *pb* acting *towards* B, for the compression in BD; *lp*, acting *from* B, the tension in the brace BF.

At the joint F, we have the polygon of forces, X*m*, *mp*, *pl*, *l*X, giving X*m* acting *from* F, for the tension in FG; and *mp* acting *towards* F, for the compression in DF.

At D, the polygon of forces is, *cb*, *bp*, *pm*, *mq*, *qc*, which gives *qc* acting *towards* D, for the compression in DE; and *mq* acting *from* D, the tension in DG.

At G, the polygon is, X*n*, *nq*, *qm*, *m*X, giving X*n* which acts *from* G, for the tension in GH; and *nq* which acts *towards* G, the compression in EG.

At the joint E, we have, *dc*, *cq*, *qn*, *nr*, *rd*, for the polygon of forces; from which we get *rd* acting *towards*

E, for the compression in EC; and *nr* acting *from* E, the tension in EH.

At H, the polygon is, X*o*, *or*, *rn*, *n*X, giving X*o* which acts *from* H, for the tension in HI, and *or* which acts *towards* H, for the compression in CH.

By repeating the figure on the upper side of the line X*o*, we get the corresponding stresses in the other half of the girder. The strong lines in the stress-diagram (fig. 75) indicate compressions, and the dotted lines indicate tensions.

In the girder shown by fig. 76, the *upper* "boom"

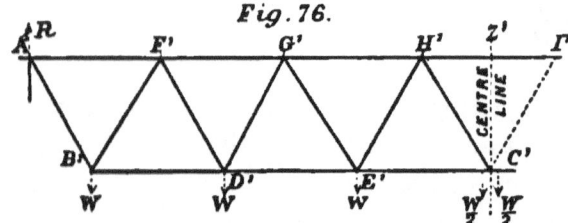

Fig. 76.

rests on the supports, and the loads, W, are supposed to be carried on the lower "boom" at B', D', E', C', &c.; the stresses in the different parts will therefore be the reverse of those in the former case (fig. 74).

To form the stress-diagram (fig. 77), take, as before, *a*X*e* to represent on scale 4 W; X*e* and X*d* being each ½ W; *ab*, *bc*, *cd*, each equal to W. Then X*a* represents the reaction, R, at A', and is equal to 3½ W. Draw horizontal lines through X, *d*, *c* and *b*, as X*o*', *dr*', *cq*', *bp*'. Draw *al*' parallel to A'B' (fig. 76) meeting X*o*' in *l*'; draw *l*'*p*' parallel to B'F", meeting *bp*' in *p*'; draw *p*'*m*' parallel to D'F", meeting X*o*' in *m*'; draw *m*'*q*' parallel to D'G', meeting *cq*' in *q*'; draw *q*'*n*' parallel to E'G' meeting X*o*' in *n*'; draw *n*'*r*' paiallel to E'H', meeting dr' in r'; draw $r'o'$ parallel to C'H', meeting Xo' in the point o'.

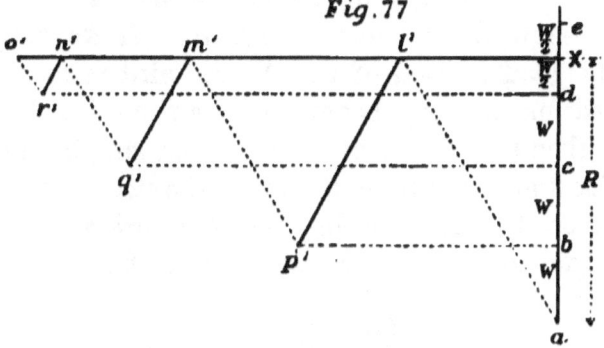

Fig. 77

Proceeding then exactly as in the former case (figs. 74, 75) we find that $l'a$ represents the tension in A'B', $m'p'$ that in D'F', $n'q'$ that in E'G', $r'o'$ that in C'H'. The line $p'l'$ represents the compression in B'F', $q'm'$ that in D'G', $r'n'$ that in E'H'. The horizontal line Xl' represents the compression in A'F', Xm' that in F'G', Xn' that in G'H', Xo' that in H'I'. The horizontal line bp' represents the tension in B'D', cq' that in D'E', and dr' that in E'C'.

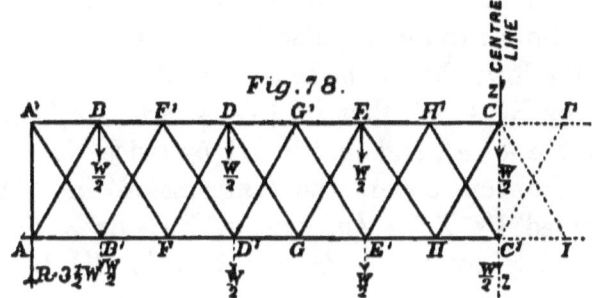

Fig. 78.

75. LATTICE GIRDER.—The girder shown by fig. 78 is formed by the combination of the two "Warren"

girders (figs. 74 and 76), the cross braces being riveted
together where they intersect. The letters A, B, F, D,
&c. show the bracing of fig. 74, and the letters A′, B′, F′,
D′, &c. show the bracing of fig. 76. If we suppose
half the load on each of the Warren girders to be distributed on the top boom, and half on the bottom
boom, then the resultant stress along any bar of the
"lattice" girder will be the sum of the stresses brought
to bear on it by the load in the single girders.

In order to form the stress-diagram (fig. 79) take a

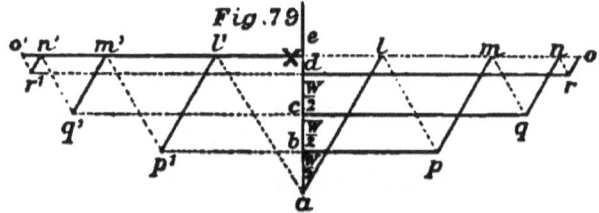

vertical line ae equal to half ae in figs. 75 and 77, or
representing 2 W. Make Xe, Xd each equal to ¼ W,
ab, bc, and cd each equal to ½ W. Draw the diagram
on the right-hand of ae exactly similar to fig. 75, only
to half its scale ; and draw the left-hand diagram
exactly similar to fig. 77, also to half its scale.

For the lower boom the tension in AF will be represented by Xl + bp', that in FG by Xm + cq', that in
GH by Xn + dr', and that in HI by twice Xo.

For the top boom, the compression in A′F′ is
represented by Xl' + bp, that in F′G′ by Xm' + cq,
that in G′H′ by Xn' + dr, and that in H′I′ by twice
Xo'.

The stresses in the braces of the lattice beam will be
the same in kind as they are in the Warren beams,

only the amounts will be reduced one-half. Thus, al, pm, qn, ro, are the compressions in the bars AB, FD, GE, HC, respectively, of fig. 78; $l'p'$, $m'q'$, $n'r'$, the compressions in the bars B'F', D'G', E'H', respectively. The lines lp, mq, nr, represent the tensions in the bars BF, DG, and EH, respectively; the lines al', $p'm'$, $q'n'$, and $r'o'$, the tensions in the bars A'B', F'D', G'E', and H'C', respectively.

CHAPTER VIII

ARCHES.

76. PRINCIPLE OF THE ARCH.—The object which the builder has in view when constructing an arch, is that it shall carry a wall or other load over an opening, the width of which is called the " span " of the arch. The outline, or " intrados," of the arch may be flat or curved, and it is generally formed of three or more blocks of stone called "voussoirs," which are cut in a wedge form, the joints of which should be perpendicular to the tangents of the curve forming the intra-

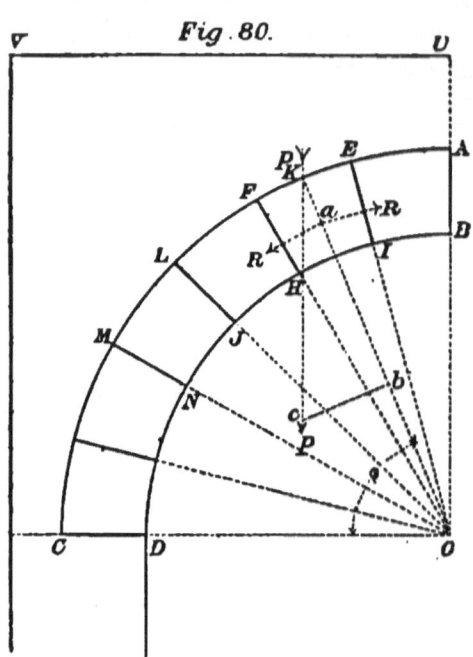

Fig. 80.

dos. If the intrados is an arc of a circle, as in fig. 80, then the lines of the joints of the voussoirs, as AB, EI, FH, &c. will, if produced, meet in the centre,

O, from which the circle is struck, forming triangular "wedges," as AOE, EOF, &c. The joints EI, FH, &c. form "inclined planes" down which the adjacent voussoirs tend to slide. Before therefore proceeding to discuss the "Theory of the Arch," we propose to consider the principles of the two above-named "Mechanical powers," with which the principle of the arch is intimately connected, namely, the "inclined plane" and the "wedge."

77. THE INCLINED PLANE.—Let AB (fig. 81) represent the section of a smooth plane surface inclined at the angle BAC (= θ) to the horizontal line AC. Draw the vertical line BC, which is called the *height* of the plane; AB being the *length* of the plane, and AC its *base*. Suppose a heavy block of stone whose weight is W, to be placed on the plane; then a reaction R will be produced acting at right angles to the inclined plane.

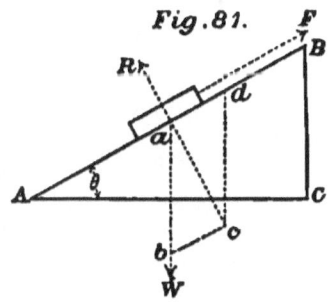

Fig. 81.

If the surfaces of the plane and load are perfectly smooth where they are in contact, then the load will slide down the plane unless held back by some force F acting up the plane. We shall have then three forces keeping the load in equilibrium, namely, its own weight W acting vertically downwards and parallel in direction to BC; the reaction R acting perpendicularly to AB; and the force F, which keeps it from sliding, acting parallel to AB. Let the vertical line *ab* represent W, and drawing *bc* parallel to AB to meet *ac* per-

pendicular to AB, and *cd* parallel to *ab*; we have by
the triangle of forces (**4**) *bc* or *ad* representing the
force F acting up the plane and preventing the load
from sliding down by the action of W, *ac* representing
the reaction R at right angles to the plane. It will be
evident that the triangle *abc* is similar to the triangle
ABC, and the angle *bac* = angle BAC = θ; consequently we find the following relations between W,
R and F—

$$F : W = bc : ab$$
$$= BC : AB;$$

therefore, $\quad F = W \times \sin. \theta$

$$R : W = ac : ab$$
$$= AC : AB;$$

therefore, $\quad R = W \times \cos. \theta$

$$F : R = bc : ac$$
$$= BC : AC;$$

therefore, $\quad F = R \times \tan. \theta \qquad . \qquad . \quad (129)$

As however we never can obtain perfect smoothness
in the surfaces in contact, it follows that the load W
will not begin to slide down the plane until a certain
amount of inclination has been attained by the plane,
on account of the " friction " which takes place between
the surfaces in contact arising from slight irregularities
or roughness which exists on the smoothest surfaces
and which interlock one with the other. The angle at
which sliding will begin depends therefore on the condition of the surfaces in contact, being less for smooth
surfaces than for rough ones; and by gradually increasing the angle θ it can be ascertained experimentally

when sliding begins. Suppose we put a for the value of θ when the load begins to slide, and F for the force of friction which has hitherto kept it from sliding, then we have from equation (129)—

$$F = R \times \tan. a \qquad . \qquad . \qquad . \qquad . \qquad (130)$$

Now it is found by experiment that the friction between two surfaces in contact varies according to the pressure of one against the other, and therefore in the present case it is proportional to the reaction R. For the same kind of surface the force F of friction is proportional to R, or is equal to R multiplied by a constant for which we put the Greek letter μ, and call it the "coefficient of friction"; we have then—

$$F = \mu \times R \qquad . \qquad . \qquad . \qquad . \qquad . \qquad (131)$$

But by equation (130) we had, $F = R \times \tan. a$; consequently we find $= \tan. a$.

The value of μ can therefore be obtained experimentally for various kinds of surface by finding the angle, a, at which they begin to slide down an inclined plane. Thus, when two blocks of hewn stone are placed in contact without mortar, the angle a at which sliding begins is about 33°, the tangent of which is ·65, or the "coefficient of friction" in this case is $\mu = $ ·65. When the stones are freshly bedded in mortar the friction is increased, and $a = 37°$, or $\mu = $ ·75; hence it follows that when an arch is being built of stone, the voussoirs will require no support from the centering until the joints make an angle of 37° with the horizontal.

If the angle of inclination, θ, of the inclined plane

N

had been so high as to cause the heavy body to slide rapidly down it, and this angle were gradually decreased during the sliding, the velocity would be diminished until at length the load was brought to a standstill. By observing the angle at which the load ceased to slide we find that it exactly coincides with the angle, a, at which sliding commences when the angle θ is on the increase; consequently a is called the "angle of repose," and may be taken either as that at which sliding ceases or at which it begins, according as the plane is being lowered or raised.

78. THE WEDGE.—Suppose the isosceles triangle ABC (fig. 82) to represent the section of a " wedge," formed of some hard and incompressible material, which is driven between the surfaces of two solids by means of a force P acting upon its back BDC. Let the line AD bisect the angle BAC of the wedge, the angle BAD being equal to CAD and called θ. Then the pressure P will act in the direction of DA, and will produce reactions R and R_1 at right angles to AB and AC, whose directions ba and ca will meet in the point a on the line AD. In the first place let us suppose that the surfaces are perfectly smooth, and that there is no friction between them. Draw bd parallel to ac, cd parallel to ab, and let ad be taken to represent the pressure P; then ab and ac

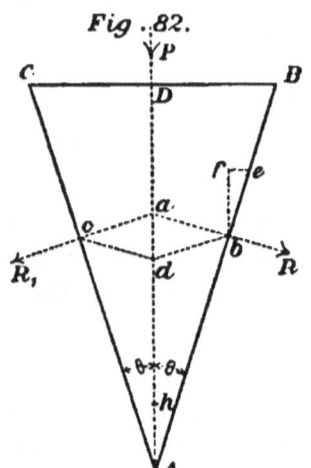

Fig. 82.

will represent the reactions R and R_1. Since AB is equal to AC, therefore ab is equal to ac, or $R_1 = R$.

From the similarity of the triangles abd and BAC, we have—

$$R : P = ab : ad$$
$$= AB : BC$$
$$= AB : 2\ BD\ ;$$

therefore,

$$R = \frac{P}{2} \times \frac{AB}{BD}$$

$$= \frac{P}{2\ \sin.\ \theta} \qquad . \qquad . \quad (132)$$

In the second place we will suppose that the friction, F, acts up and down the planes AB and AC, and parallel to them in direction; then by equation (131) we have—

$$F = \mu \times R\ ;$$

so that F can be found when μ and R are known. Take the length be to represent F on the same scale that ab represents R, or ad represents P; and draw bf parallel to AD, meeting ef parallel to BC. Then bf represents the resolved part of F which is directly opposed to the pressure produced by $\frac{1}{2}$ P on each side of the wedge; or if we call Q the total resistance to the pressure P arising from the friction on both sides of the wedge, we have—

$$Q = 2\ bf.$$

On AD take dh equal to twice bf, then $ah = P + Q$ represents the force that will just overcome the reactions R and R_1, and the friction of the surfaces in

contact.   Then since the triangle *bef* is similar to
ABD, we have—

$$Q = 2 \; bf = 2 \; be \times \frac{bf}{be}$$

$$= 2 \; F \times \frac{AD}{AB}$$

$$= 2 \; F \times \cos. \; \theta$$

$$= 2 \; \mu \; . \; R \times \cos. \; \theta$$

from equation (132), $= \mu \; . \; P \times \cot. \; \theta$.   (133)

The wedge then is kept in equilibrium by the force
P + Q, the reactions R and R_1, and the effect on its
sides of the friction $\mu \; . \; R$ and $\mu \; . \; R_1$. The friction on
the sides may, however, act either upwards or down-
wards, according as the pressure acts in the opposite
direction.   For while friction opposes the force P in
pushing in the wedge, it on the other hand opposes
any attempt in the opposite direction to withdraw the
wedge; and consequently the wedge, is held in its
place by the action of friction on the sides.   And since
ad or P represents the resolved part of R + R_1 in the
direction parallel to AD; and *dh*, or Q, that of the
friction in the same direction; it follows that so long
as Q is greater than P the wedge will not be pushed
backwards by the reactions R and R_1 after the pressure
P is removed.

79. APPLICATION TO THE ARCH.—Referring to the
fig. 80, we see that an arch consists of an assemblage of
truncated wedges called " voussoirs," as AE, EF, &c.,
the directions of the joints AB, EI, &c. meeting in the
centre O, when the intrados is an arc of a circle.
These wedges are kept in their position by means of the

pressure of the load upon their backs, called the "surcharge," and by the reactions and friction of the surfaces in contact with each other, just as in the case of the wedge (**78**). Take any one of these voussoirs, as EF, and let p be the vertical pressure on its back, the resultant of which acts at K. Since, however, the direction of the force p is inclined to the centre line, OK, it is only its resolved part in the line KO that has to be considered as the driving force on the back, corresponding to the force P in the wedge (**78**). If then we put ϕ for the angle which OK makes with the horizontal OD, we have—

$$P = p \cdot \sin \cdot \phi.$$

Take the line Kc to represent p, and draw bc at right angles to OK; then Kb represents the component of p acting down OK, or

$$Kb = Kc \cdot \sin \cdot \phi = P \qquad . \qquad (134)$$

Let R_1, R, be the reactions of the two sides, meeting in the point a; $\mu \cdot R$ the friction; then we can determine R in the manner previously shown in the case of the wedge (**78**); and the resistance of friction is found as in equation (133); putting θ for the angle KOE, and Q for the resistance of the two sides, we have—

$$Q = \mu \cdot P \cdot \cotan \cdot \theta$$

$$= \mu \cdot p \cdot \sin \cdot \phi \cdot \cotan \cdot \theta \cdot \qquad . \qquad (135)$$

And since the force Q acts with P to prevent the reaction 2R from pushing back the voussoir, consequently the voussoir will remain in its place so long as 2R is less than P + Q.

If p and θ are the same at each voussoir it will be seen from equation (134) that P and Q decrease as the angle ϕ decreases, or as the voussoirs approach the springing; being greatest when $\phi = 90°$, or at the "key-stone," and least when $\phi = 0°$, in a semicircular arch. From this we see the necessity of increasing the value of p as we get further away from the crown of the arch, otherwise it will have a tendency to break up by the rising, or pushing back, of the voussoirs at the haunches. The reaction R, and consequently the friction F, is however greater in the lower voussoirs than in the higher ones, owing to the weight of the upper ones pressing upon the lower.

If there is a heavy load on the crown of the arch and an insufficient load on the haunches, the arch will break up by the falling in of the crown, the joints AB and EI opening at B and I, while those at LJ and MN open at M and L; thus throwing all the pressure on the outer edges of the voussoirs, and causing them to crumble to pieces.

It will be seen from equation (132) that the reaction, R, of the sides of the voussoirs increases as sin. θ diminishes, or as the angle which the joints make with each other is lessened. Also from equation (133) we see that the force Q increases with the increase of the cotangent of θ or with the decrease of the angle θ. Hence it will be evidently an advantage to make the arch consist of a large number of narrow voussoirs, rather than a small number of wide ones, as the former will be able to carry a heavier load.

Since the reaction, R, measures the pressure which the surface of the bed of the stone has to sustain, we must take care that it does not exceed the limit of safety for the resistance of the material to a crushing force. If $b . d$ is the area in inches of the bed, and S is the load per square inch that may be safely borne, which is about one-tenth of the crushing-weight, we must have R not greater than $S . b . d$. If Yorkshire stone is used the value of S is about 760 lbs.; if Portland stone, S is 390 lbs., and for Bath stone S is 150 lbs. It will be evident that by increasing the area of the bed of each voussoir, or its depth d, we increase the strength of the arch itself, or its usefulness for supporting a heavy load.

For example, we will suppose an arch of 1 foot breadth of soffit has to support a wall 20 feet high, the back of each voussoir being 12 inches; to find the necessary depth d of the voussoirs, b being 1 foot. Take the keystone as bearing a load of 20 cubic feet of brickwork weighing 1 cwt. per cubic foot; then we have—

$$P = 20 \text{ cwt.} = 2240 \text{ lbs.}$$

Let $\theta = 6°$, the arch being semi-circular; then we have sin. $\theta = \cdot 10453$; and from equation (132)

$$R = \frac{P}{2 \sin. \theta}$$

$$= \frac{1120}{\cdot 10453} = 11,000 \text{ lbs. nearly.}$$

Also we have—

$$R = S . b . d$$

$$= 12 \, S . d, \text{ since } b = 12 \text{ ins.};$$

therefore, $$d = \frac{11000}{12 \, S}.$$

If we put S = 150, for Bath stone, we find—

$$d = \frac{11000}{1800} = 6{\cdot}1 \text{ ins.}$$

The radius of the arch in this case is 4 feet, or the span 8 feet.

80. JOINT OF RUPTURE.—Suppose we have a semi-circular arch, as CAL (fig. 83), resting on two supports

Fig. 83.

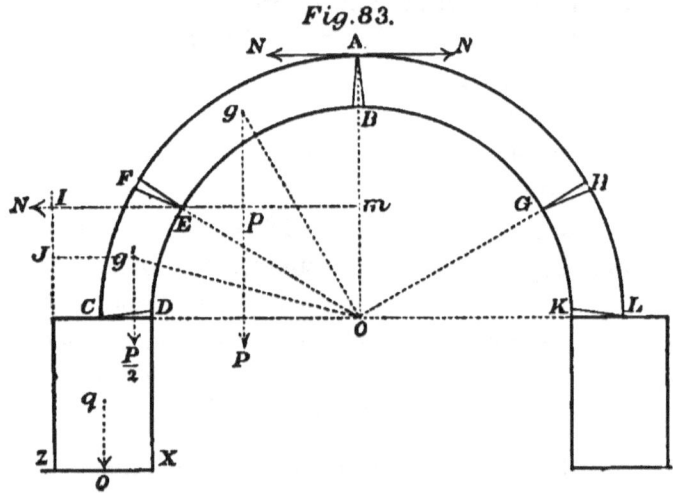

at CD and KL. Then, from what has been previously shown (**79**), if the haunches at F and H are insufficiently loaded, the voussoirs will have a tendency to rise at F and H, being pushed back by the pressure of the upper part of the arch. The arch in this case will give way by the falling in of the crown, the joint at AB opening at B, so that the whole pressure comes upon the edge at A. The haunches will rise at F and H by the opening of the joints EF and GH at F and H, whereby the whole of the pressure is thrown upon the edges at E and G. Let N represent the mutual reaction of the two halves of the arch meeting at AB; then at the instant of rupture the force N acts horizontally at the edge A, and in order to ascertain the conditions of equilibrium in the arch, we must find the position of the joint EF, where the thrust, N, of the other half of the arch will have its greatest effect; that joint being called the "joint of rupture."

Let O be the centre of curvature from which all the joints of the arch radiate; let P be the weight of any portion of the arch between AB and the joint EF; then P will act vertically at the centre of gravity, g, of the arch, and its *moment* about E will be, P × Ep. The *moment* of N about E is, N × Am; and in order that there may be equilibrium in the arch, we must have—

$$N \times Am = P \times Ep\,;$$

or,
$$N = P\frac{Ep}{Am} \qquad . \qquad . \quad (136)$$

In order to find the "joint of rupture," or that at which the value of N is greatest, we must express these quantities in terms of the angle, θ, which OEF makes with the vertical OB; and by calculating N for several values of θ, we find what value of θ makes N greatest.

It is necessary in the first place to ascertain the position, g, of the centre of gravity of the arch ABEF, which we do by means of equation (31), namely,

$$Og = \frac{4}{3}\frac{R^3 - r^3}{R^2 - r^2} \times \frac{\sin.\frac{\theta}{2}}{\theta}$$

where OA = R, OB = r, and θ is the angle BOE.

Then we have—

$$Ep = Em - mp$$

$$= r \cdot \sin \cdot \theta - Og \cdot \sin \cdot \frac{\theta}{2}.$$

$$= r \cdot \sin \cdot \theta - \frac{4}{3} \frac{R^3 - r^3}{R^2 - r^2} \cdot \frac{\left(\sin \cdot \frac{\theta}{2}\right)^2}{\theta} \quad . \quad (137)$$

$$Am = OA - Om$$

$$= R - r \cdot \cos \cdot \theta \quad . \quad . \quad (138)$$

Taking one foot for the thickness of the arch, and making δ represent the weight per cubic foot of the material; we have—

$$P = (R + r)(R - r) \frac{\theta}{2} \cdot \delta$$

$$= \delta \frac{\theta}{2}(R^2 - r^2) \quad . \quad . \quad . \quad (139)$$

Then by combining the equations (137, 138, 139), we obtain the value of N for any given values of θ, R and r, by means of equation (136).

In order to find the value of θ which makes N greatest, we take the case of R = 12, r = 10, and calculate N for various values of θ. Take θ = 56°; or, arc θ = ·9774, see table (21).

Then, $\sin \cdot \theta = \cdot 820$, $\cos \cdot \theta = \cdot 5592$, $\sin \cdot^2 \frac{\theta}{2} = \cdot 2204$; and we find

$$N = 11 \cdot 127 \times \delta.$$

Now take θ = 58°; or, arc θ = 1·0123, $\sin \cdot \theta$ = ·848, cos. $\theta = \cdot5299$, $\sin.^2 \dfrac{\theta}{2} = \cdot2354$; from which we find—

$$N = 11\cdot161 \times \delta.$$

Let $\theta = 60°$; or, arc $\theta = 1\cdot0472$, sin. $\theta = \cdot866$, cos. $\theta = \cdot5$, $\sin.^2 \dfrac{\theta}{2} = \cdot25$, from which

$$N = 11\cdot17 \times \delta.$$

Let $\theta = 62°$; or, arc $\theta = 1\cdot0821$, sin. $\theta = \cdot883$, cos. $\theta = \cdot4695$, $\sin.^2 \dfrac{\theta}{2} = \cdot2653$, and we get

$$N = 11\cdot155 \times \delta.$$

Let $\theta = 64°$; or, arc $\theta = 1\cdot117$, sin. $\theta = \cdot8988$, cos. $\theta = \cdot4384$, $\sin.^2 \dfrac{\theta}{2} = \cdot2808$,

$$N = 11\cdot106 \times \delta.$$

Hence it appears that the value of N is greatest when $\theta = 60°$, its values diminishing as θ either increases or decreases. The joint EF which makes 60° with the vertical may therefore be considered to be the " angle of rupture."

81. STABILITY OF THE ARCH.—Having found the " joint of rupture," we can now proceed to determine the necessary strength of the pier or abutment of given height to resist the thrust of the arch. By means of equation (136) we obtain the value of the horizontal thrust N, and since the *moments* of N and P balance about the point E, we can by the " transposition of couples," which has been previously demonstrated (**11**), consider N and P as acting at the point E, without

alteration of direction or value. Let ZX be the base of
the pier which supports the arch, h its height DX, t
its required thickness ZX ; then we have to equate the
moments of P and N acting at E, of the weight $\dfrac{P}{2}$ of
the arch between CD and EF acting at g', and of the
weight Q of the pier acting at its centre of gravity q ;
all these *moments* to be taken about the outer edge Z
of the pier.   From equation (139) we have (when θ
$= 60°$)—

$$P = \delta \times \cdot 5236 \ (R^2 - r^2) \qquad . \qquad (140)$$

$$Ep = \cdot 866 \, r - \cdot 3183 \left(\frac{R^3 - r^3}{R^2 - r^2} \right) \qquad . \quad (141)$$

$$Am = R - \frac{r}{2} \qquad . \qquad . \qquad (142)$$

$$N = P \times \frac{Ep}{Am}.$$

The *moment* of P, acting at E, about Z, is—

$$P \times EI = P \ (r + t - r \cdot \sin. \theta)$$
$$= P \ (t + \cdot 134 \, r).$$

The *moment* of N, acting at E, about Z, is—

$$N \times IZ = N \cdot \left(h + \frac{r}{2} \right).$$

The *moment* of $\frac{1}{2}$ P, at g', about Z, is—

$$\tfrac{1}{2} P \times Jg' = \tfrac{1}{2} P \ (t + r - Og' \cdot \cos. \ DOg')$$
$$= \tfrac{1}{2} P \left(t + r - \cdot 37 \, \frac{R^3 - r^3}{R^2 - r^2} \right).$$

The *moment* of Q, acting at q, about Z, is—

$$Q \times \frac{t}{2} = \frac{\delta}{2} h \cdot t^2.$$

Equating the *moment* of N with all the other *moments*, we obtain the " Equation of Equilibrium," namely,

$$N \left(h + \frac{r}{2} \right)$$

$$= P (t + \cdot 134\, r) + \frac{P}{2} \left(r + t - \cdot 637\, \frac{R^2 - r^3}{R^2 - r^2} \right)$$

$$+ Q \times \frac{t}{2} \qquad . \qquad . \quad (143)$$

The value of N is found from the equations (140, 141, 142) ; and if the arch and abutment are of similar material we can omit δ from the equations, or consider $\delta = 1$.

For example, let R = 12 ft., r = 10 ft., h = 10 ft, then we find from equation (140), P = 23·038, $\frac{1}{2}$ P = 11·519 ; from equation (141) Ep = 3·394 ; from equation (142) Am = 12 − 5 = 7 ;

$$\therefore \quad N = \frac{23 \cdot 038 \times 3 \cdot 394}{7} = 11 \cdot 17.$$

The moment of N at E, about Z, is—

$$11 \cdot 17 \times 15 = 167 \cdot 55 ;$$

the moment of P at E, about Z, is—

$$23 \cdot 038 (t + 1 \cdot 34) = 23 \cdot 038\, t + 30 \cdot 87 ;$$

the moment of $\frac{1}{2}$ P at g', about Z, is—

$$11 \cdot 519 (t - \cdot 54) = 11 \cdot 519\, t - 6 \cdot 22 ;$$

the moment of Q at q, about Z, is—

$$Q\frac{t}{2} = 5\,t^2.$$

Then the "equation of equilibrium" becomes—

$$167{\cdot}55 = 23{\cdot}038\,t + 11{\cdot}519\,t + 30{\cdot}87 - 6{\cdot}22 + 5\,t^2,$$

or,
$$t^2 + 6{\cdot}9\,t - 28{\cdot}6 = 0;$$

$$t = 3 \text{ ft.}$$

The equation (143) however only gives the thickness t of the pier that will just produce *equilibrium* between the forces, and in order to find the thickness that will give *stability* to the structure we must multiply N by 2 in the foregoing equation; we then have the "Equation of Stability," which in this example becomes

$$335{\cdot}12 = 34{\cdot}557t + 24{\cdot}65 + 5t^2$$

$$\text{or, } t^2 + 6{\cdot}9t - 62{\cdot}1 = 0;$$

$$t = 5 \text{ ft.}$$

The expression for P given in equation (140) shows that P, and consequently N, is proportional to the *square* of the span of the arch, so that if the span is doubled the thrust is increased four-fold; and if trebled the thrust is increased nine times, and so on; as long as the proportions of the several parts remain the same. Thus the thrust of similar arches having 10 feet, 20 feet, and 30 feet span, will be in the proportion of 1, 4, and 9.

82. LOADED ARCH.—In the foregoing investigations we have considered the arch as only sustaining its own weight; but as arches are usually employed to carry a load or "surcharge," we have now to take this into consideration. We will, for the sake of simplicity, consider that the top of the surcharge is level, as KLM (fig. 84), and that the weight per cubic foot is the same as that of the arch and pier. The force P will now be repre-sented by the figure EFLMB, acting at G, its centre of gravity. The exact determination by analysis of the posi-tion of G involves complicated for-mulæ, which we can dispense with by cutting out the figure EFLMB in cardboard and sus-pending it from two of its angles with a. plumbline, when the intersec-tion of the two directions of the plumbline will give

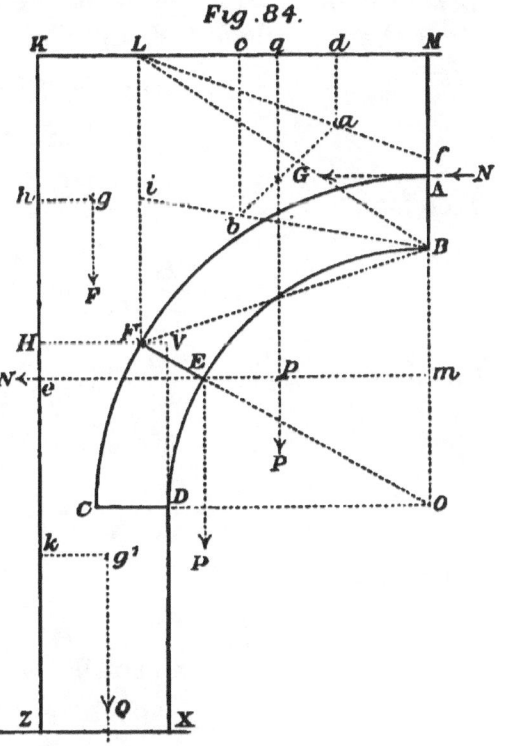

Fig. 84.

the point G, as before described (19). We can, however, find it approximately, and also the value of P, by drawing the line BF and taking the trapezium FLMB to represent the above figure. Then by dividing the trapezium into two triangles by the line BL, we can find (17) the centres a and b of each triangle; then joining ab we find G by the proportion

$$Gb : Ga = BM : FL,$$

as before described (18). Draw the vertical Gp cutting the horizontal Em in p; then we can find the length mp or Mq in the following manner. Divide ML into three equal parts at the points c and d, and draw the verticals bc, ad, and Gq. Then we have

$$cq : dq = Gb : Ga$$
$$= BM : FL;$$

therefore,
$$\frac{cq + dq}{dq} = \frac{FL + BM}{FL};$$

or,
$$\frac{dq}{cd} = \frac{FL}{FL + BM}, \text{ and } cd = \frac{ML}{3};$$

therefore,
$$dq = \frac{ML}{3} \times \frac{FL}{FL + BM}.$$

Then,
$$mp = Mq = Md + dq$$
$$= \frac{ML}{3}\left(1 + \frac{FL}{FL + BM}\right) \quad (144)$$

Then we have—

$$Ep = Em - mp$$
$$= r \cdot \sin. 60° - mp$$
$$= \cdot 866\, r - mp \qquad . \qquad . \quad (145)$$
$$Am = OA - Om$$
$$= R - \frac{r}{2} \qquad . \qquad . \qquad . \quad (146)$$

$$P = \frac{BM + FL}{2} \times LM \times \delta \quad . \quad (147)$$

$$N = P\,\frac{Ep}{Am} \qquad . \qquad . \quad (148)$$

When the angle BOE is 60°, and we put k for the height AM of the surcharge, we can find the values of BM, FL, and LM, in terms of R, r, and k—

$$BM = AB + AM$$
$$= R - r + k$$
$$FL = OM - OF \times \cos. 60°$$
$$= k + R - \frac{R}{2}$$
$$= k + \frac{R}{2}$$
$$ML = OF \times \sin. 60°$$
$$= ·866\ R.$$

For example, let $R = 12$, $r = 10$, $k = 5$, then we have, BM = 7, FL = 11, ML = 10·392.

$$P = 5·196 \times 18 . δ = 93·528 . δ,\ \text{from equation (147)}$$
$$mp = 3·464 \times \frac{20}{18},\ \text{from equation (144)}$$
$$= 5·58;$$
$$Ep = 8·66 - 5·58 = 3·08,\ \text{from equation (145)}$$
$$Am = 12 - 5 = 7,\ \text{from equation (146).}$$

Then by equation (148) we have—

$$N = \frac{93·528 \times 3·08}{7} \times δ = 41·152 . δ.$$

In order to determine the thickness t of the pier of given height h, we take the *moments* of N and P, as acting at E, about the point Z; and the pier can be taken as consisting of the two rectangles HVXZ, called Q, and

o

HFLK, called F, whose *moments* about Z are $Q \times \dfrac{t}{2}$
and $F \times gh$. Let $h = 10 = DX$.

$$KL = MK - ML = t + r - \cdot 866\ R = t - \cdot 302,$$
$$F = KL \times FL = 11\,t - 4\cdot312,$$
$$gh = \tfrac{1}{2}\,KL = \tfrac{1}{2}\,(t - \cdot 392).$$

Then the *moment* of F about Z is—

$$F \times gh = 5\cdot5\ (t^2 - \cdot784\,t + \cdot154)$$
$$= 5\cdot5\,t^2 - 4\cdot31\,t + \cdot847$$

$$HZ = h + \frac{R}{2} = 16,$$

$$Q \cdot \frac{t}{2} = HZ \times \frac{t^2}{2} = 8\,t^2.$$

The *moment* of N, acting at E, and taken about Z, is—

$$N \times Zc = N\left(h + \frac{r}{2}\right) = 41\cdot152 \times 15 = 617\cdot28.$$

The *moment* of P, at E, taken about Z, is—

$$P \times Ec = P\ (t + r - \cdot 866\,r) = 93\cdot528\ (t + 1\cdot34)$$
$$= 93\cdot528\,t + 125\cdot33.$$

The *equation of equilibrium* is therefore—

$$N \times Zc = P \times Ec + Q \times \frac{t}{2} + F \times gh \qquad (149)$$

which becomes in this case—

$$617\cdot28 = (93\cdot528 - 4\cdot31)\,t + (5\cdot5 + 8)\,t^2$$
$$+ 125\cdot33 + \cdot847,$$

or, $$13\cdot5\,t^2 + 89\cdot218\,t - 491 = 0$$

which reduces to the equation—

$$t^2 + 6\cdot6\,t - 36\cdot4 = 0,$$

the solution of which is, $t = 3\cdot55$.

To determine t for *stability* we have to put 2 N for N in equation (149); which gives—

$$1234\cdot56 = 89\cdot218\,t + 13\cdot6\,t^2 + 126\cdot177,$$

or,

$$t^2 + 6\cdot6\,t - 82\cdot1 = 0$$

$$\therefore t = 6\cdot35.$$

If we put W for the pressure on the joint EF, for 1 foot width of arch, we have—

$$W = P.\cos. 30^\circ + N.\sin. 30^\circ$$

$$= (93\cdot528 \times \cdot863 + 41\cdot152 \times \cdot5)\,\delta$$

$$= 101\cdot57 \,.\, \delta$$

Suppose the material used to be Bath stone, weighing 120 lbs. to the cubic foot; then we have—

$$W = 12,188 \text{ lbs.}$$

The safe load which Bath stone will sustain is 150 lbs. on the square inch, and as the area of the bed is 2 feet or 288 square inches, the safe load amounts to 43,200 lbs. which is $3\frac{1}{2}$ times that which the stone has to support; so that the depth of the voussoirs might safely be reduced to 7 inches at EF, and to half that at the crown. The pressure on the springing CD is 15,000 lbs., which requires that CD should be not less than $8\frac{1}{3}$ inches. There are, however, other matters to be considered in fixing the depth of the arch, which will be seen hereafter (**83**).

The following examples of the application of the

above equations show how the thickness of pier for different spans and height of pier can be calculated.

Example 1.—Let $R=6$, $r=5$, $k=3$, $h=10$; then $R - r = 1$, BM $= 4$, FL $= 6$, ML $= 5·196$, KL $= t - ·196$, HZ $= 13$.

From equation (147), $P = \dfrac{4 + 6}{2} \times 5·196 . \delta = 25·98 . \delta$

From equation (144), $mp = 1·732 \times \dfrac{16}{10} = 2·77$.

From equation (145), $Ep = 4·33 - mp = 1·56$.
From equation (146), $Am = 6 - 2·5 = 3·5$.

Therefore, $\quad N = P \dfrac{Ep}{Am} = \dfrac{25·98 \times 1·56}{3·5} \delta = 11·58 . \delta$;

$$F = KL \times FL = 6 \, (t - ·196);$$
$$F \times gh = 3 \, (t^2 - ·392t + ·04);$$
$$Q \times \frac{t}{2} = HZ \times \frac{t^2}{2} = 6·5 \, t^2;$$

$Ze = 10 + 2·5 = 12·5$; $Ee = t + 5 \times ·134 = t + ·67$.

Then the *Equation of Stability* when pier and arch are of same material, is, $2 \times 11·58 \times 12·5 = 25·98 \, (t + ·67) + 3 \, (t^2 - ·392 \, t + ·04) + 6·5 \, t^2$; which reduces to, $t^2 + 2·61t - 28·63 = 0$; from which we find, $t = 4·2$.

Example 2.—$R = 10$, $r = 8$, $k = 4$, $h = 10$; then $R - r = 2$, BM $= 6$, FL $= 9$, ML $= 8·66$, KL $= t - ·66$, HZ $= 15$, $Ze = 14$, $Ee = t + 1·072$.

$$P = \frac{6 + 9}{2} \times 8·66 . \delta = 65 . \delta;$$

$$mp = \frac{8·66}{3} \times \frac{24}{15} = 4·62;$$

$$\mathrm{E}p = 6{\cdot}928 - 4{\cdot}62 = 2{\cdot}308\,;$$

$$\mathrm{A}m = 10 - 4 = 6\,;$$

$$\mathrm{N} = \frac{65 \times 2{\cdot}308}{6} = 25\,.\,\delta\,;$$

$$\mathrm{F} \times gh = \frac{9}{2}\,(t - {\cdot}66)^2 = 4{\cdot}5\,(t^2 - 1{\cdot}32\,t + {\cdot}44)$$

$$\mathrm{Q} \times \frac{t}{2} = 15 \times \frac{t^2}{2} = 7{\cdot}5\,t^2$$

Then the *equation of stability* becomes—

$$50 \times 14 =$$

$$65\,(t + 1{\cdot}072) + 4{\cdot}5\,(t^2 - 1{\cdot}32\,t + {\cdot}44) + 7{\cdot}5\,t^2$$

or,

$$t^2 + 5\,t - 52{\cdot}36 = 0$$

$$t = 5{\cdot}15.$$

Example 3.—Let $\mathrm{R} = 12$, $r = 10$, $k = 5$, $h = 10$; then $\mathrm{R} - r = 2$; $\mathrm{BM} = 7$, $\mathrm{FL} = 11$, $\mathrm{ML} = 10{\cdot}392$, $\mathrm{KL} = t - {\cdot}392$, $\mathrm{HZ} = 16$, $\mathrm{Z}e = 15$, $\mathrm{E}e = t + 1{\cdot}34$.

$$\mathrm{P} = 93{\cdot}528\,.\,\delta$$

$$mp = 5{\cdot}58,\ \mathrm{E}p = 3{\cdot}08,\ \mathrm{A}m = 7$$

$$\mathrm{N} = \frac{93{\cdot}528 \times 3{\cdot}08}{7}\,.\,\delta = 41{\cdot}152\,.\,\delta$$

$$\mathrm{F} \times gh = 5{\cdot}5\,(t^2 - {\cdot}784\,t + {\cdot}154)$$

$$\mathrm{Q}\,.\,\frac{t}{2} = 8\,t^2$$

Then the *equation of stability* becomes—

$$1234{\cdot}56 = 89{\cdot}218\,t + 13{\cdot}5\,t^2 + 126{\cdot}177$$

or,

$$t^2 + 6{\cdot}6\,t - 82{\cdot}1 = 0$$

$$t = 6{\cdot}35.$$

Example 4.— Let $\mathrm{R} = 16$, $r = 14$, $k = 7$, $h = 15$;

then $R - r = 2$, $BM = 9$, $FL = 15$, $ML = 13.856$,
$KL = t + .144$, $HZ = 23$, $Ze = 22$, $Ec = t + 1.876$;

$$P = 166.27 \cdot \delta$$

$$mp = 7.5, \quad Ep = 12.124 - 7.5 = 4.624,$$
$$Am = 16 - 7 = 9.$$

$$N = \frac{166.27 \times 4.624}{9} = 85.423 \cdot \delta$$

$$F \times gh = \frac{15}{2}(t + .144)^2 = 7.5(t^2 + .288\,t + .021)$$

$$Q \cdot \frac{t}{2} = 11.5\,t^2.$$

The *equation of stability* becomes—

$$3759 = 19\,t^2 + 168\,t + 312,$$

or,
$$t^2 + 9\,t - 181 = 0$$

$$t = 9.65.$$

Example 5.—Let $R = 20$, $r = 17$, $k = 10$, $h = 20$;
then $R - r = 3$, $BM = 13$, $FL = 20$, $ML = 17.32$,
$KL = t - .32$, $HZ = 30$, $Ze = 28.5$, $Ec = t + 2.278$

$$P = 285.8$$

$$mp = 9.27, \quad Ep = 14.72 - 9.27 = 5.45, \quad Am = 11.5$$

$$N = \frac{285.8 \times 5.45}{11.5} \cdot \delta = 135.44 \cdot \delta$$

$$F \times gh = 10(t - .32)^2 = 10(t^2 - .64\,t + .102)$$

$$Q \times \frac{t}{2} = 15\,t^2.$$

The *equation of stability* is—

$$135.44 \times 57 = \quad .$$

$$285.8(t + 2.278) + 10(t^2 - .64\,t + .102) + 15\,t^2,$$

or, $t^2 + 11\cdot176\,t - 282\cdot7 = 0$,

$$t = 12\cdot1.$$

Example 6.—Let $R = 24$, $r = 20$, $k = 12$, $h = 25$; then $R - r = 4$, $BM = 16$, $FL = 24$, $ML = 20\cdot78$, $KL = t - \cdot784$, $HZ = 37$, $Ze = 35$, $Ee = t + 2\cdot68$;

$$P = 415\cdot6 \,.\, \delta$$

$$mp = 11\cdot09, \ Ep = 17\cdot32 - 11\cdot09 = 6\cdot23,$$

$$Am = 24 - 10 = 14$$

$$N = \frac{415\cdot6 \times 6\cdot23}{14}\,\delta = 185 \,.\, \delta$$

$$F \times gh = 12\,(t - \cdot784)^2 = 12\,(t^2 - 1\cdot568\,t + \cdot615)$$

$$Q \,.\, \frac{t}{2} = 18\cdot5\,t^2.$$

The *equation of stability* becomes—

$$185 \times 70 = 30\cdot5\,t^2 + 396\cdot8\,t + 1106\cdot4,$$

or, $t^2 + 13\,t - 388 = 0$,

$$t = 14\cdot25.$$

The following Table gives the results obtained in the foregoing examples, the material of the arch and pier being supposed to be the same in all cases, so that the value of δ can be omitted.

R.	r.	k.	h.	t.
6	5	3	10	4·20
10	8	4	10	5·15
12	10	5	10	6·35
16	14	7	15	9·65
20	17	10	20	12·10
24	20	12	25	14·25

The following method of approximating to the calculation of the thrust of a loaded arch, will be found more simple in practice, although perhaps not quite so accurate as the foregoing; the result obtained will however be sufficiently exact for the purposes of the architect.

Fig. 85.

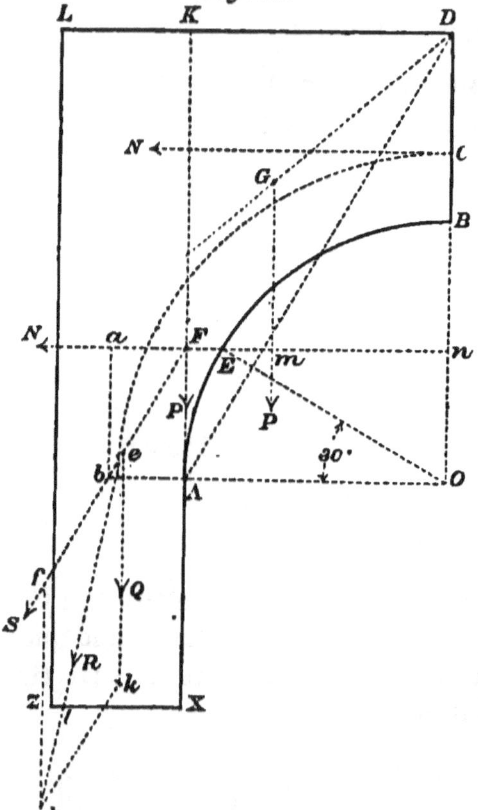

Let AB (fig. 85) be the *intrados* of the semi-arch, DKL the load line, AX the height of the pier, XZ its thickness. Draw the line AD, and suppose the triangle ADK to represent (very nearly) the area of the figure EBDKF, and let G be the centre of gravity of the triangle (17). The radius OE is drawn at an angle of 30° with the horizontal line OA, O being the centre of the arch. Through E draw the horizontal line FE*mn*; the point F being on the vertical KAX. We will consider F to be the point about which the *moments* of P at G and N at C balance one another;

N being the horizontal thrust at C, P the weight of the triangle ADK acting at G. Then, as before, we have—

$$P \times Fm = N \times Cn$$

or,

$$N = P \, \frac{Fm}{Cn}.$$

which determines the value of N.

We can now take N and P as acting horizontally and vertically at the point F, and equate their *moments* about Z with the *moment* of the weight (Q) of the pier whose sectional area is LKXZ, the force Q acting down the centre of the pier, as *ck*. Putting *t* for the thickness XZ of the pier, we have for *equilibrium*—

$$N \times FX = P \cdot t + Q \cdot \frac{t}{2}$$

And for *stability*,

$$2 N \times FX = P \cdot t + Q \cdot \frac{t}{2}$$

For example, let OA = 10, BC = $1\frac{1}{2}$; CD $\times$ $2\frac{1}{2}$, AX = 10, FX = 15, Q = 24 t, P = $\dfrac{14 \times 10}{2}$ = 70, Fm = $\dfrac{10}{3}$, Cn = $\dfrac{13}{2}$.

Then we have—

$$N = 70 \, \frac{\frac{10}{3}}{\frac{13}{2}} = 36$$

For *equilibrium*, the equation is—

$$36 \times 15 = 70 \, t + 12 \, t^2$$

or, $t^2 + 6t - 45 = 0$

$t = 4\cdot4.$

For *stability*, the equation is—

$72 \times 15 = 70t + 12t^2$

or, $t^2 + 6t - 90 = 0$

$t = 7$, nearly.

We can also determine the stability of the structure *geometrically*, as follows: Take Fa to represent on any scale the value of N, or 36 in the above example. FA and ab to represent on the same scale the value of P, or 70. Draw the diagonal Fb, which we call S, the resultant of N and P acting at F. Let e be the point where a vertical through the middle of the pier cuts the diagonal Fb, assuming some thickness, as 5, for the pier. Take ek to represent on the above scale the value of Q, the weight of the pier LKXZ, or its area : Q = 24t = 120, in this case. Complete the parallelogram efhk by making ef = Fb = kh; then the diagonal eh represents the resultant R of the forces N, P, and Q, and will cut the base ZX at a point l within the base.

In this way we can determine the amount of stability possessed by the structure for any given thickness of the pier. If the point l lies *outside* Z, the structure will be overthrown; if it falls *at* Z, there will only just be equilibrium; and for stability the point l should be at least one-fourth of ZX *within* the base, or Zl should be not less than ¼ ZX.

83. LINE OF PRESSURES.—The stability of an arch and its abutments can be determined by geometrical

methods in the following manner. Let ABCD (fig.
86) represent the half of a semi-circular arch having a
surcharge MK, and a supporting pier DZ. From O

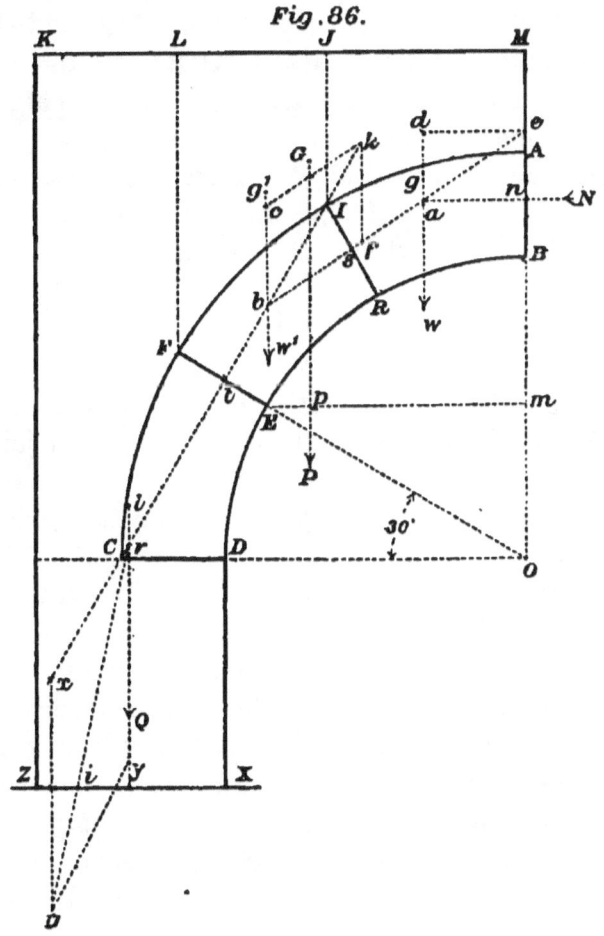

Fig. 86.

the centre of curvature draw OEF, making the angle
EOD equal to 30° with the horizontal line ODC; then
EF is the "joint of rupture" as previously determined

(80). Divide ABEF into two equal voussoirs with a common joint RI; draw the verticals IJ and FL. Find G, the centre of gravity of the arch ABEF and its surcharge, and also the weight P which will act at G, by the methods previously given (82). Also find the centre of gravity g of the voussoir ABRI and its surcharge, and the centre g' of IREF and its surcharge; and call w and w' the weights of these two parts respectively, acting at g and g'. Draw the horizontal line Em, and the vertical Gp; then find the value of the horizontal thrust N by the method given above, namely—

$$N = P \frac{E p}{A m}.$$

When the arch is in a condition of stability the horizontal pressure N will act at the centre n of the joint AB.

Suppose the line na to represent on any convenient scale the force N, and the line ad the weight of the part JMBRI, acting through the centre g; take ne equal to ad and draw the diagonal ea, which will represent the resultant of the two forces. Produce ea to meet the joint RI in s, and also to meet a vertical bc through g' in the point b. The points n and s are called "centres of resistance." Take bc to represent the weight of the second part acting through g', and make bf equal to ae; draw the vertical fk equal to bc, and draw the diagonal kb, producing it to meet the joint EF in t, and also to meet the vertical through l, the centre of gravity of the abutment, in the point r; then t is another "centre of resistance," and kb represents the resultant of the forces on the second voussoir.

A curved line drawn through the "centres of resistance," *n*, *s*, *t*, &c., is called the "line of pressures," and in order to secure the stability of the arch it is essential that this curve should lie entirely within the depth of the arch, and should in no place come nearer to the *extrados* or *intrados* than one-fourth of the depth of the arch.

To determine the stability of the abutment, draw the vertical *ry*, through *l*, the centre of gravity, of the abutment, and let *ry* represent Q on the same scale that *an* represents N; produce *tr* to *x*, making *rx* equal to *kb*; draw the vertical *xu* equal to *ry*; then the diagonal *ru* represents the resultant of all the forces acting on the pier. If the point *i*, where *ru* cuts the base of the pier, lies within the base ZX, the structure will be in a condition of stability, but if it falls either at Z or outside of it, the pier will be pushed over by the arch. The distance Z*i* should be at least one-fourth of ZX in order to secure stability to the structure.

When applying this method in practice, the arch should be drawn on a large scale and divided into several voussoirs, so as to get as many "centres of resistance" as possible.

An example of the application of the geometrical method will be found in a paper on "Vaulting" by T. H. Eagles, read before the Royal Institute of British Architects on June 1, 1874.

84. ARCADES.—It is a common occurrence in Architecture to have a number of arches arranged in a row, two adjoining arches being made to spring from the same pier, so as to form an "Arcade." When the

arches are equal in span and carry an equal load, their
horizontal thrusts will counterbalance one another, so
that the strength of the piers need not be more than
sufficient to resist the vertical load or crushing weight
of the superstructure ; except in the case of the abutments at the two extremities of the arcade, which
must be made sufficiently strong to resist the horizontal thrust N of the end arches, as calculated in the
foregoing examples (**82**).   When however it happens,
as is frequently the case, that the span of the arches
varies, then the thrust of the larger arches will only be
partially counteracted by that of the smaller ones;
and as we have seen (**81**) that the horizontal thrust increases as the square of the span in arches of similar
construction, it is evident that if one arch is double
the span of that next to it, the pier which supports
them both must be made strong enough to resist three-
fourths of the thrust of the larger arch, as the thrust
of the smaller one is only one-fourth that of the larger.
If one arch is 10 feet span and the next one is 14 feet
span, or in like proportion, the larger will have just
double the thrust of the former, so that only one-half of
its thrust is balanced by the thrust of the smaller arch,
and the other half of the thrust must therefore be sustained by the supporting pier.

The necessity of making the end piers of an arcade
stronger than the intermediate ones was pointed out
long ago in the 6th book of Vitruvius, showing that
the thrust of an arch formed of wedge-shaped voussoirs
was well understood in his days.   He says " itemque
quæ pilatim aguntur ædificia, cum cuneorum divisioni-
bus coagmentis ad centrum respondentibus fornices

concluduntur, extremæ pilæ in his latiores spatio sunt faciundæ, uti vires eæ habentes resistere possint, cum cunei ab oneribus parietum pressi per coagmenta ad centrum se prementer extrudunt incumbas. Itaque si angulares pilæ erunt spatiosis magnitudinibus, continuendo cuneos firmitatem operibus præstabunt."

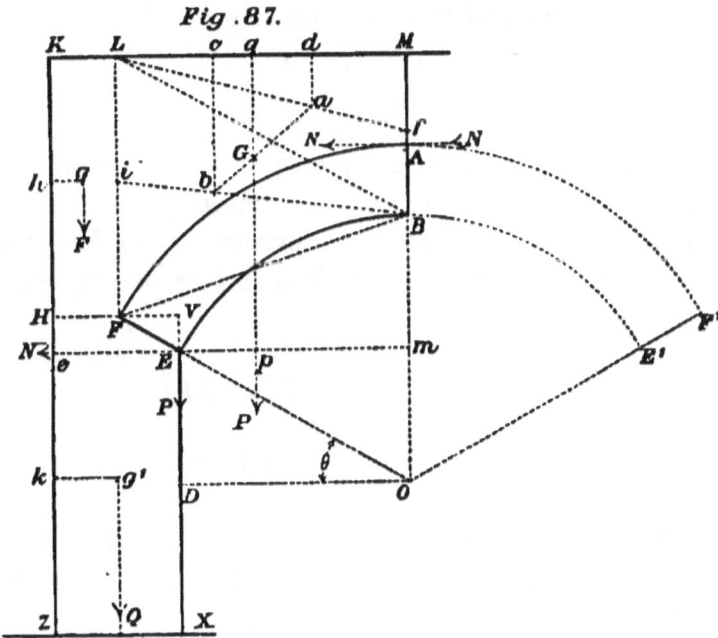

Fig. 87.

85. SEGMENTAL ARCH.—An arch which is formed by an arc of a circle less than a semi-circle is called "segmental," as EFABE'F' (fig. 87), which subtends at O an angle less than two right angles. If the angle EOD, or θ, which the joint EF makes with the horizontal line OD, is less than or equal to 30°, then the oint which makes 30° with OD is the "joint of rupture," as in the case of the semi-circular arch (**82**). When θ is greater than 30° the springing joint or " skewback " EF will be the "joint of rupture," and we proceed to find P and N in the same way as before (**82**). We can get their values approximately by drawing BF, FL, and considering P to be the area of the trapezium FLMB, as in the case of a semi-circle. Bisect FL in i and BM in f, draw Bi and Lf, and take $af = \frac{1}{3} Lf$, $bi = \frac{1}{3} Bi$; join a and b; then take

$$bG : aG = BM : FL.$$

Then G is the centre of gravity of the figure BFLM. Draw the verticals bc, ad, Gq, meeting the line of surcharge MK. Draw the horizontal line Em, and the vertical Gp; then $mp = Mq$, and as before, equation (144),

$$mp = \frac{ML}{3} \left(1 + \frac{FL}{BM + FL} \right)$$

$$P = \frac{BM + FL}{2} \times ML \times \delta$$

$ML = R \cdot \cos \theta$, $FL = MA + R - R \cdot \sin \theta = k + R(1 - \sin \theta)$, $BM = k + R - r$, $Ep = Em - mp = r \cdot \cos \theta - mp$.

$$Am = OA - OM = R - r \cdot \sin \theta$$

$$N = P \frac{Ep}{Am} \cdot$$

For example, let $\theta = 45°$, $R = 12$ ft., $r = 10$ ft., $k = 5$ ft., $h = EX = 10$ ft.

Then, $\sin \theta = \cos \theta = \cdot 707$, $ML = 8\cdot484$, $FL = 8\cdot52$, $BM = 7$.

$$mp = 2 \cdot 828 \times \frac{24 \cdot 04}{15 \cdot 52} = 4 \cdot 446$$

$$\mathrm{E}p = 7 \cdot 07 - 4 \cdot 45 = 2 \cdot 62, \quad \mathrm{A}m = 12 \times 7 \cdot 07 = 4 \cdot 93$$

$$\mathrm{P} = 4 \cdot 242 \times 15 \cdot 52 \,.\, \delta = 65 \cdot 83 \,.\, \delta$$

$$\mathrm{N} = \frac{65 \cdot 83 \times 2 \cdot 62}{4 \cdot 93} \, \delta = 35 \,.\, \delta$$

To find the requisite thickness t of the pier, we have, as before, to suppose P and N to act at E, and to take their *moments* about Z. We have also the *moment* of the rectangle FHKL, which we call F, acting at g, and of the rectangle HVXZ, which we call Q, acting at g'.

$$\mathrm{F} = \mathrm{KL} \times \mathrm{FL} = \mathrm{FL} \times (t + r \,.\, \cos. \,\theta - \mathrm{ML})$$
$$= 8 \cdot 52 \,(t - 1 \cdot 414); \text{ and } gh = \tfrac{1}{2} \mathrm{KL}; \text{ so that,}$$

$$\mathrm{F} \times gh = 4 \cdot 26 \,(t - 1 \cdot 414)^2 = 4 \cdot 26 \,(t^2 - 2 \cdot 828 \, t + 2)$$
$$\mathrm{Q} = \mathrm{HZ} \times t = (h + (\mathrm{R} - r) \sin. \theta) \, t = 11 \cdot 414 \, t; \text{ and}$$

$$kg' = \frac{t}{2}; \, \mathrm{Q} \times kg' = 5 \cdot 707 \, t^2.$$

Then the *equation of stability* is—

$$2 \, \mathrm{N} \times h = \mathrm{P} \times t + \mathrm{F} \times gh + \mathrm{Q} \times kg',$$

which becomes in this example—

$$700 = 65 \cdot 83 \, t + 9 \cdot 967 \, t^2 - 12 \cdot 05 \, t + 8 \cdot 52$$

or,
$$t^2 + 5 \cdot 4 \, t - 69 \cdot 2 = 0$$
$$t = 6 \cdot 05 \text{ feet.}$$

The span of the arch is $2 \, r \,.\, \cos. \, \theta$, which is 14·14 feet in this case.

The thrust of segmental arches of equal span but of different radii of curvature will be found to vary nearly

r

as the square of the radius r of the intrados, if all the other dimensions remain the same. Thus if in the foregoing example we keep the span 14·14 or the half span 7·07 = r . cos. θ, and double the value of r, we have cos. θ = ·3535, sin. θ = ·93544, and by calculation we find N = 131 . δ, or 3¾ times the value of N obtained above, when r = 10.

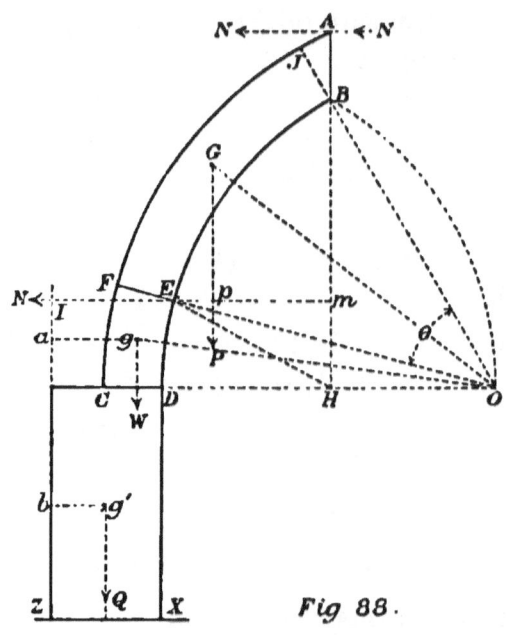

Fig 88.

86. Pointed Arch.—There are many varieties of the "pointed" or "Gothic" arch, varying from the high pitch of the "lancet" arch down to the low pitch of the "Tudor" or "four-centred" arch. We shall here consider the commonest form of pointed arch called the "equilateral" arch, in which the radius OB (fig. 88) makes an angle of 60° with the horizontal OD; or the radius, r, is the span of the arch.

We will first suppose the arch to be without sur-charge and to rest on a pier at CD, and that as in the semi-circular arch previously considered (**80**) it is about to break up by the opening of the joint AB at B, and of the joint EF at F. Then the horizontal pressure, N, acts at A, and we have to determine the position of the "joint of rupture" by taking *moments* of N and P about E, and find what is the value of the angle BOE, or θ, which makes $N = P \dfrac{Ep}{Am}$ a maximum.

We can, without material error, consider P to be the area of the arch EFJB, omitting the small triangle ABJ, and let G be its centre of gravity whose position is determined by the equation (31), namely,

$$OG = \frac{4}{3} \; \frac{R^3 - r^3}{R^2 - r^2} \times \frac{\sin. \frac{\theta}{2}}{\theta}.$$

Also we have, $\qquad P = (R^2 - r^2)\dfrac{\theta}{2} \times \delta,$

where OE $= r$, OF $=$ R. Draw the vertical line ABH, and the horizontal line Em; also the vertical line Gp. Then we have—

$$Ep = r \cdot \cos. (60° - \theta) - OG \cdot \cos. \left(60° - \frac{\theta}{2}\right)$$

$$Am = AB + BH - Hm$$

$$= \frac{R - r}{\cos. 30°} + r \cdot \sin. 60° - r \cdot \sin. (60° - \theta)$$

$$= 1{\cdot}155 \, (R - r) + {\cdot}866 \, r - r \cdot \sin. (60° - \theta)$$

$$N = P \frac{Ep}{Am}.$$

Putting $R = 12$, $r = 10$, we find $P = 22 \times \theta \cdot \delta$,

$$OG = 22 \cdot 06 \frac{\sin. \frac{\theta}{2}}{\theta},$$

$$E_p = 10 \cdot \cos. (60° - \theta) - OG \cdot \cos. \left(60° - \frac{\theta}{2}\right)$$

$$A_m = 10 \cdot 97 - 10 \sin. (60° - \theta).$$

We have now to calculate N for several values of θ, in order to find which value of θ makes N greatest.

The values of " arc θ " and $\frac{\sin. \frac{\theta}{2}}{\theta}$ can be found from the table previously given (**21**).

Take $\theta = 40°$, then " arc θ " $= \cdot 69813$, $\frac{\sin. \frac{\theta}{2}}{\theta} =$ $\cdot 4890$; and we find $P = 15 \cdot 359 \cdot \delta$, $A_m = 7 \cdot 55$, $E_p = 1 \cdot 1181$;

$$N = \frac{15 \cdot 359 \times 1 \cdot 1181}{7 \cdot 55} \cdot \delta = 2 \cdot 274 \cdot \delta.$$

Let $\theta = 42°$, " arc θ " $= \cdot 73304$, $\frac{\sin. \frac{\theta}{2}}{\theta} = \cdot 48889$; then we have—

$$P = 16 \cdot 127 \cdot \delta, \quad A_m = 7 \cdot 88, \quad E_p = 1 \cdot 1286,$$

$$N = \frac{16 \cdot 127 \times 1 \cdot 1286}{7 \cdot 88} \cdot \delta = 2 \cdot 31 \cdot \delta.$$

Let $\theta = 44°$, " arc θ " $= \cdot 76794$, $\frac{\sin. \frac{\theta}{2}}{\theta} = \cdot 48782$;

then—

$$P = 16·895 . \delta, \quad Am = 8·21, \quad Ep = 1·1323,$$

$$N = \frac{16·895 \times 1·1323}{8·21} . \delta = 2·33 . \delta.$$

Let $\theta = 45°$, "arc θ" = ·7854, $-\dfrac{\sin. \frac{\theta}{2}}{\theta} = ·48725$;

then—

$$P = 17·279 . \delta, \quad Am = 8·38, \quad Ep = 1·1317,$$

$$N = \frac{17·279 \times 1·1317}{8·38} . \delta = 2·33346 . \delta.$$

Let $\theta = 46°$, "arc θ" = ·80285, $\dfrac{\sin. \frac{\theta}{2}}{\theta} = ·48668$;

then—

$$P = 17·6627 . \delta, \quad Am = 8·55, \quad Ep = 1·128,$$

$$N = \frac{17·6627 \times 1·128}{8·55} . \delta = 2·33 . \delta.$$

Let $\theta = 48°$, "arc θ" = ·83776, $\dfrac{\sin. \frac{\theta}{2}}{\theta} = ·4855$;

then—

$$P = 18·4307 . \delta, \quad Am = 8·891, \quad Ep = 1·1171,$$

$$N = \frac{18·4307 \times 1·1171}{8·891} . \delta = 2·316 \delta.$$

Hence it appears that the maximum value of N is obtained when $\theta = 45°$, and we can take the joint EF which makes 15° with the horizontal as the "joint of rupture." If we draw HE from H the middle point of the span, we find that the angle EHD is nearly 30°, when EOD is 15°. In any other form of pointed arch the position of the "joint of rupture" can be found approximately by drawing HE at an angle of 30° with the horizontal.

For an equilateral arch of any other dimensions, we can find P and N from the following equations:—

$$P = \cdot 3927 \, (R^2 - r^2) \, . \, \delta \qquad . \quad (150)$$

$$OG = \cdot 640 \, \frac{R^3 - r^3}{R^2 - r^2} \qquad . \qquad . \quad (151)$$

$$Ep = \cdot 96593 \, r - \cdot 79335 \times OG \quad . \quad (152)$$

$$Am = 1 \cdot 154 \, R - \cdot 547 \, r \qquad . \quad (153)$$

$$N = P \, \frac{Ep}{Am} \qquad . \qquad . \quad (154)$$

In order to determine the thickness, t, of the abutment necessary to resist the thrust of such an arch, we proceed as before, in the semi-circular arch (**81**), to take N and P as acting horizontally and vertically at E; and equate their *moments* about Z with the *moments* of the lower part CDEF of the arch and of the pier DZ. Put W for the area of CDEF, then in this arch we have—

$$W = \tfrac{1}{3} P,$$

Also, to determine the position of g the centre of CDEF, we have, from equation (31)—

$$Og = \frac{4}{3} \, \frac{R^3 - r^3}{R^2 - r^2} \times \cdot 49858$$

$$= \cdot 665 \, \frac{R^3 - r^3}{R^2 - r^2} \qquad . \qquad . \quad (155)$$

$$ag = r + t - Og \, . \, \cos. \, 7\tfrac{1}{2}° = t + r - \cdot 99144 \, Og.$$

In the present example where $R = 12$, $r = 10$, we find

$$Og = 11, \quad ag = t - \cdot 906, \quad W = \frac{P}{3} = \frac{17\cdot279}{3}\, \delta$$

$= 5\cdot7597\ \delta$. Let h, the height DX of the pier, be taken as ten feet; then the *moment* of N about Z is

$$N \times IZ = N\,(h + r\,.\,\sin.\ 15°) = N\,(h + \cdot2588\ r)\,;$$

and in this case,

$$N \times IZ = 2\cdot33340 \times 12\cdot588\,.\,\delta = 29\cdot373\,.\,\delta.$$

The *moment* of P, at E, about Z, is $P \times EI$, or

$$P\,(t + r - r\,\cos.\ 15°) = P\,(t + \cdot03407\,r).$$

Also,

$$P \times EI = 17\cdot279\,(t + \cdot3407) = (17\cdot279\ t + 5\cdot887)\ \delta$$

The *moment* of W, acting at g, about Z, is—

$$W \times ag = \frac{P}{3}\,(t - \cdot906) = 5\cdot7597\,(t - \cdot906)\ \delta$$

$$= (5\cdot7597\ t - 5\cdot217)\ \delta.$$

The *moment* of Q, the area of the pier, is—

$$Q \times bg' = h\frac{t^2}{2} = 5\ t^2\,.\,\delta.$$

The *equation of stability* is therefore (**18**),

$$58\cdot746 = 17\cdot279\ t + 5\cdot887 + 5\cdot7597\ t - 5\cdot217 + 5\ t^2$$

or, $\qquad\qquad t^2 + 4\cdot61\ t - 58\cdot078 = 0\,;$

whence, $\qquad\qquad t = 5$ feet.

87. SURCHARGED POINTED ARCH.—We have now to consider the Gothic arch as loaded with a wall level

at the top, as MK (fig. 89). Let the arch be equilateral as before, having O for its centre, OD = r =
the span, OC = R. Draw OEF making the angle
EOD = 15°, and draw the vertical FL. Then, as in
the case of the semi-circular arch, we can approximate

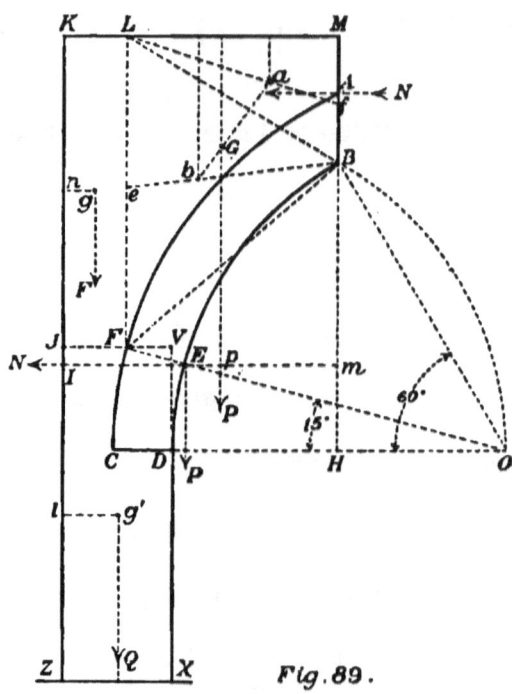

Fig. 89.

to the area of FLMBE, or P, by drawing BF and
taking P to represent the area of the trapezium
BFLM, of which G is the centre of gravity. Draw the
vertical Gp meeting the horizontal line Em in p; then,
as before, we find

$$mp = \frac{ML}{3}\left(1 + \frac{FL}{BM + FL}\right);$$

$$P = \frac{BM + FL}{2} \times ML \times \delta;$$

$$BM = MA + AB = k + 1{\cdot}155\,(R - r)$$

$$FL = BM + r \,.\, \sin. 60^\circ - R \,.\, \sin. 15^\circ$$
$$= BM + {\cdot}866\,r - {\cdot}259\,R;$$

$$ML = R \,.\, \cos. 15^\circ - r \,.\, \cos. 60^\circ = {\cdot}96593\,R - {\cdot}5r;$$

$$Em = r \,.\, \cos. 15^\circ - r \,.\, \cos. 60^\circ = {\cdot}46593r;$$

$$Ep = Em - mp.$$

By equation (153)—

$$Am = AB + Bm = 1{\cdot}154R - {\cdot}547r$$

By equation (154)—

$$N = P\,\frac{Ep}{Am}.$$

For example, let $R = 12$, $r = 10$, $k = 5$; then,
$BM = 7{\cdot}31$, $FL = 12{\cdot}862$, $ML = 6{\cdot}59116$,

$$P = 20{\cdot}172 \times 3{\cdot}296\,\delta = 66{\cdot}486 \,.\, \delta,$$

$$mp = 3{\cdot}603,\ Em = 4{\cdot}66,\ Ep = 1{\cdot}057,\ Am = 8{\cdot}38$$

$$N = \frac{66{\cdot}486 \times 1{\cdot}057}{8{\cdot}38}\,\delta = 8{\cdot}386 \,.\, \delta.$$

To find the thickness t of the pier DZ whose height
is h, we proceed as before to take *moments* of the forces
about Z. Put F for the area of the rectangle KLFJ,
acting at g its centre of gravity; then F × gn is its
moment about Z. Let Q be the area or weight of the
rectangle JVXZ, whose centre is g'; and the *moment*
of Q is $Q \times \dfrac{t}{2}$.

$$F \times gn = FL \times \frac{KL^2}{2} = \frac{FL}{2}(MK - ML)^2$$

$$= \frac{FL}{2}(t + r - \cdot 96393\ R)^2,$$

$$= 6\cdot431\ (t - 1\cdot592)^2 \cdot \delta$$
$$= (t^2 \times 6\cdot431 - 20\cdot476\ t + 16\cdot297)\ \delta,$$

$$Q \times \frac{t}{2} = (h + R \sin. 15°)\ \frac{t^2}{2} \cdot \delta = (h + \cdot259\ R)\frac{t^2}{2} \cdot \delta$$

$$= 6\cdot554\ t^2 \cdot \delta.$$

The *moment* of N, at E, taken about Z, is N × ZI, or, N ($h + r$ sin. 15°) = N ($h + \cdot259\ r$) = 105·57 . δ.

The *moment* of P, at E, taken about Z, is P × EI, or
$$P(r + t - r \cdot \cos. 15°) = P(t + \cdot034\ r)$$
$$= (66\cdot486\ t + 22\cdot605)\ \delta.$$

Then the *equation of stability* (**81**) is

$$211\cdot14 = 66\cdot486\ t + 22\cdot605 + 6\cdot431t^2 - 20\cdot476\ t$$
$$+ 16\cdot297 + 6\cdot554\ t^2$$

or, $t^2 + 3\cdot54\ t - 13\cdot25 = 0.$

Whence we find, $t = 2\cdot3$ feet; when the arch and pier are built of similar materials.

Example 2.—Let R = 16, r = 14, k = 7, h = 12;

then BM = 9·31, ML = 8·46, FL = 17·27,
$$KL = t - 1\cdot456 .$$

P = 112·44 . δ, mp = 4·6523, Ep = 1·871, Am = 10·8

$$N = \frac{112\cdot44 \times 1\cdot871}{10\cdot8}\ \delta = 19\cdot5 . \delta;$$

$$F \times gn = 8\cdot63 \, (t - 1\cdot456)^2 \cdot \delta$$
$$= 8\cdot63 \, (t^2 - 2\cdot912 t + 2\cdot12) \cdot \delta$$

$$N \times ZI = 19\cdot5 \times 15\cdot626 \cdot \delta = 304\cdot7 \cdot \delta$$

$$P \times EI = 112\cdot44 \, t \cdot \delta + 53\cdot52 \cdot \delta$$

$$Q \times lg' = 8\cdot07 \, t^2 \cdot \delta$$

The *equation of stability* is

$$609\cdot4 = 112\cdot44 \, t + 53\cdot52 + 8\cdot63 \, t^2 - 25\cdot13 \, t + 18\cdot3$$
$$+ 8\cdot07 \, t^2$$

which can easily be reduced to

$$t^2 + 5\cdot23 \, t - 32\cdot16 = 0,$$

from which we get, $t = 3\cdot65$.

Example 3.—Let $R = 20$, $r = 17$, $k = 10$, $h = 15$.
Then we have, BM $= 13\cdot465$, ML $= 10\cdot82$, FL $= 23$;

$$P = 197\cdot276 \cdot \delta, \; mp = 5\cdot887, \; Ep = 2\cdot03,$$
$$Am = 13\cdot785;$$

$$N = \frac{197\cdot276 \times 2\cdot03}{13\cdot785} \cdot \delta = 27\cdot75 \cdot \delta;$$

$$F \times gn = 11\cdot5 \, (t - 2\cdot32)^2 \cdot \delta$$
$$= 11\cdot5 \, (t^2 - 4\cdot64 \, t + 5\cdot34) \cdot \delta$$

$$N \times ZI = 27\cdot75 \, (15 + \cdot259 \times 17) \cdot \delta = 538\cdot4 \cdot \delta$$

$$P \times EI = 197\cdot276 \, t \cdot \delta + 114\cdot43 \cdot \delta$$

$$Q \times lg' = 20\cdot18 \, \frac{t^2}{2} \cdot \delta = 10\cdot1 \, t^2 \cdot \delta$$

The *equation of stability* is therefore

$$1\cdot077 = 197\cdot3 \, t + 114\cdot4 + 11\cdot5 \, t^2 - 53\cdot3 \, t + 61\cdot4$$
$$+ 10\cdot1 \, t^2$$

which reduces to

$$t^2 + 6\cdot67t - 41\cdot7 = 0.$$

from which we obtain, $t = 3\cdot92$ feet.

Example 4.—Let $R = 24$, $r = 20$, $k = 12$, $h = 20$;
$R - r = 4$, $BM = 16\cdot62$, $FL = 27\cdot72$, $ML = 13\cdot18$,
$$KL = t - 3\cdot182;$$

$P = 306 \cdot \delta$, $mp = 7\cdot134$, $Ep = 2\cdot184$, $Am = 16\cdot76$;

$$N = \frac{306 \times 2\cdot184}{16\cdot76} \cdot \delta = 39\cdot87 \cdot \delta;$$

$$F \times gn = 13\cdot86 \, (t - 3\cdot182)^2 \cdot \delta$$
$$= 13\cdot86 \, (t^2 - 6\cdot364t + 10\cdot12) \cdot \delta$$

$$N \times ZI = 39\cdot87 \times 25\cdot18 \cdot \delta = 1004 \cdot \delta,$$
$$P \times EI = 306 \, t \cdot \delta + 208 \cdot \delta,$$

$$Q \times lg' = 26\cdot22 \, \frac{t^2}{2} \cdot \delta = 13\cdot11 \, t^2 \cdot \delta$$

The *equation of stability* becomes

$$2008 = 306 \, t + 208 + 13\cdot86 \, t^2 - 88\cdot3 \, t + 140\cdot3$$
$$+ 13\cdot11 \, t^2$$

which can be reduced to

$$t^2 + 8 \, t - 65 = 0;$$

therefore we have, $t = 5$ feet.

The following Table gives the results obtained in the foregoing examples, the arch and pier being supposed to be built of similar materials.

R.	r.	k.	h.	t.
12	10	5	10	2·30
16	14	7	12	3·65
20	17	10	15	3·92
24	20	12	20	5·00

88. TUDOR ARCH.—We now propose to investigate the thrust of an arch which is usually drawn by the architect by means of two different lengths of radius and from four different centres. The height MB (fig. 90) of this kind of arch is generally less than half

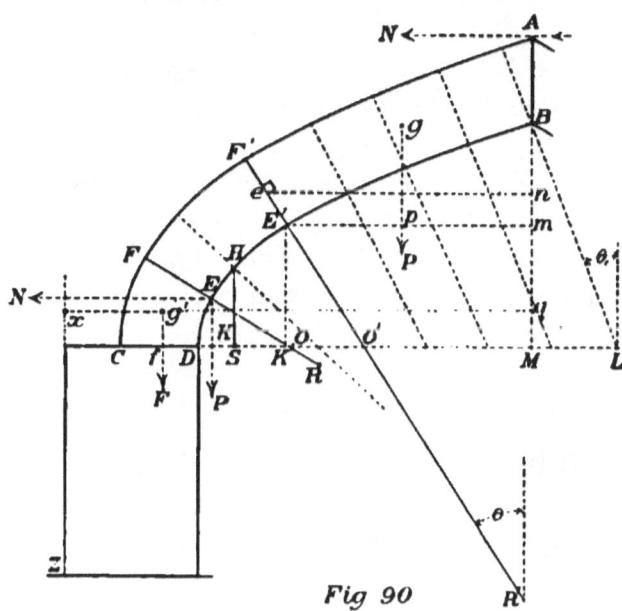

Fig 90

the span, and the mathematical curve which most nearly approaches it in form is that known as the Parabola, for modes of drawing which the reader is referred to the author's "Practical Geometry." Let DEB be a part of a parabola which nearly represents the half of a Tudor arch, S being the focus of the curve, and the ordinate HS = 2DS. Let DM, the half span of the arch, be 9 times DS, MB = 6 times DS, according to the property of the curve, namely, that the square of the ordinate MB equals 4 times DS multiplied by DM the abscissa. Or, if we put DS $= m$, x any abscissa measured from D, y any ordinate belonging to the abscissa x, we have,

$$y^2 = 4\,m\,.\,x.$$

There will be a similar half-arch turned the opposite way abutting against the joint AB, and producing the thrust N at A, the two half-arches together making an obtuse angle at the crown A. The *normal* at any point, as E', can be easily drawn by letting fall the vertical E'K' and measuring K'O' = 2DS = $2m$; then E'O' is the *normal*, or perpendicular to the tangent at the point E'. The *radius of curvature* at any point of the curve lies in the *normal*, and its length, R'E' or ρ, is found from the equation

$$\rho = \frac{2\,m}{\sin.^3\theta},$$

where θ is the angle which the normal makes with the vertical. The line K'O' is called the "subnormal," and its length is the same for all points on the curve. Also

$$O'E'\,.\,\sin.\,\theta = O'K' = 2\,m,$$

or,
$$O'E' = \frac{2\,m}{\sin.\,\theta};$$

$$E'K' = O'E' \times \cos.\,\theta = 2\,m\,.\,\text{cotan.}\,\theta;$$

$$DK' = \frac{(E'K')^2}{4\,m} = m\,.\,\text{cotan.}^2\theta;$$

$$O'R' = \rho - O'E' = \frac{2\,m}{\sin.^3\theta} - \frac{2\,m}{\sin.\,\theta} = 2\,m\,\frac{\text{cotan.}^2\theta}{\sin.\,\theta};$$

$$MO' = MD - (O'K' + DK')$$
$$= 9\,m - 2\,m - m\,.\,\text{cotan.}^2\theta = m\,(7 - \text{cotan.}^2\theta).$$

For the sake of convenience we shall assume in this case that m, or DS, is 1 foot, and omit m altogether from the equations, DM being 9 feet, BM being 6 feet, and the depth of the voussoirs CD, EF, or E'F', 2 feet. Take any small area e between E' and F', and let R'$e = r$; then O'$e = r -$ O'R',

also, $O'e \sin. \theta = (R'e - R'O') \sin. \theta$
$$= r \sin. \theta - 2 \cotan.^2 \theta.$$

Draw the horizontals mE' and ne; then—

$$ne = O'e \,.\, \sin. \theta + \mathrm{MO'}$$
$$= r \,.\, \sin. \theta - 2 \cotan.^2 \theta + 7 - \cotan.^2 \theta$$
$$= 7 + r \,.\, \sin. \theta - 3 \cotan.^2 \theta.$$

Let g be the centre of gravity of the arch between E'F' and AB, of which the area or weight is represented by P ; let fall the vertical gp meeting mE' in p. Take ML $= 2$DS or 2 feet, and draw the normal BL making the angle θ_1 with the vertical ; then θ_1 is found to be 18° 26', and " arc θ_1" $= \cdot 3143$. If we take a very small area or weight at any point e on the arch, and call it dP, or the " differential " of P, then we have—

$$d\mathrm{P} = r \,.\, dr \,.\, d\theta$$

where dr and $d\theta$ are the corresponding differentials of r and θ. Therefore we have—

$$\mathrm{P} = \iint r \,.\, dr \,.\, d\theta$$

the limits of integration being $r = \rho$ and $r = \rho + 2$, $\theta = \theta_1$ and $\theta = \theta$. Expressing ρ in terms of θ and integrating between these limits, we find for the value of P—

$$P = 21 \cdot 9862 + 2\,\theta - 2\,\frac{\cos.\,\theta}{\sin.^2\theta} + 2\,\log.\,\text{cotan.}\,\frac{\theta}{2}\,.\,(156)$$

Also by the principles of the centre of gravity, we have—

$$mp = \frac{\displaystyle\iint (7 + r\,.\,\sin.\,\theta - 3\,\cot.^2\theta)\,r\,.\,dr\,.\,d\theta}{P}.$$

Then by integrating this between the same limits as before we obtain—

$$mp \times P = 116 \cdot 426 + \frac{\cos.\,\theta}{\sin.^4\theta} + 20\,\theta - \frac{8}{3}\,\cos.\,\theta$$

$$- 2\,\text{cotan.}\,\theta - 18 \cdot 5\left(\frac{\cos.\,\theta}{\sin.^2\theta} + \log.\,\text{cotan.}\,\frac{\theta}{2}\right)\,.\quad(157)$$

which being divided by the value of P obtained from equation (156) gives the value of mp. Also we find—

$$pE' = 9 - \text{cotan.}^2\,\theta - mp \qquad . \qquad . \quad (158)$$

$$AM = 8 \cdot 108 - 2\,\text{cotan.}\,\theta \qquad . \qquad . \quad (159)$$

Then, as before, since N and P are supposed to balance about the point E', we have—

$$N = P\,\frac{pE'}{Am} \qquad . \qquad . \quad (160)$$

We have now to ascertain the angle θ; or the point E, at which N is greatest as found from equation (160). This can only be done by calculating the foregoing equations for several values of θ.

Let $\theta = 54°$, or "arc θ" $= \cdot 94248$, cos. $\theta = \cdot 58779$, cotan. $\theta = \cdot 72654$, cotan.$^2\,\theta = \cdot 52786$, log. cotan. $\dfrac{\theta}{2} =$

·67427, $\dfrac{\cos.\theta}{\sin.^4\theta} = 1·3721$, $\dfrac{\cos.\theta}{\sin.^2\theta} = ·89805$. Then,
$P = 20·72652 . \delta$, $mp = 5·0437$, $E'p = 3·42844$,
$Am = 6·65492$; and we get—

$$N = \frac{20·72652 \times 3·42844}{6·65492} \delta = 10·678 . \delta.$$

Let $\theta = 56°$; then we find $P = 21·05 . \delta$,
$E'p = 3·4344$, $Am = 6·759$; giving—

$$N = 10·697 . \delta$$

Let $\theta = 58°$; then $P = 21·357 . \delta$, $E'p = 3·436$,
$Am = 6·858$;
$$N = 10·700 . \delta$$

Let $\theta = 60°$; then $P = 21·649 . \delta$, $E'p = 3·4342$,
$Am = 6·9533$;
$$N = 10·692 . \delta$$

Let $\theta = 62°$; then $P = 21·9272 . \delta$, $E'p = 3·4289$,
$Am = 7·0446$;
$$N = 10·673 . \delta$$

From which it appears that as we approach to or recede
from $\theta = 58°$, the value of N diminishes; hence we
may assume that for this case the point E, where N is
a maximum, is that at which the normal makes an
angle of 58° with the vertical, and this will therefore
be the "joint of rupture."

To find the point E where the normal EO makes 58°
with the vertical, take $DK = ·39046$, $KO = 2$.

Putting F for the area or weight of the lower part of
the arch between EF and CD, we have to integrate the

q

equations for P and mp, given above, between the limits $\theta = 58°$ and $\theta = 90°$; from which we obtain—

$$F = 3\cdot7707 \ . \ \delta$$

and the distance qg' is obtained by the integral for mp, taken between the same limits, giving—

$$qg' = 9\cdot907,$$

therefore $\mathrm{D}f = qg' - \mathrm{MD} = 9\cdot907 - 9 = \cdot907$; and $g'x = t - \mathrm{D}f = t - \cdot907$. The *moment* of F about Z is—

$$\mathrm{F} \times g'x = \mathrm{F} \ (t - \mathrm{DF}) = 3\cdot7707 \ t \ . \ \delta - 3\cdot42 \ . \ \delta.$$

Putting Q for the area or weight of the pier whose height is h and thickness t, we have for the *moment* of Q about Z—

$$Q \times \frac{t}{2} = \tfrac{1}{2} h \ . \ t^2 \ . \ \delta.$$

The *moment* of N, acting horizontally at E, taken about Z, is—

$$N \ (h + \mathrm{KE}) = \mathrm{N} \ (h + 2 \ \text{cotan.} \ 58°)$$
$$= 10\cdot7 \ (h + 1\cdot24974) \ \delta \ ;$$

The *moment* of P, acting vertically at E, taken about Z, is—

$$\mathrm{P} \ (\mathrm{DK} + t) = \mathrm{P} \ (\cdot39 + t) = 21\cdot357 \ t \ . \ \delta + 8\cdot33 \ . \ \delta$$

Taking $h = 10$, we have for the *equation of stability* (**81**), $241 = 21\cdot357 \ t + 8\cdot33 + 3\cdot7707 \ t - 3\cdot42 + 5 \ t^2$ which reduces to—

$$t^2 + 5 \ t - 47\cdot2 = 0 \ ;$$

from which we obtain, $t = 4\cdot8$.

It appears therefore that for a Tudor arch without surcharge having a span of 18 feet, and a height at the centre of 6 feet or one-third of the span, the depth of the voussoirs being 2 feet, the thickness of a pier 10 feet high should be 4·8 feet in order to ensure stability.

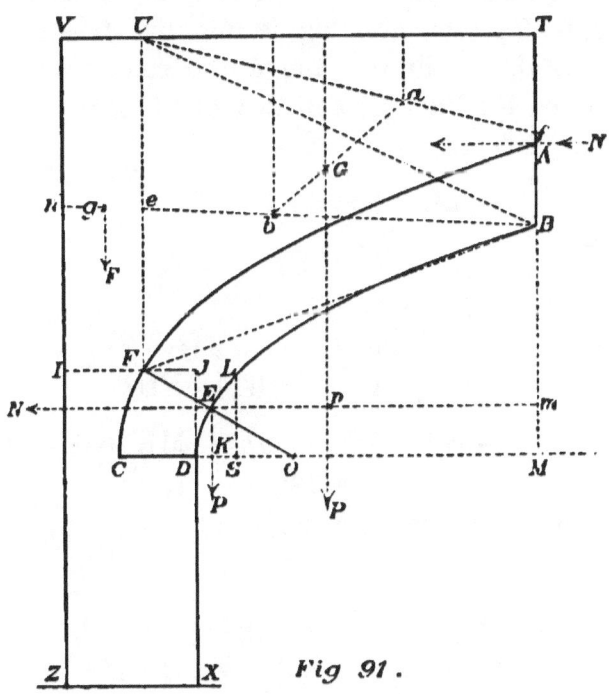

Fig 91.

89. SURCHARGED TUDOR ARCH.—Let us now suppose that the parabolic arch, whose "joint of rupture" we found above (**88**), has a surcharge whose height AT (fig. 91) above the crown we call k, the top of the surcharge being taken as level. Let EF be the "joint of rupture" as above determined, making 58° with the vertical; DM = 9, BM = 6, CD = EF = 2, DK =

·39, AT $= 3 = h$, BT $= 5\cdot108$, OE $= 2\cdot358$, EK $= 1\cdot24974$.

If P represents the area or weight of the arch and its surcharge EFUTB, acting at G its centre of gravity, we can approximately find P and mp by assuming as in the former cases (**82**) (**87**) that the trapezium FUTB represents P; then dividing it into two triangles by the line BU, we find a and b the centres of these triangles on the lines Uf and Be; and G is determined from the proportion—

$$aG : bG = \text{FU} : \text{BT}.$$

Then,
$$P = \frac{\text{BT} + \text{FU}}{2} \times \text{TU};$$

$$mp = \frac{\text{TU}}{3}\left(1 + \frac{\text{FU}}{\text{BT} + \text{FU}}\right);$$

$$Ep = m\,\text{E} - mp = \text{MK} - mp, \quad \text{MK} = 9 - \text{DK}$$
$$= 8\cdot6,$$

$$\text{BT} = 5\cdot108, \quad \text{FU} = \text{MT} - (\text{OE} + 2)\,.\,\sin.\ 32'$$
$$= 8\cdot799,$$

$$\text{TU} = \text{MK} + 2\,.\,\cos.\ 32° = 10\cdot296,$$

$$\text{VU} = 9 + t - \text{TU} = t - 1\cdot296.$$

From which we find, P $= 71\cdot593$, $mp = 5\cdot93$,

$$Ep = 8\cdot6 - 5\cdot93 = 2\cdot67, \quad \text{A}m = 6\cdot858,$$

$$N = P\,\frac{Ep}{\text{A}m} = 27\cdot873.$$

Let the height h of the pier be 10, and F the area or weight of the rectangle VUFI, Q that of the rectangle

IJXZ; then the *moment* of F, acting at g, taken about Z, is—

$$F \times gn = \frac{FU}{2} \times UV^2$$

$$= \frac{FU}{2} (MD + t - TU)^2$$

$$= 4\cdot4 (t - 1\cdot296)^2 = 4\cdot4\ t^2 - 11\cdot4\ t + 7\cdot38.$$

The *moment* of Q, acting at its centre of gravity, and taken about Z, is—

$$Q \times \frac{t}{2} = (h + OF . \sin. 32^\circ) . \frac{t^2}{2} = 6\cdot16\ t^2.$$

The *moment* of N, acting horizontally at E, and taken about Z, is—

$$N (h + EK) = 27\cdot873 \times 11\cdot25 = 313\cdot56.$$

The *moment* of P, acting vertically at E, and taken about Z, is—

$$P (DK + t) = 71\cdot593 (t + \cdot39) = 71\cdot593\ t + 28\cdot637.$$

Then the *equation of stability* is (**81**)—

$$627\cdot12 = 71\cdot593\ t + 28\cdot637 = 11\cdot4\ t + 7\cdot38 + 10\cdot6\ t^2,$$

which can be reduced to —

$$t^2 + 5\cdot7\ t - 55\cdot76 = 0;$$

$$t = 5\cdot15.$$

If the actual thrust, N, of the arch is required for one ·foot width, we must multiply the value of N obtained above by δ the weight of a cubic foot of the material used in the arch and surcharge. The *normal*

OE can be easily drawn at any point E of the arch, by dropping the vertical EK on the horizontal line MD, and taking KO = 2 DS, where S is the *focus* of the curve. The joints of the voussoirs should always be in the directions of the *normals*.

In the foregoing example we have taken DS as 1 foot, but for any similar arch of different span the same formulæ will apply, by multiplying all the linear dimensions by DS, or m as it is generally called. The *subnormal* KO is always $2m$, and the length of the *radius of curvature*, ρ, at any point E, is—

$$\rho = \frac{2\,m}{\sin.^3 \theta},$$

where θ is the angle which ρ makes with the vertical. For example, if we make MD = 12 instead of 9, then DS or $m = \frac{4}{3} = 1\cdot33$, BM = 8, EF = $\frac{8}{3} = 2\cdot67$, $k = 4$, and so on for the other dimensions. The weight P and the horizontal thrust N must however be multiplied by m^2, which in this case would be $\frac{16}{9} = 1\cdot778$.

If we *reduce* the span and make MD = 6, BM = 4, $k = 2$; then $m = \frac{2}{3} = \cdot667$, $m = \frac{4}{9} = \cdot444$.

If the weight, δ, of a cubic foot of the arch is 120 lbs. then we have for one foot breadth of arch and pier when the span is 18 feet, and the height 6 feet—

$$N = 27\cdot873 \times 120 = 3,345 \text{ lbs.},$$

When the span is 12 feet, and height 4 feet,

$$N = \cdot444 + 3345 = 1,485 \text{ lbs.},$$

When the span is 24 feet, and height 8 feet,

$$N = 1\cdot778 \times 3345 = 5,947 \text{ lbs.}$$

90. ELLIPTIC ARCH.—The thrust of an elliptic arch can be obtained approximately by comparison with a semi-circular arch of equal span. Let ABDC (fig. 92)

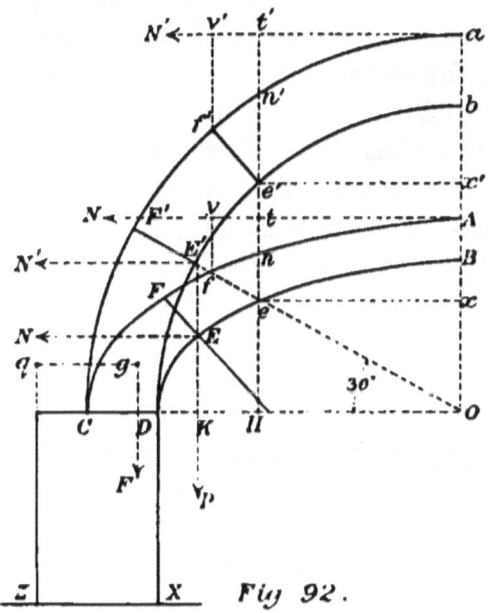

Fig 92.

be a semi-elliptic arch, OD and OB being the semi-axes of the *intrados*, OC and OA the semi-axes of the *extrados.* Draw the semi-circular arch CD *ab,* where the depth *ab* is to the depth AB as OD to OB; then—

$$OA : Oa = OB : OD.$$

Draw any vertical, as H*t'*, cutting the two arches in the points *e, n, e', n'* ; then—

$$He : He' = OB : OD.$$

If we draw the horizontal lines *ex,* and *e'x'*, we have

$$Ax : ax' = OB : OD;$$

or, the vertical distances of e and e' below the vertices
A and a are in the constant ratio of OB to OD. The
segments ABen and $abe'n'$ have the same horizontal
dimensions, namely ex and $e'x'$, while their vertical
dimensions are in the proportion of OB to OD. Consequently, the areas of these segments are in the same
proportion of OB to OD, and their centres of gravity are
in the same vertical line. Let P be the area or weight
of ABen acting at its centre of gravity, p its lever arm
with respect to the vertical Ht', N the corresponding
horizontal thrust of the elliptical arch at A, and q its
lever arm with respect to the line ex; then, as before,
we have—

$$N = P \frac{p}{q}.$$

Let N', P', p', q', be the same for the corresponding
quantities of the circular arch; then we have also—

$$N' = P' \frac{p'}{q'},$$

therefore, $\quad p = p', \ P = P'' \times \dfrac{OB}{OD}, \ q = \dfrac{OD}{OB} \times q',$

$$N = N',$$

or the thrust of both arches is the same.

Hence it appears that the thrust of an elliptical arch
may be considered as nearly equal to that of a circular
arch of the same span, the depth at the crown of the
circular arch being to that of the elliptic arch in the
proportion of the *major-axis* of the ellipse to its *minor-
axis*; the loading being the same in both cases. This
circular arch is called

    " The equivalent circular arch "

of the ellipitic arch whose thrust it is required to determine.

In the investigation of the thrust of a circular arch (**80**) it was shown that the joint E'F' which makes 30° with the horizontal, is the "joint of rupture," or that joint at which N' is a maximum. Draw the vertical E'K cutting the intrados of the elliptic arch in E, draw the *normal* HEF, then EF will be the "joint of rupture" in the elliptic arch.

For example, let OD = 2 OB, then P' = 2 P, $q = 2q'$; or, if OD = 10, OB = 5, AB = 1, then $ab = 2$; and the circular arch is the same as in the example previously given (**81**), where R = 12, $r = 10$, P' = δ (R² − r²) × ·5236 = 23 = 2P, $p' = 3.4$, $q' = 7$, N' = 11·2 × δ = N; P = 11·5 . δ.

To find the strength of pier necessary to resist this thrust, we must consider N and P as acting horizontally and vertically at E on the intrados of the ellipse, and take their *moments* about Z, as before. If we call F the area or weight of the lower part of the elliptic arch between CD and FE, then F = ½ P, and its centre of gravity g will be in a vertical drawn from the centre of gravity of the circular arch CDE'F'. Then the *moment* of F about Z (omitting δ), is—

$$F \times gy = \frac{P}{2}\left(t + r - ·637 \frac{R^3 - r^3}{R^2 - r^2}\right) = 5·75\,t - 3·11.$$

The *moment* of P, acting at E, and taken about Z, is—

$$P \times (t + DK) = 11·5 (t + 1·34) = 11·5\,t + 30·87.$$

The *moment* of N, acting at E, and taken about Z, is ($h = 10$)—

$$N\,(h + EK) = 11{\cdot}2 \times 12{\cdot}5 = 140.$$

The *moment* of Q, the area or weight of the pier, is—

$$Q \times \frac{t}{2} = \frac{h}{2}\,t^2 = 5\,t^2.$$

Then the *equation of stability* becomes—

$$280 = 11{\cdot}5\,t + 15{\cdot}4 + 5{\cdot}75\,t - 3{\cdot}11 + 5\,t^2,$$

or,　　　　　$$t^2 + 3{\cdot}45\,t - 53{\cdot}6 = 0,$$

$$t = 5{\cdot}78.$$

Comparing this with the thickness of pier requisite for the " equivalent circular arch " of the above dimensions (**81**), we find that $t = 5$ in the circular arch, so that is nearly one-sixth more in the elliptic arch of same span.

In applying this method to an elliptic arch of any form and dimensions, there is no necessity to draw the circular arch, as the point E can be determined on the ellipse by taking OK = ·866 × OD, and EK = ½ OB ; and the horizontal dimensions are the same as in a circular arch in which OD = r, and OC = R.

91. Surcharged Elliptic Arch.—When the elliptic arch has a surcharge MR (fig. 93), we can approximate to the value of N in the same manner as we have done for other forms of arch. Let OD be the half span, OB the height of the intrados, OC the half *major-axis*, and OA the half *minor-axis* of the extrados. With O as a centre and OD for a radius, describe the arc DE' ; and draw OE' making an angle of 30° with OD, or take KE' equal to half OD. Draw the vertical KE' cutting the intrados of the elliptic arch at E ; then, as before shown (**90**), E is the point at which the thrust N has its greatest effect.

Then OK = OE′ × cos. 30° = ·866 × OD.

If HEF is the *normal* to the curve at E, then we find

$$HK = \left(\frac{OB}{OD}\right)^{2} \times OK.$$

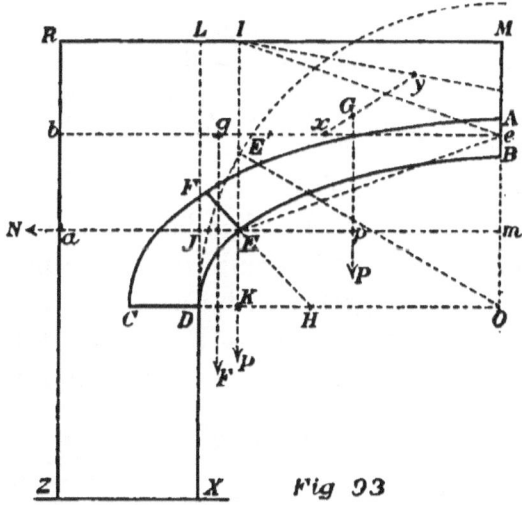

Fig 93

Take e the middle point between A and B, and draw Ee and the vertical EI; then the trapezium EIMe will represent approximately the weight P of the arch and its surcharge between the point E and the crown. Divide this trapezium, as before, into two triangles of which *x* and *y* are the centres; and take G*x* : G*y* = M*e* : EI. Then the load P will act in the vertical G*p* cutting the horizontal E*m* in *p*; consequently

$$P = \frac{EI + Me}{2} \times MI,$$

$$mp = \frac{\text{MI}}{3}\left(1 + \frac{\text{EI}}{\text{EI} + \text{M}e}\right).$$

Let F be the area or weight of the rectangle EJLI acting at g its centre; then F = EI × EJ; EI = OM − KE = OM − $\frac{\text{OB}}{2}$; EJ = ·134 × OD = DK; Me = k + $\frac{\text{AB}}{2}$; MI = OK = ·866 × OD; Ep = Em − mp = OK − mp; Am = OA − KE = OA − $\frac{\text{OB}}{2}$; bj = t + $\frac{\text{EJ}}{2}$ = t + ·067 × OD; RZ = DX + OM = h + k + OA; EK = $\frac{1}{2}$ OB.

Then the *moment* of N acting horizontally at A, and taken about E, will be in equilibrium with the *moment* of P acting at G, together with the *moment* of F acting at y. The *moment* of F about E however acts in the opposite direction to that of P, consequently we have

$$N = \frac{\text{P} \times \text{E}p - \text{F} \times \frac{\text{DK}}{2}}{\text{A}m}.$$

We can then transpose N, P, and F, to the point E and equate their *moments* about Z with that of Q the weight of pier RLXZ, as before.

For example, let OD = 10, OB = 5, OC = 12, AO = 6, k = MA = 2, h = DX = 10, AB = 1. Then we find, EI = 8 − 2·5 = 5·5, Me = 2·5, MI = 8·66, EJ = 1·34, P = 34·64, mp = 2·89 $\frac{13\cdot5}{8}$ = 4·875, Ep = 8·66 − 4·875 = 3·785, Am = 6 − 2·5 = 3·5, P × Ep = 131·11 . δ, F × $\frac{\text{DK}}{2}$ = 4·94 . δ,

$$N = \frac{131 \cdot 11 - 4 \cdot 04}{3 \cdot 5} \, \delta = 36 \, . \, \delta.$$

The *moment* of N (omitting δ), supposed to act at E, and taken about Z is,

$$N \times (h + EK) = 36 \times (10 + 2 \cdot 5) = 450.$$

The *moment* of P, acting at E, and taken about Z, is, $P \, (t + DK) = 34 \cdot 64 \, (t + 1 \cdot 34) = 34 \cdot 64 \, t + 46 \cdot 4.$

The *moment* of F, acting at E, and taken about Z, is $F \, (t + DK) = 5 \cdot 5 \times 1 \cdot 34 \, (t + 1 \cdot 34) = 7 \cdot 37 \, t + 9 \cdot 88.$

The *moment* of Q is, $Q \times \dfrac{t}{2} = \dfrac{t^2}{2} \, (h + OM) = 9 \, t^2$

Then the *equation of stability* **(81)** becomes

$$900 = (34 \cdot 64 + 7 \cdot 37) \, t + 46 \cdot 4 + 9 \cdot 88 + 9t^2$$
$$\text{or, } t^2 + 4 \cdot 67 \, t - 93 \cdot 75 = 0, \text{ and } t = 7 \cdot 625.$$

92. VAULTED ROOFS. CY-LINDRICAL.—When a roof is formed of a stone or brick vaulting, the thrust is often concentrated at certain points on the wall or abutment, and is not distributed uniformly along the wall. This kind of vaulting consists of a main rib as shown on plan by AB, ED, CF, (fig. 94) placed at right angles to the walls EF and CD, and also two (or more) diagonal ribs, as AC, AD, BF, BE, springing from the same points on the wall, so that we have three ribs all pressing on the same point. The diagram

Fig. 94.

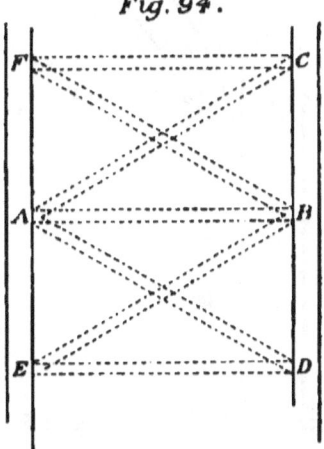

shows the ribs on plan, the diagonal ribs being here drawn as making 30° with the main transverse ribs; but in practice we find them placed at different angles, usually varying from 30° to 45°. The diagonal arches will be of greater span than the transverse arch, and consequently if the latter is semicircular, the former must be elliptical if the height is the same.

Since the angle CAB = 30°, we have AC = 1·154 × AB, and if a and b are the *semi-major* and *semi-minor axes* of the elliptic arches AC, &c., then we have

$$b : a = 1 : 1\text{·}154, \quad \text{or,} \quad \frac{b}{a} = \text{·}866.$$

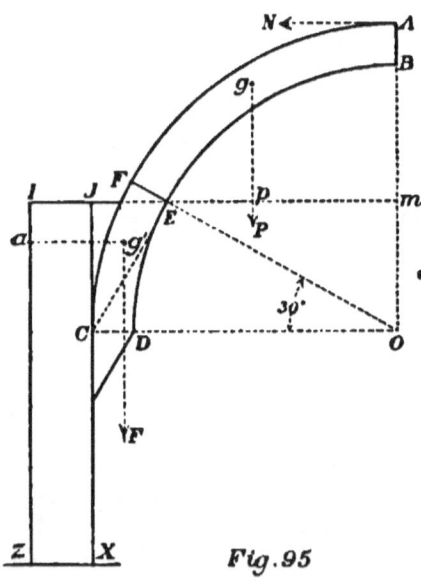

Fig.95

Let ABCD (fig. 95) represent the half section of one of the transverse ribs, resting at its springing on a corbel let securely into the wall, and filled in at the back up to the level of E, EO making 30° with the horizontal OD, and EF being the "joint of rupture" as previously determined (80). Put R for OF the radius of the extrados, r for OE the radius of the intrados, and let l be the lateral breadth of the rib. Then if P is the weight of the arch EFAB, g its centre of gravity, δ the weight of a cubic foot of the material, we have from equation (140)

$$P = \cdot 5236 \cdot \delta \cdot l \cdot (R^2 - r^2);$$

By equation (141)—

$$Ep = \cdot 866\, r - \cdot 3183 \left(\frac{R^3 - r^3}{R^2 - r^2}\right);$$

And by equation (142)—

$$Am = R - \frac{r}{2}$$

$$N = P\,\frac{Ep}{Am}$$

and N is the horizontal thrust of the arch acting at E.

For the diagonal elliptical ribs we have (90) to take N from a semi-circular arch of same span whose depth at the crown is to that of the elliptic arch in the proportion of a to b, or as $1\cdot 154 : 1$. If then $R - r$ is the depth of the elliptic arch, that of the " equivalent circular arch " is, $1\cdot 154 (R - r)$. Putting R' and r' for the radii of the " equivalent circular arch," we have

$$r' = 1\cdot 154\, r, \quad R' = r' + 1\cdot 154\,(R - r) = 1\cdot 154\, R$$

$$P' = \cdot 5236 \cdot l \cdot \delta\,(R'^2 - r'^2)$$

$$= \cdot 7 \cdot l \cdot \delta\,(R^2 - r^2)$$

$$E'p' = \cdot 866\, r' - \cdot 3183 \left(\frac{R'^3 - r'^3}{R'^2 - r'^2}\right) = r - \cdot 367 \left(\frac{R^3 - r^3}{R^2 - r^2}\right)$$

$$A'm' = R' - \frac{r'}{2} = 1\cdot 154 \left(R - \frac{r}{2}\right)$$

$$N' = P'\,\frac{E'p'}{A'm'}.$$

The thrust produced by each diagonal rib upon the

wall is the resolved part of N′ perpendicular to the wall, namely, N′ cos. 30° = ·866 N′, and for the two diagonal ribs the thrust will be twice this, or 1·732 N′. Then the total thrust, T, upon the wall at A or B (fig. 93) is

$$T = N + 1\cdot732 \, N'$$

which may be considered as acting at the level of the point E (fig. 94).

For example, let us suppose the span to be 20 feet, the depth and width of the ribs 1 foot; then $R = 11$, $r = 10, l = 1$; $P = \cdot5236 \,.\, \delta \times 21 = 11\,.\, \delta$; $Ep = 8\cdot66 - 5 = 3\cdot66, Am = 6.$

$$N = \frac{11 \times 3\cdot66}{6} \,.\, \delta = 6\cdot71 \,.\, \delta$$

$$P' = 14\cdot7 \,.\, \delta, \ E'p' - 4\cdot2, \ A'm' = 6\cdot92,$$

$$N' = \frac{14\cdot7 \times 4\cdot2}{6\cdot92} \,.\, \delta = 8\cdot916 \,.\, \delta$$

$$T = (6\cdot71 + 1\cdot732 \times 8\cdot916) \, \delta = 22\cdot15 \, \delta.$$

To determine the thickness, t, of the wall, we have to take the *moments* about Z of T, P and 2P′ acting at E; T acting horizontally, P and 2P′ acting vertically. Also we have the *moment* of F the weight of the lower part JEDC of the transverse arch, and 2F′ the same for the two diagonal ribs.

We can find approximately the value of F, by considering it to be the area or weight of the triangle JEC, or—

$$F = \frac{\delta}{2} \times JE \times CJ = \frac{\delta}{2}(11 - 8\cdot66) \times 5 = 5\cdot85 \,.\, \delta.$$

Also, $F' = F$, very nearly; and if g' is the centre of gravity, then $ag' = t + \frac{1}{3} JE = t + \cdot 78$.

Let Q be the weight of the wall IJXZ, whose thickness $XZ = t$, and whose length will be AF (fig. 94) or

$$AC \times \sin. 30° = \frac{AC}{2} = R' = 12\cdot7.$$

Let $h \doteq CX = 20$, then $IZ = 25$; and the *moment* of Q about Z is—

$$Q \times \frac{t}{2} = 12\cdot7 \times 25 \times \frac{t^2}{2} \cdot \delta = 158\cdot8 \cdot \delta \cdot t^2.$$

The *moment* of T is—

$$T \times IZ = 25 \times 22\cdot15 \cdot \delta = 553\cdot75 \cdot \delta.$$

The *moment* of $P + 2 P'$ is—

$$(P + 2P')(t + JE) = 40\cdot4(t + 2\cdot34)\delta = (40\cdot4t + 94\cdot5)\delta.$$

The *moment* of $F + 2 F'$ or 3 F is—

$$3 F(t + \cdot78) = 17\cdot55(t + \cdot78)\delta = (17\cdot55 t + 13\cdot69)\delta.$$

Then the *equation of stability* is—

$$1107\cdot5 = 40\cdot4 t + 94\cdot5 + 17\cdot55 t + 13\cdot69 + 158\cdot8 t^2,$$

or,

$$t^2 + \cdot365 t - 6\cdot3 = 0;$$

$$t = 2\cdot3 \text{ ft.}$$

Suppose that instead of a solid wall from E to F, we had only piers 3 feet wide at E, A, and F; to determine the thickness, t, of the pier? The value of $Q \times \dfrac{t}{2}$ will then be $3 \times 25 \times \dfrac{t^2}{2} \cdot \delta = 37\cdot5 \, t^2 \cdot \delta$; all the other *moments* remaining as before. Then the *equation of stability* is reduced to—

$$t^2 + 1\cdot545\,t - 26\cdot65 = 0;$$
$$t = 4\cdot5 \text{ ft.}$$

In the latter case, however, the quantity of walling is less than one-half what it is in the former, the proportion being 13·5 to 29·2; consequently it is more economical to have piers than a continuous wall.

In the foregoing investigation we have taken no account of the panelling or filling in between the ribs. This is usually made as light as possible, and we will assume that its thickness decreases from E towards B (fig. 94), in such a manner that the pressure on every part of the ribs is uniform, and also that the effect of the panelling is the same as if we doubled the width l of the ribs, keeping the depth $R - r$ the same as before.

In this case we have to double the values of P and P′, and therefore also of N and N′, while F may be considered as unaltered. Applying this to the foregoing example, we have $l = 2$, and all dimensions the same as before. Then we have—

$$P = 22 \cdot \delta, \; P' = 29\cdot4 \cdot \delta, \; T = 44\cdot3\,\delta.$$

Then for a continuous wall 12·7 feet in length the *equation of stability* becomes, when reduced,

$$t^2 + \cdot62\,t - 12\cdot7, \text{ or } t = 3\cdot5 \text{ feet.}$$

For a pier 3 feet wide we have to determine t from the equation

$$t^2 + 2\cdot6\,t - 53\cdot7 = 0;$$
$$t = 6\cdot2 \text{ feet.}$$

So that the strength of the abutment must be increased in the proportion of 3 to 2, to allow for the panelling of the vault.

Instead of making the diagonal ribs elliptical, the plan is sometimes adopted of making them semicircular, and "stilting" the transverse rib so as to bring the crowns on the same level. In the foregoing case the semi-circular transverse arch will have to be stilted by ·154 r, or about a foot and a half when r = 10; the springing of the semicircle in the transverse rib being so much above that of the diagonal ribs.

93. GOTHIC VAUL-
TING. — In Gothic
vaulting the ribs are
formed of pointed
arches, as previously
described (**86**); and
we shall here assume
that the ribs, both
transverse and diago-
nal, are equilateral
arches. Let ABCD
(fig. 96) represent in
section one of the
transverse ribs, the
"joint of rupture"
EFO making 15°
with the horizontal OD, as before determined (**86**).

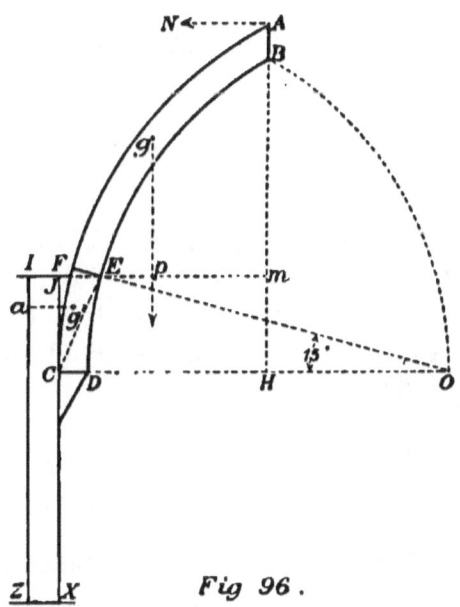

Fig 96.

We will also assume that the masonry is solid up to the level of EI, I being the top of the wall or pier. Let the plan (fig. 94) of the ribs be the same as in the case of the semicircular ribs, the diagonals making 30° with the transverse. Let R and r be the radii of the transverse rib, R′ and r′ those of the diagonals; then as before, R′ = 1·154 R, r' = 1·154 r. From equation (150) we have—

$$P = ·3927\ (R^2 - r^2)\ \delta\ .\ l,$$

where l is the width of the arch.

From equations (151, 152) we have—

$$Ep = ·966\ r - ·516\ \frac{R^3 - r^3}{R^2 - r^2},$$

And from equation (153)—

$$Am = 1·154\ R - ·547\ r$$

$$N = P \times \frac{Ep}{Am}.$$

For example, let R = 21, r = 20, l = 1; then we find, P = 16·1 . δ, Ep = 3·32, Am = 13·3, N = 4 . δ; P′ = $(1·154)^2 \times P$ = 1·331 × P, E′p' = 1·154 × Ep, A′m' = 1·154 × Am; N′ = $(1·154)^2 \times N$ = 1·331 × N = 5·324 . δ; P″ = 21·43 δ.

The *moment* of N acting at E, and taken about Z, is—
$$N\ (h + r\ .\ \sin.\ 15°) = 100·72\ .\ \delta.$$

The *moment* of N′ at E, taken about Z, is—
$$·866\ .\ N\ (h + r'\ \sin.\ 15°) = 120\ .\ \delta.$$

The *moment* of P at E, taken about Z, is—
$$P\ (t + R - r\ .\ \cos.\ 15°) = (16·1\ t + 27·1)\ \delta.$$

The *moment* of P′ about Z, is (for the diagonals)—
$$P'\ (t + R' - r'\ .\ \cos.\ 15°) = (21·43\ t + 36)\ \delta.$$

Taking F for the weight of the part of the arch below

E, and represented by the triangle CJE, whose centre is g'; the *moment* of F about Z, is

$$F (t + \cdot 5) = (5 \cdot 18\, t + 2 \cdot 59)\, \delta.$$

The *moment* of the weight Q of the wall taken as 12·7 feet in length and 20 feet high, is

$$Q \times \frac{t}{2} = 12 \cdot 7 \times 25 \cdot 8 \times \frac{t}{2} = 163\, t \cdot \delta.$$

Then the *equation of stability* becomes

$$2\ (100 \cdot 72 + 240) = 16 \cdot 1\, t + 27 \cdot 1 + 42 \cdot 86\, t + 72$$
$$+ 15 \cdot 54\, t + 7 \cdot 77 + 163\, t,$$

or,
$$t^2 + \cdot 457\, t - 3 \cdot 525 = 0;$$

$$t = 1 \cdot 65 \text{ feet.}$$

If instead of a wall there is a pier 3 feet wide, we can determine its thickness, t, from the equation

$$38 \cdot 7\, t^2 + 74 \cdot 5\, t - 574 \cdot 57 = 0,$$

or,
$$t^2 + 1 \cdot 92\, t - 14 \cdot 8 = 0;$$

$$t = 3 \text{ feet.}$$

In the above investigation the weight of the load of panelling or filling in between the ribs has been left out of account.

We will suppose, as in the case of the semi-circular vaulting (**92**), that this load is equal to the weight of the ribs themselves, so that by taking the ribs as 2 feet wide instead of 1 foot, while the depth remains the same, we have the load included with the ribs. We have then to double the values of P and P′, and also of N and N′, and the above *equation of stability*, with a wall 12·7 feet long, is,

$$1362 \cdot 88 = 32 \cdot 2\, t + 54 \cdot 2 + 85 \cdot 72\, t + 144 + 15 \cdot 54\, t$$
$$+ 7 \cdot 77 + 103\, t^2,$$

or, $\qquad t^2 + \cdot 82\, t - 7 \cdot 1 = 0;$

$$t = 2 \cdot 3 \text{ feet.}$$

For a pier 3 feet wide, the *equation of stability* is,

$$t^2 + 3 \cdot 45\, t - 30 = 0;$$

$$t = 4 \text{ feet.}$$

The pressure on the joint AB of the transverse rib when loaded will be 2 N or 8 . δ, and upon the joint EF it will be, $W = 2\,(P \cos. 30° + N . \sin. 30°) = 2\, \delta . 16 = 32 . \delta$. For Bath stone we may take $\delta = 120$ lbs., and the safe load per square inch as 150 lbs.: so that we have—

$$W = 3{,}840 \text{ lbs.}$$

The safe load, S, on a surface 12 inches square will be—

$$S = 144 \times 150 = 21{,}600 \text{ lbs.}$$

which is about $5\frac{1}{2}$ times as great as the value of W, so that we might reduce the sectional area of the ribs in that proportion. This allows a good margin for the reduction of the section of the ribs by mouldings and carvings. It is necessary however that the depth should be sufficient to allow of the " line of pressures " (83) falling within the depth of the arch.

94. Arched Iron Ribs.—Wrought-iron ribs of a semi-circular form are frequently used for carrying a roof-covering over a wide span. The ribs differ from those of stone which we have been considering, in that they are all in one piece, or of separate pieces of iron so riveted together as to make them practically one

piece, and therefore are not liable to break up like a
voussoir arch. The effect of a load, however, will be
similar in both cases, and we treat the arched iron rib
in the same manner as we did the stone rib; there
being a point E (fig. 97) on the intrados where the

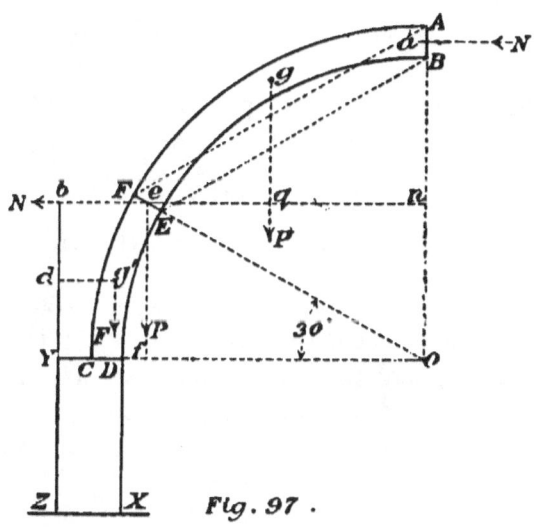

Fig. 97 .

thrust, N, will be greatest. Let ABCD (fig. 97) be the
half of a rib standing on a pier or support at CD, let
O be the centre of the circle, and draw OEF making
30° with OD; bisect EF in e, then e will be the point
at which N, supposed to act at a, the middle of the
joint AB, will have its greatest effect (80). The
covering is generally carried by means of purlins
resting upon the rib at intervals of a few feet, but it
will be sufficient to consider the load as uniformly
distributed over the whole circumference of the rib,
and, as in the Chapter on Roofs, we take this load,
including the effects of wind, snow, &c., as 66 lbs. to

the square foot of roofing. Then, if the ribs are 10 feet apart, we have 660 lbs. for the load on every foot of their circumference. If we put W for the weight of each half-rib and its load, R and r for the external and internal radii of the rib, we have—

$$W = \frac{\pi}{2} R \times 660 = 1037\, R.$$

Let P be the weight of the part between AB and EF, F that of the part between EF and CD; then—

$$P = \tfrac{2}{3} W = 691 . R$$
$$F = \tfrac{1}{3} W = \tfrac{1}{2} P = 346 . R.$$

Draw the vertical OB, and let g be the centre of gravity of the arch between AB and EF; then P acts at g in the vertical line gq, cq and An are determined by the equations (141, 142) namely—

$$cq = \cdot 866\, \frac{R + r}{2} - \cdot 3183 . \frac{R^3 - r^3}{R^2 - r^2}$$

$$an = \frac{R + r}{2} - \frac{R + r}{4} = \frac{R + r}{4}.$$

Then we have,　　　$N = P \times \dfrac{cq}{an}$,

which gives the horizontal thrust at c.

To determine the strength of the rib at EF to resist the stress upon it from the *moment* of P about c, we have to compare the *moment* P × eq with the *moment* of resistance of a beam of I section about the point c as given by equation (56)—

or,　　　P × cq with $\dfrac{5}{6\, D}$ $(b . D^3 - \overline{b - t} . d^3)$, where D and d are the depths, b the breadth and t the thickness of the rib; and the fibres at the top and bottom are subjected to a stress of 5 tons per square inch of section; all the dimensions being expressed in inches.

The resistance of the pier or support is determined as before by taking *moments* about Z, of P and N acting at e, F acting at g', and Q, the weight of the pier itself, acting at its centre.

As an example, let $R = 11$, $r = 10$, $\dfrac{R + r}{2} = 10.5$, DX $= h = 20$, $R^2 - r^2 = 21$, $R^3 - r^3 = 331$, P $= 3.4$ tons, F $= 1.7$ tons, $eq = 9.09 - 5.07 = 4$, $an = 5.25$.

$$N = \frac{3.4 \times 4}{5.25} = 2.6 \text{ tons.}$$

Transposing N and P to e, and taking their *moments* about Z, we have, for the *moment* of N about Z—

$$N\left(h + \frac{R + r}{4}\right) = 2.6 \times 25.25 = 65.65.$$

The *moment* of P about Z is—

$$P \times eb = P\ (t + Df) = P\ (t + r - .433\ (R + r)\ )$$
$$= 3.4\ t + 3.1.$$

The *moment* of F acting at g', about Z, is—

$$F \times dg' = F\left(t + r - .637\ \frac{R^3 - r^3}{R^2 - r^2}\right) = 1.7\ t.$$

Taking the length of the wall as 10 ft. and weighing $\frac{1}{20}$th of a ton per cubic foot, we have for the *moment* of Q—

$$Q \times \frac{t}{2} = 10 \times 20 \times \frac{t}{20} \times \frac{t}{2} = 5\,t^2.$$

Then the *equation of stability* becomes—

$$131 \cdot 3 = 3 \cdot 4\,t + 3 \cdot 1 + 1 \cdot 7\,t + 5\,t^2,$$

or, $t^2 + t - 25 \cdot 64 = 0\,;\ t = 4 \cdot 58$ ft.

For the resistance of the rib, we must assume certain dimensions for its section, and see whether it is strong enough for its purpose. Let $D = 12''$, $d = 11''$, $b = 6''$, $t = \frac{1}{2}''$; then we have for the safe *moment* of resistance of the section about c—

$$\frac{5}{6\,D}\,(b \cdot D^3 - \overline{b - t} \cdot d^3) = \frac{5}{72} \times 3027 = 210.$$

And for the *moment* of stress, about c, of the force P, we have—

$$P \times eq = 3 \cdot 4 \times 4 \times 12 = 163\,;$$

or, the resistance is to the stress as 210 to 163, which shows that the rib is considerably stronger than necessary for safety. If we reduce the width b to $4\frac{1}{2}$ inches, we find the safe *moment* of resistance to be 170, which is very nearly equal to the *moment* of stress.

As another example, let $R = 16$ ft., $r = 15$ ft., $h = 25$ ft., then $P = 4 \cdot 94$ tons, $F = 2 \cdot 47$ tons, $eq = 13 \cdot 42 - 7 \cdot 4 = 6$, $am = 7 \cdot 75$,

$$N = \frac{4 \cdot 94 \times 6}{7 \cdot 75} = 3 \cdot 82.$$

The *moment* of N at c, taken about Z, is—

$$N\left(h + \frac{R + r}{4}\right) = 3 \cdot 82 \times 32 \cdot 75 = 125 \cdot 1.$$

The *moment* of P, at *e*, taken about Z, is—

$$P(t + r - \cdot 433(R + r)) = 4 \cdot 94(t + 1 \cdot 58) = 4 \cdot 94\,t + 7 \cdot 8.$$

The *moment* of F, about Z, is—

$$F \times dg' = F\left(t + r - \cdot 637\,\frac{R^3 - r^3}{R^2 - r^2}\right) = 2 \cdot 47\,t + \cdot 494.$$

The *moment* of Q is—

$$\frac{25 \times 10}{20} \times \frac{t^2}{2} = 6 \cdot 25\,t^2,$$

the wall being taken as 10 feet long and 25 feet high.
Then the *equation of stability* is—

$$250 \cdot 2 = 4 \cdot 94\,t + 7 \cdot 8 + 2 \cdot 47\,t + \cdot 494 + 6 \cdot 25\,t^2$$

or,
$$t^2 + 1 \cdot 2\,t - 37 \cdot 1 = 0\,;$$

$$t = 5 \cdot 6.$$

The *moment* of stress at EF is $P \times cq = 4 \cdot 94 \times 6 \times 12 = 356$. For the *moment* of resistance when $D = 12''$, $d = 11''$, $t = \frac{1}{2}$, $b = 8''$, we find

$$\frac{5}{6\,D}(b \cdot D^3 - (b - t) \cdot d^3) = 267,$$ or resistance is to stress as 267 to 356, which shows that the strength of the rib is insufficient for safety.

Let the thickness of the metal be $\frac{3}{4}$ inch, then $D = 12$, $d = 10 \cdot 5$, $t = \cdot 75$, $b = 8$; and the *moment* of resistance is found to be 377, or rather more than the *moment* of stress.

Since the *moment* of stress diminishes from EF towards the crown, it will be evident that a saving of material may be effected by diminishing the depth of the rib towards the crown, where it need only be strong enough to resist the direct pressure of the force N.

A very considerable saving can be effected in the quantity of walling to sustain the thrust of the rib, by making the ribs rest on piers with an open space between them. In the last example suppose the pier to be 2·5 ft. long instead of 10 ft., then the *moment* of Q is 1·56 t^2; and the *equation of stability* is—

$$t^2 + 4·7\,t - 155 = 0\,;\; t = 10·3\,\text{ft.}$$

Here the amount of walling is less than half what it was in the former case, or as 26 to 56.

In the foregoing investigation no account has been taken of the *stiffness* of the iron arch, or its resistance to change of form, which is a very important element, and constitutes the chief advantage obtained by the employment of iron for this purpose; for until the arch begins to bend at the crown there will be little or no thrust upon the supporting piers. If we supposed the half-arch from EF to AB, together with the corresponding half-arch on the opposite side, to be superseded by a straight beam, we can find easily the deflexion in the middle caused by the load P at g, by the rules given in Chapter V. Let $\frac{1}{2}$ W be the load at A whose *moment* about e will be equal to the *moment* of P; then we have—

$$\frac{W}{2} \times ne = \text{P} \times ge, \quad \text{or,} \quad \text{W} = 2\,\text{P} \times \frac{ge}{ne}.$$

Then W represents the vertical pressure at A, caused by the load P at g in the two sides of the arch. For a horizontal beam whose length is $l = 2\,ne$, the deflexion (53) will be—

$$\delta = \frac{1}{48} \cdot \frac{W}{E} \cdot \frac{l^3}{I},$$

where
$$I = \frac{b \cdot D^3 - (b - t) d^3}{12}.$$

Taking the foregoing example in which $D = 12$, $d = 11$, $b = 4\frac{1}{2}$, $t = \frac{1}{2}$, $ne = 9{\cdot}09$, we find—

$$W = \frac{6{\cdot}8 \times 4}{9{\cdot}09} = 3 \text{ tons.}$$

Taking $E = 10,000$ tons, we find $\delta = {\cdot}315$ for the deflexion at the middle of a beam $18{\cdot}18$ ft. long. The deflexion allowed by Tredgold for such a beam would be ·455 inch. The deflexion of less than $\frac{1}{3}$ inch would not produce a perceptible change of form in an arch 12 inches deep, and the actual deflexion at the crown would certainly not exceed that of a horizontal beam, but would rather be less. Consequently, it would appear that the resistance of the continuous arch to bending will counteract the greater part of the horizontal thrust as calculated above, as long as the *moment* of stress does not exceed the safe *moment* of resistance. The strength of the piers will then be little more than is necessary to resist the vertical load upon the rib.

Since the resistance to deflexion increases as the cube of the depth, when other dimensions remain unaltered, it follows that by increasing the depth of the rib we can ensure any amount of *stiffness* in the arch, and thereby overcome all horizontal thrust upon the supports.

A nearer approximation to the resistance to bending can be obtained by supposing the arch to be divided at AB, and a straight beam (as shown by the dotted lines

AF and BE) to take the place of the arch between EF and AB, the length of which beam will be $\dfrac{R+r}{2}.$ The deflexion (δ) of such a beam loaded with a distributed weight W is, by equation (71)

$$\delta = \frac{1}{8} \cdot \frac{W}{E} \cdot \frac{l^3}{I},$$

where I is the *moment of inertia* of the section.

W is here represented by the resolved part of P at right angles to the length of the beam; or W = ·866 P. Putting D = 12″, d = 11″, b = 4½″, t = ½″, we have

$$I = \frac{b \cdot D^3 - (b - t)\, d^3}{12} = 204.$$

Let E = 10,000, R = 11 ft., r = 10 ft., $l = \dfrac{R + r}{2}$ = 10½ ft. = 126 ins., P = 3·4 tons, W = ·866 P = 2·944 tons;

$$\delta = \frac{1}{8} \cdot \frac{2 \cdot 944}{10000} \cdot \frac{(10 \cdot 5)^3 \times 1728}{204} = \cdot 361 \text{ inch.}$$

By Tredgold's rule (**50**) the maximum deflexion is ·25 inch, so that the above is half as much again.

Taking the same dimensions, but making b = 6″, we have

$$I = 252, \quad \delta = \cdot 292 \text{ inch.}$$

Let D = 13″, d = 12″, then in the last example we find

$$I = 307, \quad \delta = \cdot 24 \text{ inch.}$$

When D = 14″, and d = 13″, we find

$$I = 365, \quad \delta = \cdot 2 \text{ inch.}$$

CHAPTER IX

DOMES, SPIRES.

95. HEMISPHERICAL DOME.—A Dome or Cupola is an arched roof covering a circular chamber, and consists of a series of arches both vertical and horizontal. The vertical arches have all the same radius of

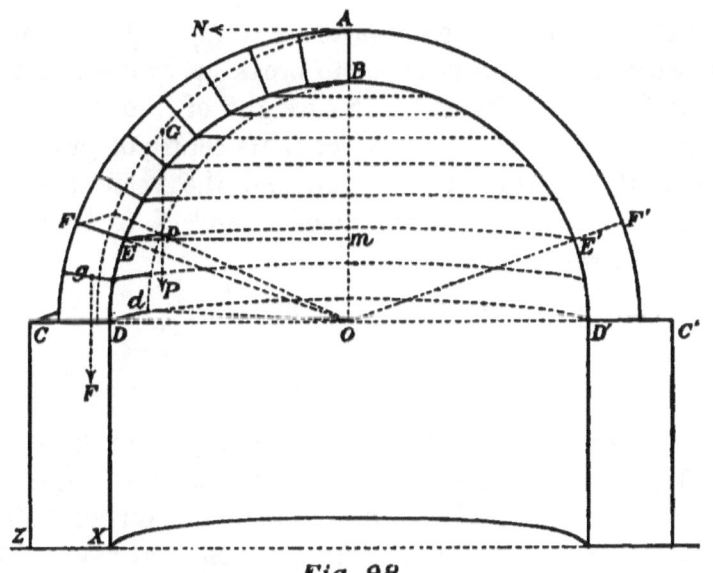

Fig 98.

curvature but diminish in width from the springing towards the vertex, and may be considered as "lunes" cut out of the hemisphere. The horizontal arches diminish in diameter from the springing towards the vertex.　In order to investigate the stability of a dome, we shall consider it as made up of a number of vertical "lunes" or ribs, as ABCD (fig. 98), whose base Dd subtends a very small angle, ϕ, at the centre O.　The horizontal arches are shown by the dotted lines as EE′, DdD′.

Let N be the horizontal pressure acting at the vertex, A, when the dome is about to break up by opening at B and F; and let P be the weight of the "lune" between EF and AB, acting at G its centre of gravity. Draw the horizontal line Em, and the vertical Gp; then, supposing the *moments* of P and N to balance about E, we have—

$$ \mathrm{P} \times \mathrm{E}p = \mathrm{N} \times \mathrm{A}m, $$

or,
$$ \mathrm{N} = \mathrm{P}\,\frac{\mathrm{E}p}{\mathrm{A}m} \qquad . \qquad . \quad (161) $$

In order to find the "joint of rupture," as in the arch **(80)**, we must express N in terms of the angle θ which any joint, as OEF, makes with the vertical; or, $\theta = \mathrm{BOE}$; and then find by calculation what value of θ gives the greatest value of N.　We will suppose for convenience that the angle DOd, or ϕ, subtends 2° at O, or the arc Dd to be $\frac{1}{180}$th part of the circle; then, "arc ϕ" = ·0349, expressed in circular measure, as in the Table previously given **(21)**.　Let OE = r, OF = R, δ = weight of 1 cubic foot of the arch and pier or "drum" which supports it.　Then we find, by means of the Integral Calculus, that—

$$ \mathrm{P} = \frac{\phi}{3}\,(1 - \cos. \theta)\,(\mathrm{R}^3 - r^3)\,.\,\delta \quad . \quad (162) $$

$$ = \cdot01163\,(1 - \cos. \theta)\,(\mathrm{R}^3 - r^3)\,.\,\delta. $$

Also we find—

$$mp = \frac{3}{8} \cdot \frac{\sin.\frac{\phi}{2}}{\frac{\phi}{2}} \cdot \frac{R^4 - r^4}{R^3 - r^3} \cdot \frac{\theta - \frac{1}{2}\sin.2\theta}{1 - \cos.\theta}$$

$$= \frac{3}{8} \cdot \frac{R^4 - r^4}{R^3 - r^3} \cdot \frac{\theta - \frac{1}{2}\sin.2\theta}{1 - \cos.\theta} \qquad . \quad (163)$$

since $\dfrac{\sin.\frac{\phi}{2}}{\frac{\phi}{2}}$ may be taken as very nearly equal to unity.

$$Ep = Em - mp = r . \sin.\theta - mp \quad . \quad (164)$$

$$Am = R - r . \cos.\theta \qquad . \qquad . \quad (165)$$

In order to find the value of θ which gives N a maximum value, we take, as in the arch, $R = 11$, $r = 10$, and calculate N by means of the foregoing equations for different values of θ. We have in this case $R^2 - r^2 = 21$, $R^3 - r^3 = 331$, $R^4 - r^4 = 4641$. When θ is 69°, we find—

$$N = \cdot 73577 . \delta$$

When θ is 70°, we find

$$N = \cdot 73607 . \delta$$

When θ is 71°, we find

$$N = \cdot 73575 . \delta.$$

We therefore conclude that the maximum value of N is obtained when the angle BOE, or θ, is 70°, the angle EOD being 20°; and that EF is the "joint of rupture" when EOD = 20°. In this case we have "arc θ" = 1·22173, sin. θ = ·93969, cos. θ = ·34202, sin. 2θ = ·64279; consequently the value of N, or the thrust of the 180th part of the dome, from equation (161) becomes—

$$N = \frac{·007192\,(R^3 - r^3)\,r - ·0039273\,(R^4 - r^4)}{R - ·34202\,r}.\,\delta \quad (166)$$

From equation (162) we have for the value of P,

$$P = ·007656\,(R^3 - r^3)\,.\,\delta \qquad . \quad (167)$$

Putting F for the weight of the part of the arch between CD and EF, we find from equation (162)

$$F = ·00398\,(R^3 - r^3)\,.\,\delta \quad . \qquad . \quad (168)$$

If Q is the weight of the portion of the "drum" which supports the "lune" ABCD, we find

$$Q = ·017453\,(2r + t)\,.\,t\,.\,h \quad . \quad (169)$$

where t is the thickness and h the height of the "drum" from the surface of the ground.

In order to determine the thickness, t, of the walls of the "drum," we equate *twice* the *moment* of N acting at E, and taken about the outer edge Z of the wall, with the *moment* of P acting at E, the *moment* of F acting at the centre of gravity of the lower part of the "lune," and the *moment* of Q. And if the wall and dome are built of similar material, we may omit the quantity δ from our equations.

The *moment* of N about Z, is—

$$N\,(h + r\,.\,\cos.\,\theta) = N\,(h + ·34202\,r) \quad . \quad (170)$$

The *moment* of P about Z is—

$$P\left(t + ·06031\ r\right) \qquad . \qquad (171)$$

The *moment* of F about Z is—

$$F\left(t + r - ·7351\frac{R^4 - r^4}{R^3 - r^3}\right) \quad . \quad (172)$$

The *moment* of Q about Z is—

$$Q\frac{(3\ r + t)\ t}{3\ (2\ r + t)} = ·0058177\ (3\ r\ t^2 + t^3)\ h \quad . \quad (173)$$

For example, let R = 11, r = 10, h = 50; then the *moment* of N is, from equation (170)—

$$39·32.$$

The *moment* of P is, from equation (171)—

$$2·534\ t + 1·53.$$

The *moment* of F is, from equation (172)—

$$1·32\ t - ·404.$$

The *moment* of Q is, from equation (173)—

$$8·7265\ t^2 + ·29\ t^3.$$

Then the " equation of stability " becomes—

$$78·64 = 2·534\ t + 1·53 + 1·32\ t - ·404 + 8·73\ t^2 + ·29\ t^3$$

which reduces to

$$t^3 + 30\ t^2 + 13·3\ t - 267 = 0;$$

from which we find, t = 2·66.

[The solution of this equation is obtained by means of " Horner's process," which will be found described in the Appendix to De Morgan's Arithmetic.]

If we put $\delta = 120$ lbs., then in the above example we find $N = 88.33$ lbs., $P = 304.08$ lbs. Suppose W to represent the pressure on the joint EF, then

$$W = N. \sin. 20° + P . \cos. 20°$$
$$= 88.33 \times .34204 + .93969 \times 304.08$$
$$= 316 \text{ lbs.}$$

If the dome is of Bath stone, the safe load is 150 lbs. per square inch, and if of brick it is 50 lbs. per square inch. Now the area, A, of the joint EF in inches is—

$$A = \frac{\phi}{2} (R^2 - r^2) . \cos. 20° \times 144$$
$$= 2.3612 (R^2 - r^2) = 49 \text{ inches.}$$

From which it appears that the load is only $\frac{1}{24}$th of the safe load in the case of Bath stone, and $\frac{1}{8}$th of the safe load for brick; so that the dome might be safely loaded to a considerable extent by a "lantern" or other heavy weight without affecting its stability.

When the dome has only its own weight to carry, and has no "lantern," the values of N and P diminish with the width as long as the internal radius is unaltered. Thus, if we make $R = 10.3$, $r = 10$, we have $R^2 - r^2 = 6$, $R^3 - r^3 = 92$, $R^4 - r^4 = 1248$; so that A is reduced by $\frac{2}{7}$ or .286, P is reduced by .278, and N is reduced by .339, or nearly one third. Hence it appears that the thickness of a dome may be reduced almost indefinitely when it has only its own weight to carry.

The thrust of the dome at the level of EF may be counter-balanced by means of a ring of wrought iron fixed tightly round it. Now, according to Rankine (*Applied Mechanics*, p. 184), "The thrust round a circular ring under an uniform normal pressure, is the product of the pressure on an unit of circumference by the radius." Consequently, if T is the total normal pressure on the ring at the level of EF, where the radius is R . cos. 20°; we have, the pressure on an unit

$$\text{of circumference} = \frac{T}{2\pi \cdot R \cdot \cos. 20°}$$

and the thrust round the ring is—

$$\frac{T}{2\pi \cdot R \cdot \cos. 20°} \times R \cdot \cos. 20° = \frac{T}{2\pi} = \cdot 16\,T \quad (174)$$

and this represents the tension in a ring of iron placed round the dome at the level of the joint EF.

In the case of a dome in which R = 11, r = 10, we have,

$$T = 180\,N = 88\cdot 33 \times 180 = 15900\,\text{lbs.} = 7\cdot 1\,\text{tons.}$$

The tension in a ring of iron is therefore, from equation (174),

$$= \cdot 16\,T = 1\cdot 136\,\text{tons.}$$

Taking the safe tensile strength of wrought iron as 5 tons per inch of section, we find that the sectional area of a belt for this dome, is $\dfrac{1\cdot 136}{5} = \cdot 227$ inches, or about $\frac{1}{4}$ of a square inch, so that $1'' \times \frac{1}{4}''$ will suffice. If this belt is fixed round the dome, the strength of the wall of the "drum" need only be sufficient to support the vertical weight of the dome, which is

$$(P + F)\,180 = 462\cdot 48 \times 180 = 83246\,\text{lbs.} = 37\cdot 3\,\text{tons.}$$

If the thickness of the drum is the same as that of the dome, or 1 foot, the area will be $\pi\,(R^2 - r^2) = 66$ feet, and the pressure will be 1261 lbs. per foot of walling. A thickness of 1 foot would suffice for a wall only 10 feet high, but for a height of 50 feet it should be at least 2 feet.

As another example, let $R = 31\cdot5$, $r = 30$, $h = 60$; then we have, $R^2 - r^2 = 93$, $R^3 - r^3 = 4280$, $R^4 - r^4 = 176049$; from equation (167), $P = 32\cdot767\;.\;\delta$; from equation (166), $N = 10\cdot924\;.\;\delta$; from equation (168), $F = 16\cdot03\;.\;\delta$.

The *moment* of N, at E, taken about Z, from equation (170), is,

$$767\cdot5.$$

The *moment* of ${}_.$P, at E, about Z, from equation (171), is,

$$32\cdot767\;.\;t + 59\cdot285.$$

The *moment* of F about Z, from equation (172), is,

$$16\cdot03\;.\;t - 3\cdot767.$$

The *moment* of Q about Z, from equation (173), is,

$$31\cdot4154\;.\;t^2 + \cdot34906\;.\;t^3.$$

The *equation of stability* is therefore (putting 2N for N) $1535 = 32\cdot767\,t + 59\cdot285 + 16\cdot03\,t - 3\cdot767 + 31\cdot4154\,t^2 + \cdot34906\,t^3$; which can be reduced to—

$$t^3 + 90\,t^2 + 140\,t - 4241 = 0,$$

from which we find, $t = 6$, nearly.

The total *normal* thrust of the whole dome at the level of the joint EF in this example, is—

$$T = 10\cdot924 \times 120 \times 180 = 234{,}078 \text{ lbs.} = 105 \text{ tons.}$$

The tensile strain on an iron belt to counteract this thrust by equation (174), is, $\cdot 16 \times 105 = 17$ tons, so that the belt should have a sectional area of $\dfrac{17}{5} =$ $3\cdot 4$ square inches; so that an iron belt $7'' \times \frac{1}{2}''$ will suffice.

In this example the area of the joint EF is 220 inches, and the pressure upon it is—

$$W = 120 \,(10\cdot924 \times \cdot34204 + 32\cdot767 \times \cdot93969)$$
$$= 4143 \text{ lbs.,}$$

which is nearly 20 lbs. on the square inch, or two-fifths of the safe pressure when the material of the dome is brick.

Suppose that in the last example the thickness is reduced to $\cdot5$, or $R = 30\cdot5$; then $R^2 - r^2 = 30$, $R^3 - r^3 = 1,365$, $R^4 - r^4 = 55,133$; $P = 10\cdot45 \cdot \delta$, $N = 3\cdot86 \cdot \delta$, $F = 5\cdot433 \cdot \delta$; $h = 60$.

The *moment* of N about Z is $271\cdot2$;

That of P about Z is $10\cdot45\,t + 18\cdot9$;

That of F about Z is $5\cdot433\,t + 1\cdot684$;

That of Q is $31\cdot4154\,t^2 + \cdot34906\,t^3$.

The *equation of stability* is therefore—

$$542\cdot4 = 15\cdot883\,t + 31\cdot4154\,t^2 + \cdot349\,t^3 + 20\cdot584,$$

which can be reduced to—

$$t^3 + 90\,t^2 + 45\cdot5\,t - 1496 = 0\,;$$

from which we find, $t = 3\cdot8$, nearly.

In this case the area of the joint EF is 71 inches, and the pressure upon it is 1,337 lbs., or 19 lbs. on the square inch. The total normal pressure round the whole of the dome at the level of the joint EF is, $T = 37.2$ tons, and the tensile strain on an iron belt by equation (174) would be $.16 \times 37.2 = 6$ tons.

It will be seen from equation (166) that for similar domes, or those in which the radii R and r are in the same proportion to one another, the thrust N varies as the *cube* of the radii. Thus, a dome whose radii are $R = 33$ and $r = 30$, will have 27 times the thrust of a dome in which $R = 11$, and $r = 10$.

When the quantity $R - r$ is small as compared with r, we can simplify the formulæ given above, by putting

$$\frac{R^4 - r^4}{R^3 - r^3} = \frac{4}{3}r \; ;$$

in which case the equation (163) becomes—

$$mp = \frac{r}{3} \times \frac{\theta - \frac{1}{2}\sin. 2\theta}{1 - \cos. \theta} = .6841\,r,$$

and the equation (166) becomes—

$$N = .00196 \frac{r\,(R^3 - r^3)}{R - .34202\,r} \cdot \delta \quad . \quad . \quad (175)$$

96. Semi-Domes.—We find in many ancient buildings, especially in the churches of the Byzantine period, that there are side-chapels, built out from the main structure, which are semicircular on plan and are covered with a domical roof. This roof is similar to that shown on fig. 97, only instead of the half-dome abutting against another half-dome, a vertical wall or arch is placed as at OA (fig. 99), against which the half-dome abuts.

The thrust of this half-dome upon the wall on which it stands, DZ, can be calculated by the formula given above, the value of N for a *lune* subtending an angle of 2° at O, being calculated either by equation (166) or equation (175); and the strength of pier or wall supporting it is found by equations (170, 171, 172, 173). Equating twice the *moment* of N to the *moments* of P, F, and Q, taken about Z, we obtain a cubic equation in terms of t, the thickness of the wall.

For example, let us take the case of the "exhedra" of Sta. Sophia at Constantinople, where R = 52 feet, r = 50 feet, and h, the height of the sustaining wall, is 75

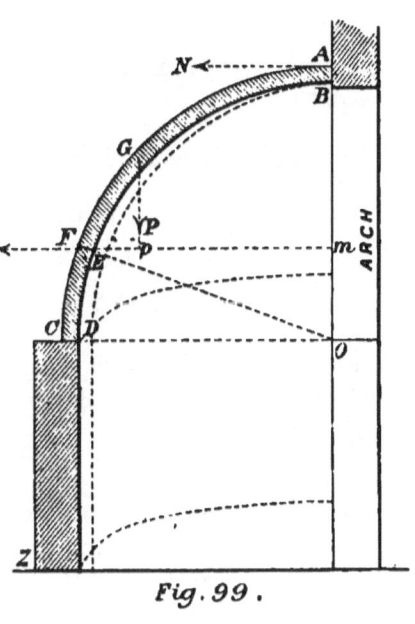

Fig. 99.

feet. We will suppose that δ, the weight of a cubic foot of the material, is 1 cwt. throughout, so that it can be omitted from the calculation. From equation (167) we have,

$$P = \cdot 007656 \, (R^3 - r^3) \, \delta = 119 \cdot 5 \, \delta.$$

From equation (175), we have,

$$N = 43 \cdot 7 \, \delta = 2 \cdot 185 \text{ tons.}$$

From equation (168), we have,

$$F = \cdot00398 \times 15608 \, \delta = 62 \cdot 12 \, \delta.$$

The *moment* of N acting horizontally at EF, taken about Z, is by equation (170)

$$43 \cdot 7 \, (75 + 17 \cdot 1) \, \delta = 4024 \cdot 77 \, \delta.$$

The *moment* of P acting vertically at E, is by equation (171)

$$119 \cdot 5 \, (t + 3 \cdot 015) \, \delta = (119 \cdot 5 \, t + 360 \cdot 3) \, \delta.$$

The *moment* of F acting vertically at the centre of gravity of the part CDEF, is by equation (172)

$$62 \cdot 12 \, (t + 1) \, \delta = (62 \cdot 12 \, t + 62 \cdot 12) \, \delta.$$

The *moment* of Q, the weight of the pier DZ, acting vertically at its centre of gravity, is by equation (173)

$$\cdot00582 \, (150 \, t^2 + t^3) \, 75 = (65 \cdot 5 \, t^2 + \cdot4365 \, t^3) \, \delta.$$

Then the *equation of stability* becomes

$$8049 \cdot 54 = \cdot4365 \, t^3 + 65 \cdot 5 \, t^2 + 181 \cdot 62 \, t + 422 \cdot 42$$

which reduces to

$$t^3 + 150 \, t^2 + 416 \, t - 17473 = 0;$$

from which we find $t = 9$ feet very nearly.

We have next to consider the thrust of the crown AB of the semi-dome against the wall or arch OA, as the vertices of all the *lunes* into which we suppose the dome to be divided meet at AB, and therefore concentrate the thrust of the half-dome on that point.

The value of N obtained by equation (175) will be

the horizontal pressure at the apex of the half-dome of each *lune*, but acting obliquely to the plane of the wall or arch OA, according to the position of the *lune* in the dome.

Therefore to obtain the pressure, R, at right angles to that plane, we must multiply N by the *sine* of the angle which it makes therewith ; or the pressure of each half of the semi-dome at the vertex is

$$\frac{1}{2} R = N \left(\sin. 1° + \sin. 3° + \ldots + \sin. 89°\right)$$

$$= N \times 28\cdot 64938$$

$$R = 2 N \times 28\cdot 64938 = 57\cdot 29876 \times N \quad . \quad (176)$$

which is the pressure of the dome at its apex at right angles to the wall or arch OA.

For example, we will apply this to the semi-dome or exhedra of Sta. Sophia, as given above. Here we have

$$N = 43\cdot 7\, \delta$$

Therefore, $R = 2504\, \delta = 125\cdot 2$ tons.

This pressure R serves the purpose of resisting the outward thrust of the central dome as calculated below (**97**).

97. DOME OF STA. SOPHIA AT CONSTANTINOPLE.—We will now apply the formula found above (**95**) to the investigation (approximately) of the thrust of the central Dome of Sta. Sophia. This dome is circular on plan and covers a square chamber by means of "pendentives" at the four angles. The dome presents on the inside the *appearance* of being nearly hemispherical, having a radius of about 55 feet ; but *constructionally* it must be considered a segmental dome, the springing of which is at the joint EF (fig. 100), the radius OE making an angle of 60° with the vertical OA, where O is the centre of curvature. The base of the dome from F to D is thickened out so as to form an abutment, the inner surface only being worked to the curve of the dome.

Referring to article (95) and taking the 180th part of the circumference, we have—

P = weight of the *lune* ABEF ;

δ = weight of a cubic foot of material ;

N = the horizontal thrust at A of the opposite *lune ;*

G, the centre of gravity of P ;

θ = the angle BOE ;

R = outer radius of the dome, as OA or OF ;

r = inner   do.   do.   as OB or OE.

Then, by equation (161) we have—

$$N = P \frac{Ep}{Am}.$$

By equation (162)—

$$P = \cdot01163 \ (1 - \cos. \ \theta) \ (R^3 - r^3) \ \delta,$$

and when $\theta = 60°$, cos. $\theta = \frac{1}{2}$. Also R = 57, r = 55. Therefore, P = $\cdot01163 \times 5 \times 18818 \ . \ \delta = 109\cdot27 \ . \ \delta$. Also, by equations (163,164)

$$Ep = 55 \times \cdot866 - \frac{3}{8} \times 74\cdot68 \frac{1\cdot047 - \cdot433}{\cdot5} = 13\cdot24.$$

By equation (165)—

$$Am = 57 - 55 \times \cdot5 = 29\cdot5.$$

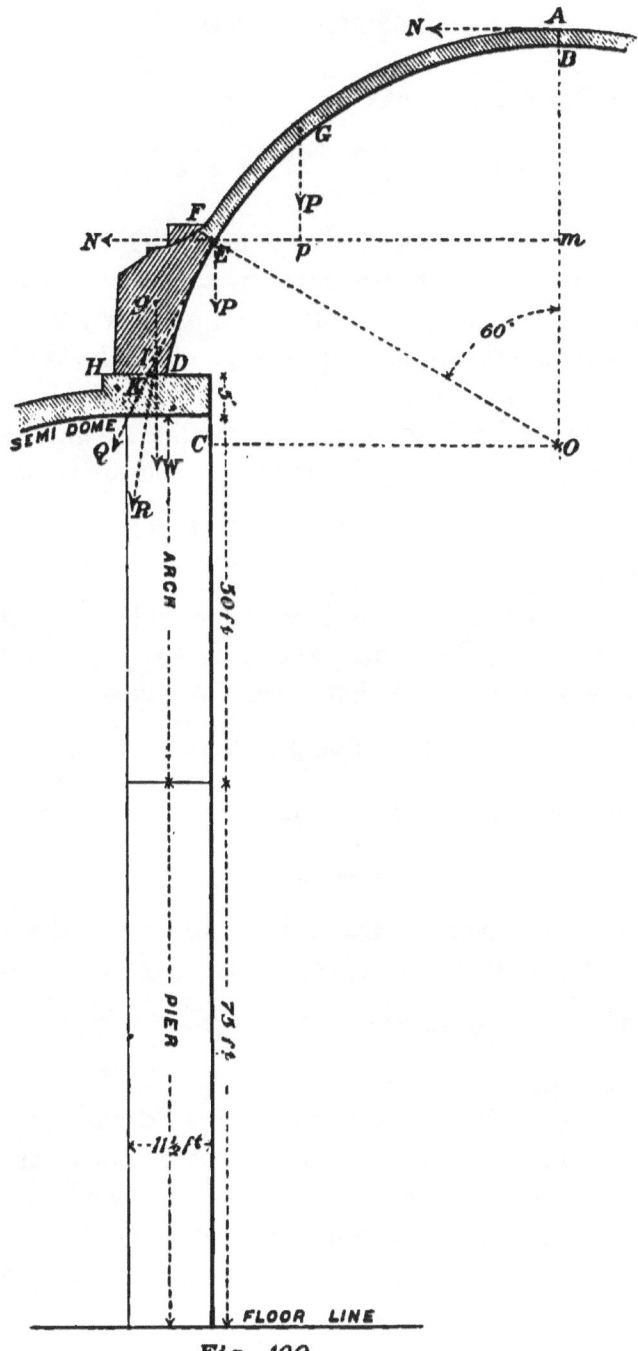

Fig. 100.

Consequently, we have—

$$N = \frac{109 \cdot 27 \times 13 \cdot 24}{20 \cdot 5} \delta = 49 \cdot \delta$$

= thrust of 180th part of the dome, at EF.

Putting $\delta = 1$ cwt., we have the thrust of the 180th part of the dome equal to 49 cwts. or nearly $2\frac{1}{2}$ tons. So that the total thrust of the dome at the level of EF is $180 \times 2\frac{1}{2}$ or 450 tons.

The radius Em is 46 feet, therefore the circumference at E is 289 feet, and the thrust of the dome at the level of EF is $\frac{450}{289}$ or $1 \cdot 56$ tons per foot length of circumference.

If a ring or belt of iron is placed round the dome at F to counteract the thrust, and T is the total normal pressure on the ring at that level, we have—

$$T = 180 \ N = 450 \text{ tons.}$$

And by equation (174) the tensile stress in the belt is—

$$\cdot 16 \ T = 72 \text{ tons.}$$

Taking 5 tons per square inch of section as the safe strain, we find that the belt must have a sectional area of $\frac{72}{5} = 14 \cdot 4$ square inches, or be $14\frac{1}{2}$ inches wide by 1 inch thick.

The weight, W, of the portion of the "drum" or base of the dome, FH, corresponding to the 180th part of the circumference, may be roughly estimated at $12\frac{1}{2}$ tons, acting vertically through its centre of gravity, g. Compounding this with the resultant, Q, of the forces

P and N at E, we get IR as the direction of the resultant R of these forces, making an angle of about 8° with the vertical at I.   The forces N, P, Q, W, and R, are very nearly in the proportions of 5, 11, 12, 25, and 36.

For the resistance offered by the abutments to this oblique thrust, we have on each of two sides of the chamber, a semi-dome of 50 feet radius; the vertex of which abuts at H, and acts as a flying buttress.   We have also the arch HD, 100 feet span, whose weight will also tend to counter-balance the thrust.

On the two other sides we have four solid piers 18 feet × 25 feet carrying arches 70 feet span.   These ought to be sufficient to resist outward thrust and also to bear the crushing weight of the whole of the superstructure.

It has been shown (96) that the thrust of the semidome against the arch forming one side of the chamber carrying the large dome, amounts to 125 tons; and we have seen above that the total thrust of the large dome is 450 tons.   If we suppose one-fourth of this to press against each of the four walls, we have $112\frac{1}{2}$ tons for the pressure against each; so that the above thrust of the half-dome, namely, 125 tons, will counterbalance this thrust on two sides.

98. SURCHARGED DOME. — Domical vaults are frequently used to carry a heavy surcharge, in cases where great strength is required in the roof of a circular chamber, as in bomb-proof powder magazines. Suppose the horizontal line MK (fig. 101) to represent the top of the surcharge, ABCD the section of the dome whose thickness is $R - r$; and let $AM = k$.

EF is the "angle of rupture" as before determined
(95), the radius OE making 20° with the horizontal

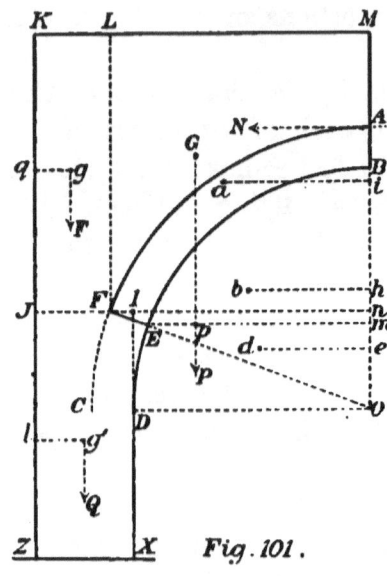

Fig. 101.

OD. Draw the horizon-
tal lines, F*n* and E*m*,
and let G be the centre
of gravity of a piece,
whose weight is P, cut
out of the dome by two
vertical planes making
an angle ϕ (= 2°) with
each other at the verti-
cal OM; drop a vertical
from G cutting E*m* in
p. Then P is the mass
MLFEB, or we may
assume, very nearly, P
= MLF*n* − BE*m*B ;
or

$$P = \frac{\phi}{2} \, R^2 \, (\cos. 20°)^2 \times FL - EBm$$

$$= \cdot 0154 \, R^2 \, (k + \cdot 658 \, R) \, . \, \delta - EBm$$

$$EBm = BEOB - OEmO$$

$$= \frac{\phi}{3} \, r^3 \, (1 - \cos. 70°) \, \delta$$

$$- \frac{\phi}{6} \, r^3 \, (\sin. 70°)^2 \times r \, . \, \cos. 70° \times \delta$$

$$= (\cdot 00765 \, r^3 - \cdot 001756 \, r^3) \, \delta = \cdot 0059 \, r^3 \, . \, \delta,$$

Therefore,

$$P = (\cdot 0154 \, R^2 \, (k + \cdot 658 \, R) - \cdot 0059 \, r^3) \, \delta \quad . \quad (177)$$

Let *a* be the centre of gravity of the part MLF*n*, a*i*

its distance from the axis OM ; b that of the part
BEOB, bh its distance from OM ; d that of the part
OEmO, de its distance from OM.

Then $ai = \frac{2}{3} ML = \frac{2}{3} R \, . \, \sin. \, 70^\circ = \cdot 626 \, R \, ;$

$de = \frac{r}{2} \sin. 70^\circ = \cdot 47 \, r; \; bh = \frac{3}{8} \, r \, \dfrac{\theta - \frac{1}{2} \sin. 2 \, \theta}{1 \cos. \theta} = \cdot 513 \, r,$

when $\theta = 70^\circ$.

$P \times mp = \{ \cdot 0154 \; R^2 \, (h + \cdot 658 \; R) \times \cdot 625 \; R +$
$\cdot 001756 \, . \, r^3 \times \cdot 47 \, r - \cdot 00765 \, r^3 \times \cdot 513 \, r\} \times \delta$
$= \{ \, \cdot 00964 \; R^2 \, (h + \cdot 658 \; R) - \cdot 0031 \, . \, r^4 \} \, . \, \delta \; . \quad (178)$

$$mp = \frac{P \times mp}{P},$$

$Ep = r \, . \, \sin. \, 70^\circ - mp = \cdot 93969 \, r - mp,$

$Am = R - r \, . \, \cos. \, 70^\circ = R - \cdot 342 \, r,$

$$N = P \, \frac{Ep}{Am}.$$

Since N and P balance about E, we can consider
them as acting at E parallel to their original directions,
and take their *moments* from that point about Z, the
outer edge of the supporting wall or " drum." Let F
be the mass of the part KLFJ, acting at g its centre of
gravity ; let Q be that of JIXZ acting at g' its centre
of gravity. Then to obtain the *equation of stability* we
have to equate twice the *moment* of N about Z, with
the sum of the *moments* of P, F, and Q. Let DX = h.

The *moment* of N is—

$$N \, (h + r \, . \, \sin. \, 20^\circ) = N \, (h + \cdot 342 \, r).$$

T

The *moment* of P is—

$$P\ (t + \cdot 0603\ r).$$

$$F = \frac{\phi}{2} \left\{ (r + t)^2 - R^2 \cdot (\cos.20°)^2 \right\} (k + R - R \cdot \sin.20°)\delta$$

$$= \frac{\phi}{2} \left((r + t)^2 - \cdot 883\ R^2 \right) (k + \cdot 658\ R)\ \delta,$$

$$gq = r + t - \frac{2}{3}\ \frac{(r + t)^3 - \cdot 7796\ R^3}{(r + t)^2 - \cdot 883\ R^2}$$

$$= \frac{1}{3}\ \frac{(r + t)^3 - 2\cdot 649\ R^2\ (r + t) + 1\cdot 559\ R^3}{(r + t)^2 - \cdot 883\ R^2}$$

$$F \times gq = \cdot 0058\ (k + \cdot 658\ R)\ \{(r + t)^3 - 2\cdot 65\ R^2\ (r + t)$$
$$+ 1\cdot 56\ R^3\}\ \delta,$$

which is the *moment* of F about Z.

The *moment* of Q is the same as in equation (173), namely—

$$\cdot 0058177\ (3\,rt^2 + t^3)\,h\ .\ \delta.$$

As an example of application of these formulæ, let R = 11, r = 10, h = 20, k = 4. Then we find from equation (177), P = 16·8 . δ; and, from equation (178), P × mp = 113·22 δ, mp = 6·74, Ep = 2·55, Am = 7·58;

Therefore, $$N = \frac{16\cdot 8 \times 2\cdot 55}{7\cdot 58}\ \delta = 5\cdot 65\ \delta.$$

The *moment* of N, at E, about Z (omitting δ), is—

$$5\cdot 65 \times 23\cdot 42 = 132\cdot 32.$$

The *moment* of P, at E, about Z, is—

$$16\cdot 8\,t + 10\cdot 13.$$

The *moment* of F, at *g*, about Z, is—

$$\cdot 0652\,t^3 + 1\cdot 965\,t^2 - 1\cdot 37\,t - 15\cdot 26.$$

The *moment* of Q, at g', about Z, is—

$$\cdot11635\,t^2 + 3\cdot51\,t^3.$$

The *equation of stability* becomes

$$264\cdot64 = \cdot18155\,t^3 + 5\cdot466\,t^2 + 15\cdot53\,t - 5\cdot13,$$

which can be reduced to

$$t^3 + 30\,t^2 + 85\cdot5\,t - 1485\cdot5 = 0,$$

from which we find $t = 5\cdot4$ ft., very nearly.

The pressure on the joint EF in this example is—

$$\mathrm{W} = \cdot342\,\mathrm{N} + \cdot94\,\mathrm{P} = 2127 \text{ lbs.}$$
$$= 150 \times 14\cdot18;$$

so that if Bath stone is used, the area of the joint must be 14·18 square inches. If brick is used, in which 50 lbs. per inch is the safe load, the area must be 42·54 inches. The actual area in this case is 49 inches, so that the resistance will be ample when brick is used.

99. DOME OF UNEQUAL THICKNESS. — Since the heaviest pressure is borne by the lower part of the dome, a saving in weight and material can be effected by reducing the thickness from the springing towards the crown. We will here take an example in which the depth at the crown is half that at the springing.

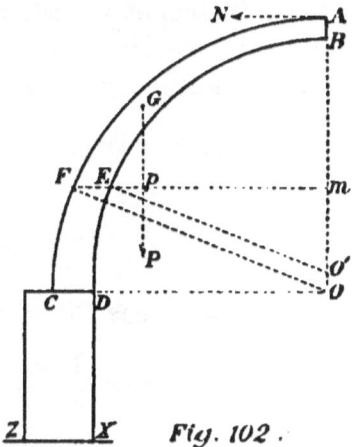

Fig. 102.

Let AC (fig. 102) be the outer curve of the dome

struck from O as a centre, BD the inner curve struck from O' as a centre ; call

$$OA = R, \; O'B = r, \; \text{then } CD = R - r, \; AB = \frac{R - r}{2}$$

Draw O'E making an angle of 70° with the vertical OA, then the "joint of rupture" will be at E. Draw the horizontal line FEm, and let P be the weight of the part of a "lune" of the dome between EF and AB. Let P_1 be the weight of the wedge AFm, P_2 that of BEm; z_1 and z_2 the horizontal distances of their centres of gravity from AO. Let G be the centre of gravity of P, mp or z, its distance from AO. Then P = $P_1 - P_2$,

$$z = \frac{P_1 z_1 - P_2 z_2}{P_1 - P_2}.$$

Let ϕ be a small angle which a small *lune* of the dome subtends at O', which, as before, we take as 2°, so that in circular measure we have " arc ϕ " = ·0349. From the geometry of the figure we find—

$$P_1 = \frac{\phi}{2} \left\{ R^2 (R - Om) - \frac{R^3 - Om^3}{3} \right\} \cdot \delta$$

$$P_2 = \frac{\phi}{2} \left\{ r^2 (r - O'm) - \frac{r^3 - O'm^3}{3} \right\} \delta$$

$$O'm = r \cdot \cos. 20° = ·342 \, r$$

$$Om = O'm + \frac{1}{2} (R - r) = ·5 \, R - ·158 \, r$$

$$r - O'm = ·658 \, r$$

$$r^3 - O'm^3 = ·96 \, r^3$$

$$R - Om = ·5 \, R + ·158 \, r$$

$$Ep = Em - mp = \cdot9397\, r - z$$

$$Am = AO - Om = R - Om = \cdot5\,R + \cdot158\,r$$

$$P_1 = \cdot0175 \left\{ R^2\,(\cdot5\,R + \cdot158\,r) - \right.$$

$$\left. R^3 - \frac{(\cdot5\,R + \cdot158\,r)^3}{3} \right\} . \delta$$

$$P_2 = \cdot0175\,(\cdot658\,r^3 - \cdot32\,r^3)\,.\,\delta = \cdot00592\,r^3\,.\,\delta.$$

If we put x for any measurement along OA, either from O or O′, then we have—

$$P_{1z_1} = \frac{\phi}{3}\,\delta\,.\int (R^2 - x^2)^{\frac{3}{2}}\,.\,dx,$$

the limits of integration being from $x = Om$ to $x = R$; when—

$$P_{1z_1} = \cdot01163 \left\{ \frac{3\,R^4}{8}\left(1\cdot5708 - \sin.^{-1}\,.\,\frac{Om}{R}\right) \right.$$

$$- \frac{Om}{4}\left(\frac{5\,R^2}{2} - Om^2\right)\sqrt{R^2 - Om^2} \left. \right\}\,.\,\delta$$

$$= \cdot01163 \left\{ \cdot458\,R^4 - \frac{Om}{4}\left(\frac{5\,R^2}{2} - Om^2\right)\sqrt{R^2 - Om^2} \right\}\,.\,\delta$$

$$P_{2z_2} = \frac{\phi}{3}\,\delta\,.\int (r^2 - x^2)^{\frac{3}{2}}\,.\,dx,$$

the limits of integration being from $x = O'm$ to $x = r$.

$$P_{2z_2} = \cdot01163 \left\{ \frac{3\,r^4}{8}\left(1\cdot5708 - \sin.^{-1}\,.\,\frac{O'm}{r}\right) \right.$$

$$- \frac{O'm}{4}\left(\frac{5\,r^2}{2} - O'm^2\right)\sqrt{r^2 - O'm^2} \left. \right\}\,.\,\delta$$

$$= \cdot003101\,r^4\,.\,\delta$$

$$N = P\,\frac{Ep}{Am}.$$

For example, let $R = 11, r = 10$; then $O'm = 3\cdot42$; $Om = 5\cdot5 - 1\cdot58 = 3\cdot92$; $P_1 = 7\cdot58 \cdot \delta$, $P_2 = 5\cdot92 \cdot \delta$, $P = P_1 - P_2 = 1\cdot66 \cdot \delta$.

$P_1 z_1 = 44\cdot34 \cdot \delta$, $P_2 z_2 = 31\cdot01 \cdot \delta$, $P_1 z_1 - P_2 z_2 = 13\cdot33 \cdot \delta$

$$mp = z = \frac{13\cdot33}{1\cdot66} = 8,$$

$$Ep = 9\cdot397 - 8 = 1\cdot397$$

$$Am = 5\cdot5 + 1\cdot58 \doteqdot 7\cdot08$$

$$N = \frac{1\cdot66 \times 1\cdot397}{7\cdot08} \cdot \delta = \cdot327 \cdot \delta.$$

Putting F for the weight of the lower portion of the *lune* between EF and CD, we can approximate to its value by taking it to equal the area of the section at EF plus that of the section at CD, multiplied by half Om; or we have—

$$F = \frac{\phi}{2}(R^2 - r^2)(1 + \cos.^2 20°)\frac{Om}{2} \cdot \delta$$

$$= \cdot016475 (R^2 - r^2)(\cdot5 R - \cdot158 r) \cdot \delta.$$

The lever arm of F about Z may be taken as very nearly—

$$r + t - \frac{R + r}{2} = t - \frac{R - r}{2}$$

or, the *moment* of F about Z, is—

$$F\left(t - \frac{R - r}{2}\right).$$

If Q is the weight of the part of the wall DZ, $Q \cdot q$ its *moment* about Z is found by equation (173), and is—

$$Q \cdot q = \cdot0058177 (3 r t^2 + t^3) h \cdot \delta.$$

The *moment* of N acting at E, about Z, is—

$$N (h + Om) = N (h + \cdot 5\,R - \cdot 158\,r).$$

Putting $h = 50$, we have for the *moment* of N in the above example—

$$N (h + Om) = 19 \cdot 27 \cdot \delta.$$

The *moment* of P acting at E, about Z, is—

$$P (r + t - r \times \cdot 9397) = P (t + \cdot 0603\,r)$$
$$= (1 \cdot 66\,t + 1) \cdot \delta, \text{ in this example.}$$

The *moment* of F about Z is—

$$F'\left(t - \frac{R - r}{2} \right) = \left(1 \cdot 356\,t - \cdot 678 \right) \cdot \delta.$$

The *moment* of Q is—

$$Q \cdot q = (8 \cdot 7265\,t^2 + \cdot 29\,t') \cdot \delta.$$

Then the *equation of stability* becomes
$$38 \cdot 54 = 1 \cdot 66\,t + 1 + 1 \cdot 356\,t - \cdot 678 + 8 \cdot 7265\,t^2 + \cdot 29\,t'$$
which reduces to

$$t^3 + 30\,t^2 + 10 \cdot 4\,t - 131 \cdot 8 = 0;$$

from which we find, $t = 1 \cdot 88$.

Comparing this dome with the dome of uniform thickness and same diameter (**95**), we see that the horizontal thrusts are in the proportion of 39 : 88 ; or the thrust of the dome of varying thickness is less than half what it is in the dome of uniform thickness ; also that the thickness required for the walls supporting them is nearly in the proportion of 2 to 3, where the height is the same.

100. GOTHIC DOME.—This term is applied to a dome whose section is a Gothic or pointed arch, as ABCD (fig. 103), where O is the centre of curvature

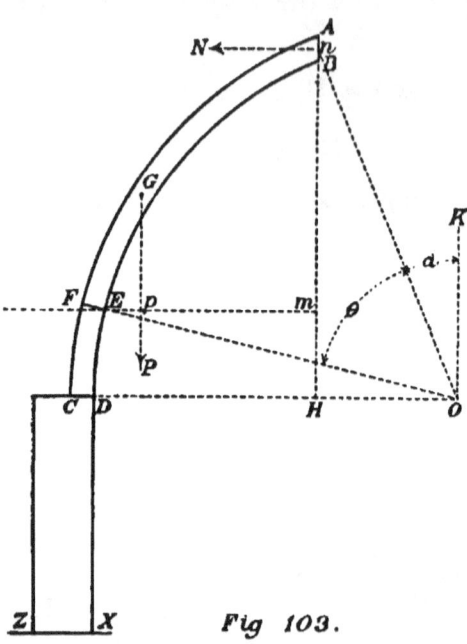

Fig 103.

for one half of the section, AH being the axis or centre line of the dome, and DH the half-span. Let n be the middle point of the vertical joint AB; then the form of arch varies according to the angle, a, which On makes with the vertical OK. If the section is an equilateral arch (**86**), then $a = 30°$; and in the great dome of the Cathedral at Florence we find $a = 22\frac{1}{2}°$. We will call OC or OF $= R$, OD or OE $= r$; then O$n = \frac{1}{2}(R + r)$, CD $=$ EF $= R - r$. We will take as before a small *lune* cut out of the dome and subtending an angle $\phi = 2°$ at the centre O; in circular measure "arc ϕ" $= \cdot 0349$. Let N, the horizontal thrust, act at the point n, and we have in the first place, as before, to find the joint EF at which N is greatest. Let G be the centre of gravity of the part of the *lune* between AB and EF whose weight is P, and let OE make the angle θ with On. Then we have to express N in terms of θ, and find what value of θ makes N a maximum. The value of P is determined from the Integral

$$P = \phi \cdot \delta \cdot \iint r^2 (\sin.(a + \theta) - \sin. a) \, dr \cdot d\vartheta \quad (179)$$

where the limits of integration are r and R, and $\theta = 0$ to $\theta = 0$; so that

$$P = \cdot 01163 \, (R^3 - r^3) \, (\cos. a - \cos. (a + \theta) - \theta \cdot \sin. a) \, \delta$$

$$mp = \frac{\sin. \frac{\phi}{2}}{\frac{\phi}{2}} \cdot \frac{\iint r^2 (\sin.(a + \theta) - \sin. a)^2 \, dr \cdot d\vartheta}{\iint r^2 (\sin.(a + \theta) - \sin. a) \, dr \cdot d\theta}$$

and since ϕ is small, we can put $\dfrac{\sin. \frac{\phi}{2}}{\frac{\phi}{2}} = 1$; and we

have

$$mp = \frac{3}{4} \cdot \frac{R^4 - r^4}{R^3 - r^3} \times$$

$$\left\{ \frac{\theta}{2} (1 + 2 \sin.^2 a) + \tfrac{1}{4} (\sin. 2 a - \sin. 2 (a + \theta)) - 2 \sin. a (\cos. a - \cos. (a + \theta)) \right\}$$

$$\div \left\{ \cos. a - \cos. (a + \theta) - \theta \cdot \sin. a \right\}$$

$$= \frac{3}{16} \cdot \frac{R^4 - r^4}{R^3 - r^3} \times$$

$$\left\{ 2 \theta (1 + 2 \sin.^2 a) + \sin. 2 a - \sin. 2 (a + \theta) - 8 \sin. a (\cos. a - \cos. (a + \theta)) \right\}$$

$$\div \left\{ \cos. a - \cos. (a + \theta) - \theta \cdot \sin. a \right\}$$

$$E_p = r \left(\sin. (a + \theta) - \sin. a \right) - mp$$

$$mn = \frac{1}{2} (R + r) \cos. a - r . \cos. (a + \theta)$$

$$N = P . \frac{E_p}{mn}.$$

We will apply these formulæ to the case where $a = 22\frac{1}{2}°$, or "arc a" = ·3927, sin. a = ·38268, cos. a = ·92388, sin. $2a$ = cos. $2a$ = ·70711, sin.² a = ·14645, $1 + 2$ sin.² a = 1·2929.

$$P = ·01163 (R^3 - r^3) \{·9239 (1 - \cos. \theta) - ·3827 (\theta - \sin. \theta)\} \delta$$

$$mp = \frac{3}{16} \frac{R^4 - r^4}{R^3 - r^3} \times$$

$$\left\{ 2·5858 \, \theta + ·707 \, (1 - \cos. 2\theta - \sin. 2\theta) \right.$$
$$\left. - 2·828 \, (1 - \cos. \theta) - 1·172 \sin. \theta \right\}$$
$$\div \left\{ ·9239 \, (1 - \cos. \theta) - ·3827 \, (\theta - \sin. \theta) \right\}$$

$$E_p = r \{·9239 \sin. \theta - ·3827 \, (1 - \cos. \theta)\} - mp$$

$$mn = ·4619 \, (R + r) - (·9239 \cos. \theta - ·3827 \sin. \theta) \, r.$$

In order to find the greatest value of N, we put $R = 11$, $r = 10$, and calculate the above formulæ for different values of θ; then we find that N is greatest when $\theta = 54°$, or, $a + \theta = 76\frac{1}{2}°$, so that OE makes $13\frac{1}{2}°$ with the horizontal line OC. In this case we have $\theta = ·94248$, cos. $\theta = ·5878$, sin. $\theta = ·80902$, sin. $2\theta = ·95106$, cos. $2\theta = - ·30902$.

$$P = ·003837 \, (R^3 - r^3) \, \delta$$

$$mp = ·32756 \frac{R^4 - r^4}{R^3 - r^3}$$

$$N = \frac{\cdot 002262 \, (R^3 - r^3) \, r - \cdot 001258 \, (R^4 - r^4)}{\cdot 46194 \, R + \cdot 22849 \, r}.$$

If we put F for the weight of the part of the *lune* between EF and CD acting at its centre of gravity; we find from the equation (179), by integrating from $\theta = 54°$ to $\theta = 90° - a$, that

$$F = \cdot 01163 \, (R^3 - r^3) \, \{\cos.(a + \theta) - (1\cdot5708 - a - \theta) \sin. a\} \, \delta$$

which in the present case becomes

$$F = \cdot 001667 \, (R^3 - r^3) \, . \, \delta.$$

The lever arm of F about Z may be taken as

$$t - \frac{R - r}{2}$$

where t is the thickness of the wall of the "drum;" or the *moment* of F about Z is

$$F\left(t - \frac{R - r}{2}\right).$$

The forces P and N being supposed to act at E, the *moment* of N about Z, (putting h for the height of the drum) is

$$N \, (h + r \, . \, \cos. \, (a + \theta)) = N \, (h + \cdot 23345 \, r).$$

The *moment* of P about Z, is

$$P \, (t + r - r \, . \, \sin. \, (a + \theta)) = P \, (t + \cdot 02763 \, r).$$

If Q is the weight of the portion of the wall of the "drum" corresponding to the *lune* ABCD, acting at its centre of gravity, q its lever arm about Z; then we have for the *moment* of Q, from equation (173), putting s for the half-span,

$$Q \, . \, q = \cdot 0058177 \, (3st^2 + t^3) \, h \, . \, \delta.$$

The half span HD of the dome is

$$s = HD = r - \frac{1}{2} (R + r) \sin. a$$

$$= r - \cdot19134 (R + r), \text{ when } a = 22\frac{1}{2}°.$$

Suppose, for example, that the span is 20 feet and the thickness, CD or EF, is 1 foot, then $R = r + 1$, HD $= 10 = s$, $s = 10 = r - \cdot19134 (2r + 1)$

$$r = \frac{10 \cdot 19134}{\cdot 61732} = 16 \cdot 5;$$

therefore $R = 17 \cdot 5$. Let $h = 50$ feet. Then we find $P = 3 \cdot 327 . \delta$, $N = \cdot 64334 . \delta$, so that the thrust at E is less in this dome than in the hemispherical one of equal span and thickness, in the proportion of about 7 to 8. The *moment* of N, at E, about Z, is 35; that of P is $3 \cdot 328 t + 1 \cdot 517$. The *moment* of F is $1 \cdot 446 t - \cdot 723$, and that of Q is $\cdot 291 (t^3 + 30 t^2)$.

The *equation of stability* becomes

$$70 = 3 \cdot 328 t + 1 \cdot 517 + 1 \cdot 446 t - \cdot 723 + \cdot 291 (t^3 + 30 t^2)$$

or, $t^3 + 30 t^2 + 16 \cdot 5 t - 239 = 0;$

therefore, $t = 2 \cdot 416$ feet.

If we put $\delta = 120$ lbs., then the value of N is 77·2 lbs. and multiplying this by 180 we have 13,876 lbs. or 6·2 tons, for the horizontal thrust of the entire dome at the level of the joint EF. If an iron belt is placed round the dome at the level of F, we find by equation (174) that its sectional area should be ·16 × 6·2 ÷ 5, or one-fifth of an inch, if we allow 5 tons per inch as the safe stress on wrought iron; so that a belt 1″ × ⅕″

would entirely counteract the outward thrust of the dome at this point.

We will now apply these formulæ to a circular dome of the same diameter as the octagonal dome of Florence, the diameter of which is 130 feet; and will take the thickness at 6 feet. We find in this case that $r = 107$ feet, and $R = 113$ feet; $R + r = 220$, $R^2 = 12,769$, $r^2 = 11,449$, $R^3 - r^3 = 217,854$, $R^4 - r^4 = 31,967,760$. We will suppose that the dome stands upon a solid circular wall 175 feet high. We have then

$$P = \cdot 003837 \times 217854 \cdot \delta = 836\, \delta$$

$$N = \frac{\cdot 002262 \times 107 \times 217854 - \cdot 001258 \times 31967760}{\cdot 46194 \times 113 + \cdot 22849 \times 107}\, \delta$$
$$= 162\cdot 5 \cdot \delta.$$

The *moment* of N about Z is

$$N\,(h + \cdot 23345\, r) = 162\cdot 5 + 200 \cdot \delta = 32500\, \delta.$$

The *moment* of P about Z is

$$P\,(t + \cdot 02763\, r) = (836\, t + 2462)\, \delta.$$

The *moment* of F about Z is

$$F\left(t - \frac{R - r}{2}\right) = \left(363\, t - 1089\right)\, \delta.$$

The *moment* of Q about Z is

$$Q \cdot q = 1\cdot 018\,(321\, t^2 + t^3)\, \delta.$$

The *equation of stability* becomes

$$65000 = 836\, t + 2462 + 363\, t - 1089$$
$$+ 1\cdot 018\,(321\, t^2 + t^3)$$

which reduces to

$$t^3 + 321\ t^2 + 1119\ t - 62500 = 0\ ;$$

the solution of which is $t = 12\cdot12$ feet.

Putting $\delta = 120$ lbs., we have $N = 19,500$ lbs. as the horizontal thrust at F for the 180th part of the dome; and the total thrust of the dome is, $T = 3,510,000$ lbs. or 1,567 tons. Multiplying this by $\cdot16$, equation (174), and dividing by 5, we get 50 square inches for the sectional area of an iron belt placed round the dome about the level of F, that will counteract the thrust; or a belt 2 inches thick and 25 inches wide would be required for this purpose.

The radius of this dome is $6\frac{1}{2}$ times that of the first example where the span was 20 feet, and the cube of $6\frac{1}{2}$ is 275. We find from the formulæ that the thrusts of the two domes are in the ratio of $\cdot64334$ to $162\cdot5$, or as 1 to 252, which is nearly as the cube of their diameters.

It will be evident from the foregoing formulæ that the position of the "joint of rupture" EF must vary with the value of the angle a, which is nothing in a hemispherical dome, and is 30° in a dome whose section is an equilateral pointed arch, in which latter case we find OE makes an angle of 11° with the horizontal line OC. By substituting the angle 30° for a and 49° for θ in the expressions for P, N, and F, we can calculate their values for a dome whose section is an equilateral arch. In this case we find the following equations,

$$P = \cdot00288\ (R^3 - r^3)\ \delta$$

$$N = \frac{\cdot001387\ (R^3 - r^3)\ r - \cdot000407\ (R^4 - r^4)}{\cdot43302\ R + \cdot24222\ r}\ \delta.$$

The lever arm of P about Z is

$$t + \cdot 01837\ r.$$

The lever arm of N about Z is

$$h + \cdot 1908\ r.$$

The *moment* of F about Z is

$$\cdot 001103\ (R^3 - r^3)\ (t + s) - \cdot 000407\ (R^4 - r^4).$$

The *moment* of Q is the same as before.

101. GOTHIC DOME WITH LANTERN.— When a dome is used to form the roof of such an edifice as the Cathedral of Florence, it is usually surmounted by an ornamental lantern, the weight of which must be taken into consideration when investigating the thrust of the dome upon its supporting walls. Let IKL (fig. 104) represent the section of such

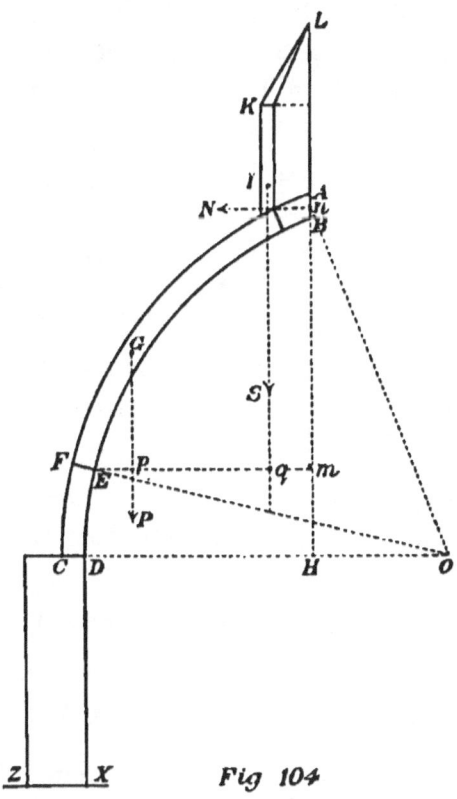

Fig 104

a lantern, and S the weight of 180th part of it acting in the direction Iq cutting Em in q. Then in finding the outward pressure, N, at E, we must equate the *moment* of N about E with the sum of the *moments* of P and of S; P being the weight of the 180th part of the dome between AB and EF; so that we have

$$N \times mn = P \times Ep + S \times Eq,$$

or

$$N = \frac{P \times Ep + S \times Eq}{mn}$$

Taking the case of the Florentine dome, we will suppose the total weight of the lantern to be 100 tons, and put $S = 10 \cdot \delta$; also $mq = 10 \cdot 5$; $R = 113$, $r = 107$; $Em = \cdot 58969 \, r = 63 \cdot 1$; $Eq = Em - mq = 52 \cdot 6$.

$$mn = \cdot 92388 \frac{R + r}{2} - \cdot 23345 \, r = 76 \cdot 65; \quad P = 836 \cdot \delta.$$

$$mp = \cdot 32756 \frac{R^4 - r^4}{R^3 - r^3} = 48 \cdot 07.$$

$$Ep = Em - mp = 63 \cdot 1 - 48 \cdot 07 = 15 \cdot 03.$$

$$N = \frac{836 \times 15 \cdot 03 + 10 \times 52 \cdot 6}{76 \cdot 65} \, \delta = 171 \, \delta.$$

This is only $\frac{1}{20}$th more than the thrust of the same dome *without* a lantern, which we found above (**99**) to be $162 \cdot 5 \cdot \delta$. In this case the *equation of stability* becomes

$$68400 = 1 \cdot 018 \, (t^3 + 321 \, t^2) + 1200 \, t + 2501 - 1089,$$

or,

$$t^3 + 321 \, t^2 + 1188 \, t - 65803 = 0;$$

$$t = 12 \cdot 38 \text{ft}.$$

Without a lantern we found that $t = 12 \cdot 12$ ft., so that its weight has comparatively very little effect upon the thrust of the dome. If we reduce the thickness of the dome one-half, we find that the thrust caused by the weight of the lantern is increased in proportion to that caused by the weight of the dome itself. Putting $R - r = 3$ft. instead of 6ft., we have

$$P = \cdot003837 \, (110^3 - 107^3) \, \delta = 406 \cdot 6 \cdot \delta.$$

$$mp = \cdot32756 \, \frac{R^4 - r^4}{R^3 - r^3} = 47 \cdot 4.$$

$$Ep = 63 \cdot 1 - 47 \cdot 4 = 15 \cdot 7.$$

$$mn = 100 \cdot 24 - 25 = 75 \cdot 24.$$

Without the weight of a lantern, we have

$$N = \frac{406 \cdot 6 \times 15 \cdot 7}{75 \cdot 24} \, \delta = 84 \cdot 8 \, \delta.$$

With a lantern of the above weight, we have

$$N = \frac{406 \cdot 6 \times 15 \cdot 7 + 10 \times 52 \cdot 6}{75 \cdot 24} \cdot \delta = 91 \cdot 8 \, \delta;$$

or the thrust *with* a lantern is in this case $\frac{1}{12}$th more than *without* it. The total outward thrust of this dome *with* the lantern amounts to 885 tons, so that an iron belt placed round it to counteract the thrust would require to be by equation (174), $\cdot 16 \times 885 \div 5$, or 28·32 square inches in section; and if it were 2 inches thick its width must be $14\frac{1}{8}$ inches.

The area of the joint EF in this example is

$$A = \frac{\phi}{2} \, (R^2 - r^2) \, \cos. \, 13\tfrac{1}{2}° = 11 \cdot 036 \text{ft.} = 1589 \text{ square}$$

inches.

The pressure on the joint is

$$W = N \, . \, \sin. \, 13\tfrac{1}{2}° + (P + S) \, . \, \cos. \, 13\tfrac{1}{2}° = 426 \cdot 51 \, \delta.$$

Putting $\delta = 120$lbs., we find $W = 51,181$lbs., or 32·2lbs. on every square inch of the joint. Since the

u

safe load for brick is 50lbs. per inch, we see that we might safely reduce the thickness of the dome still further.

102. CONICAL DOME.—In this, which is the strongest form of dome, the maximum value of N is at the base CD (fig. 105), or CD may be considered as the "joint of rupture," and the *moments* of P and N balance about the outer edge C. Let *a* be the angle which the slant side BD makes with the vertical OB; let *dn*, whose length is *l*, be the centre line of the thickness of the dome, OA the axis or central line of the dome. Then we take, as before, a small slice of the dome equal to 180th of the whole, and put P for its weight acting at G its centre of gravity; N the horizontal thrust acting at *n* the middle point between A and B. We have then

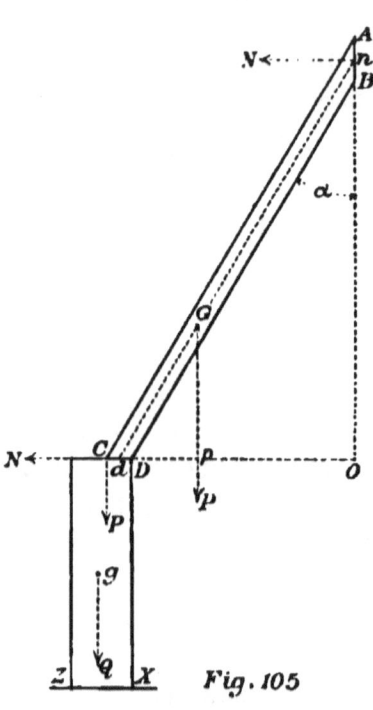

Fig. 105

$$On = l \cdot \cos a, \; Od = l \cdot \sin a, \; OC = R, \; OD = r,$$
$$Od = \tfrac{1}{2}(R + r).$$

Let the vertical from G cut the line OD in *p*, then

$$Op = \tfrac{2}{3} Od = \tfrac{1}{3}(R + r), \; Cp = R - Op = \tfrac{1}{3}(2R - r)$$

$$P = \frac{\pi}{180} \left\{ R^2 \frac{OA}{3} - r^2 \frac{OB}{3} \right\} \delta$$

$$= \cdot 00582 \, (R^3 - r^3) \, \delta \, . \, \cot. \, a$$

$$On = \tfrac{1}{2} \, (R + r) \, . \, \cot. \, a$$

$$N = P \frac{Cp}{On}$$

$$= \cdot 00388 \, \frac{R^3 - r^3}{R + r} \, (2 \, R - r) \, \delta,$$

which is independent of the angle a, and depends only on the values of R and r.

For example, let $R = 11$, $r = 10$, $a = 5°$, $\tfrac{1}{2} \, (R + r)$ $= 10\cdot 5$, cot. $a = 11\cdot 43$, $l = \dfrac{10\cdot 5}{\sin. \, a} = \dfrac{10\cdot 5}{\cdot 08716} = 120\cdot 47$.

$$P = 22 \, . \, \delta, \; N = \cdot 734 \, \delta.$$

To determine the thickness t of the wall which we suppose to be circular on plan, we take P and N as acting at C parallel to their original directions, and equate the *moment* of twice N, taken about Z, with that of P and of Q the weight of the wall. The *moment* of Q is given by equation (173), and if we put $h = 100$ for the height of the wall, we have

$$Q \, . \, q = \cdot 582 \, (t^3 + 30 \, t^2) \, \delta.$$

The *moment* of P is

$$P \, (t - (R - r)) = (22 \, t - 22) \, \delta.$$

The *moment* of N is

$$N \times h = 73\cdot 4 \, . \, \delta.$$

The *equation of stability* is

$$146\cdot 8 = \cdot 582 \, (t^3 + 30 t^2) + 22 \, t - 22,$$

which reduces to

$$t^3 + 30\ t^2 + 38\ t - 290 = 0\ ;$$

from which we get, $t = 2 \cdot 46$.

The outer dome of St. Paul's Cathedral in London is formed of a framework of timber,* supported upon a conical brick dome 18 inches thick, in which $a = 23°$, and the radius R is 50ft., $r = 48 \cdot 37$ft., $R - r = 1 \cdot 63$ft.,

$$R + r = 98 \cdot 37,\ 2R - r = 51 \cdot 63,\ R^3 - r^3 = 11912,$$

$$\text{cot. } 23° = 2 \cdot 35585,$$

$$P = \cdot 00582 \times 11912 \times 2 \cdot 35585\ \delta = 163 \cdot 33\ \delta.$$

$$N = \frac{\cdot 00388 \times 11912 \times 51 \cdot 63}{98 \cdot 37}\ \delta = 23 \cdot 73\ \delta.$$

Putting $\delta = 120$lbs., we have $N = 2847 \cdot 6$lbs., and the total outward thrust of the dome is 228 tons.

To find the thickness t of a solid wall 200 feet high that will support this dome, we have for the moment of N,

$$N \times h = 4746 \ . \ \delta.$$

The *moment* of P is

$$P\ (t - 1 \cdot 63) = (163 \cdot 33\ t - 266 \cdot 2)\ \delta$$
$$Q \ . \ q = \cdot 0058177\ (3\ r\ t^2 + t^3)\ h \ . \ \delta$$
$$= 1 \cdot 16354\ (t^3 + 145 t^2) \ . \ \delta.$$

The *equation of stability* is

$$9492 = 1 \cdot 16354\ (t^3 + 145\ t^2) + 163 \cdot 33\ t - 266 \cdot 2,$$

which reduces to

$$t^3 + 145\ t^2 + 140\ t - 8387 = 0\ ;$$

$$t = 7\text{ft., nearly.}$$

---

* See the author's edition of *Tredgold's Carpentry*, p. 144.

103. OCTAGONAL SPIRE.—
This form of spire may be
considered as a case of the
" Conical Dome " (**102**), only
having eight flat sides instead
of being circular on plan. The
plan of the spire is shown by
fig. 106, where IKLM is one
side of the base, OD the radius
of the inscribed circle of the
inner octagon, OC that of the
outer octagon. Let OC = R,
OD = r,

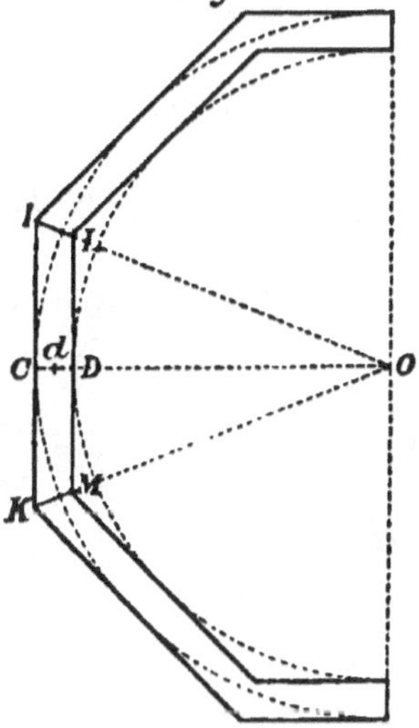

Fig. 106.

Od = ½ (R + r) ; IK =
S, LM = s ; then 'we have,

$$S = ·8286 \text{ R}, s = ·8286 \, r.$$

If we call A the area of IKLM, we have

$$A = \tfrac{1}{2} R . S - \tfrac{1}{2} r s = ·4143 (R^2 - r^2).$$

We can, with very slight error, consider d to be the
centre of gravity of the trapezium IKLM.

Let a be the angle which two opposite sides
make with each other at the vertex of the spire; nd
= l (fig. 105); then the centre of gravity, G, of one
side may be considered to be in the line nd (fig. 105).

$$O d = l . \sin. a; \quad O n = l . \cos. a = \tfrac{1}{2} (R + r) . \cot. a ;$$
$$l = \frac{O d}{\sin. a} = \frac{R + r}{2 \sin. a}.$$

Let P be the weight of one side acting at G, N the
horizontal thrust at the vertex; then we can take—

$$P = \tfrac{1}{2} A . l . \delta = \frac{\cdot 1036}{\sin. a} (R^2 - r^2)(R + r)\, \delta,$$

$$dp = \tfrac{1}{3} Od, \quad Cd = R - Od, \quad Op = \tfrac{1}{3}(R + r),$$

$$Cp = R - Op = \frac{2\,R - r}{3}.$$

Taking *moments* about C of P and N, we have

$$N \times On = P \times Cp;$$

or, $$N = P\frac{Cp}{On} = \frac{\cdot 069}{\cos. a}(R^2 - r^2)(2\,R - r)\,\delta.$$

To determine the thickness of the supporting wall, we proceed as before (**102**) to equate twice the *moment* of N, acting at CD, about Z the base of the wall, with the *moment* of P at C, and the *moment* of Q at *g*, Q being the weight of the wall forming one side of the octagonal tower.

$$Q = \cdot 4143\,((r + t)^2 - r^2)\,h\,.\,\delta, \text{ where } h = DX.$$
$$= \cdot 4143\,(2rt + t^2)\,h\,.\,\delta.$$

The *moment* of N about Z, is $N \times h$.

The *moment* of P about Z is $P\,(t - (R - r))$.

The *moment* of Q is $Q \times \dfrac{t}{2}$, very nearly.

Therefore the *equation of stability* becomes

$$2\,N\,.\,h = P\,(t - (R - r)) + Q \times \frac{t}{2}$$

For example, let $R = 11$, $r = 10$, $h = 100$, $a = 5°$, $\cos. a = \cdot 99619, \sin. a = \cdot 08716, l = 120\cdot 47, R + r = 21$, $R - r = 1$, $R^2 - r^2 = 21$; $2\,R - r = 12$;

$$P = \frac{\cdot 1036}{\cdot 08716}\,.\,441\,\delta = 524\,\delta.$$

$$N = \frac{\cdot 069}{\cdot 99619} \cdot 252\,\delta = 17\cdot 56\,\delta \,.$$

$$Q = 41\cdot 43\ (t^3 + 20\,t)\,.\,\delta.$$

The *equation of stability* is therefore

$$3512 = 524\,t - 524 + 20\cdot 715\ (t^3 + 20\,t^2)$$

which reduces to

$$t^3 + 20\,t^2 + 25\cdot 3\,t - 195 = 0\,;$$
$$t = 2\cdot 44.$$

[The foregoing investigations are based upon papers by the author, which were read before the "Royal" Society in 1866, and published in the "Proceedings" No. 85, 1866; and subsequently in the "Civil Engineer and Architect's Journal" for February and March, 1868, "On the Stability of Domes."]

CHAPTER X.

BUTTRESSES.—SHORING.—RETAINING WALLS.—FOUNDATIONS.

104. Buttresses.—It often happens in the construction of buildings that the horizontal thrust, as of a roof or vault, is concentrated on a few points of the outer walls, while the intermediate parts have little if any thrust to sustain. We have seen this to be the case in vaulted roofs (**92**) (**93**), where the ribs of the vaulting cause the whole thrust to be borne at the points from which they spring. This is also the case with roof-trusses where there is no tie-beam, as in the Hammer-beam roof (**65**).

Fig. 107.

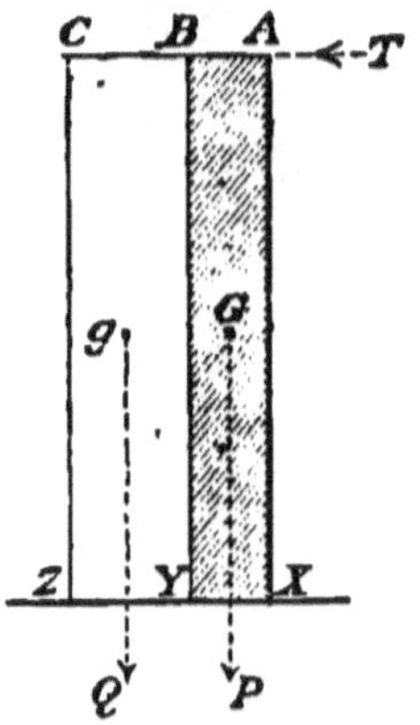

In such cases it is evidently more advantageous and economical to increase the strength of the wall at the points where the thrust is greatest, rather than making the whole wall of a sufficient thickness to resist the thrust. The method usually adopted where there is sufficient space outside the walls, is to build out a mass of masonry called a "buttress."

Thus, let ABXY (fig. 107) be the vertical section of the wall having a horizontal thrust, T, as its summit, and let BCYZ be a "buttress" built out from the wall and of the same height. The plan of the wall and buttress is shown by *abcd* (fig. 108). Suppose P to represent the weight of the portion of the wall from centre to centre of the buttresses, and to act at its centre G. We will first suppose the buttress to have a uniform projection *x* for its whole height; *t* being the thickness and *h* the height of the wall, *l* its length from centre to centre of the buttresses. Let *b* be the breadth *ab* (fig. 108) of the buttress, and Q its weight acting at its centre *g*. Then, in order that there may be equilibrium between the forces T, P and Q, their *moments* about the outer edge Z of the buttress must balance ; or the *equation of equilibrium* is

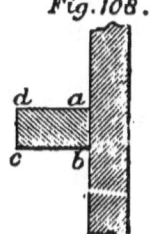

Fig. 108.

$$T \times h = P\left(\frac{t}{2} + x\right) + Q \cdot \frac{x}{2}.$$

And the *equation of stability* is

$$2\,T \times h = P\left(\frac{t}{2} + x\right) + Q \cdot \frac{x}{2} \qquad . \quad (180)$$

giving a quadratic equation in terms of *x*, by which we can determine the projection *x* of the buttress, for a given thickness *b* and thrust T. Or, if *x* and *b* are given, we can find the amount of thrust that can safely be put on the wall and buttress.

Take, for example, the case of a wall which has to sustain the thrust of a vaulted roof, as in a previous example (**92**) where T = 44·3 δ = 5,316 lbs., when δ = 120 lbs. ; let *h* = 20 feet, *l* = 10 feet, *t* = 1·5 feet, *b* = 2 feet. Then we have—

$$P = h \times l \times t \times \delta$$
$$= 300 \, \delta$$
$$Q = h \times b \times x \, . \, \delta$$
$$= 40 \, x \, . \, \delta$$

Then the *equation of stability* (180) becomes

$$40 \times 44{\cdot}3 = 300 \, (\tfrac{3}{4} + x) + 20 \, x^2$$

or, $$88{\cdot}6 = 15 \, x + 11{\cdot}25 + x^2,$$

or, $$x^2 + 15 \, x - 77{\cdot}35 = 0;$$

$$x = 4 \text{ feet.}$$

Now suppose that x is given us 3 feet, all the other dimensions being as above; to find what thrust, T, can be safely put upon the top of the wall. The equation (180) becomes

$$40 \times T = 300 \times 3\tfrac{3}{4} + 20 \times 9$$
$$T = 32{\cdot}65 \, \delta = 3{,}918 \text{ lbs.}$$

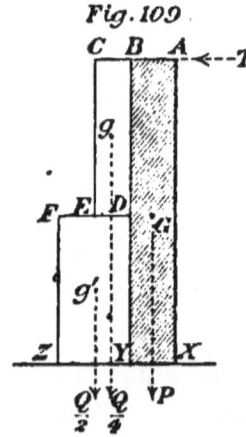

Fig. 109

Let us now reduce the projection of the upper half of the buttress by one-half, (fig. 109) making BC = $\frac{1}{2} x$, CE = $\frac{1}{2} h$; then the weight of the part BCED is $\frac{1}{4}$ Q, and its *moment* about Z is $\frac{1}{4}$ Q $\times \frac{3}{4} x$; the weight of the part FDYZ is $\frac{1}{2}$ Q, and its *moment* about Z is $\frac{1}{2}$ Q $\times \frac{1}{2} x$; or the *moment* of the weight of the buttress is

$$\left(\tfrac{3}{8} Q + \tfrac{1}{4} Q \right) \frac{x}{2} = \tfrac{7}{8} Q \times \frac{x}{2},$$

which is one-eighth less than in the former case (fig. 107), while the weight is reduced by one-fourth.

The *equation of stability* (180) becomes in this case

$$2\,T \times h = P\left(\frac{t}{2} + x\right) + \frac{7}{8}\,Q \times \frac{x}{2},$$

which in the example above given is

$$88\cdot6 = 15x + 11\cdot25 + \cdot875x^2,$$

or,

$$x^2 + 17x - 88\cdot4 = 0\,;$$

$$x = 4\cdot18.$$

Hence it appears that very little stability is lost by the reduction of the buttress, but a considerable saving in material is effected.

Now suppose the buttress to be triangular in section, as BYZ (fig. 110); then its weight is $\frac{1}{2}$ Q, and its *moment* about Z is $\frac{1}{2}$ Q $\times$ $\frac{2}{3}$ x, the weight acting at g the position of which is found as previously described (**17**); so that the *moment* of Q about Z is reduced by one-third, while the weight Q is reduced by one-half. The *equation of stability* (180) in this case becomes

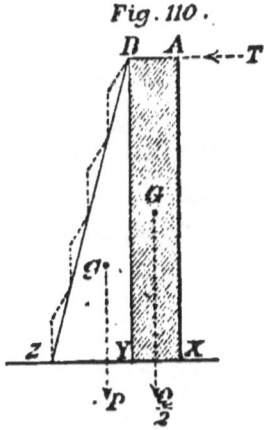

Fig. 110.

$$2\,T.\,h = P\left(\frac{t}{2} + x\right) + \frac{2}{3}\,Q \times \frac{x}{2},$$

which in the above example gives

$$x^2 + 22\cdot5\,x - 116 = 0\,;$$

$$x = 4\cdot32.$$

From this it appears that the triangular form of buttress is the most economical, being the most effective for a given weight of material. Such a buttress is generally built in steps as shown by the dotted lines.

The *equation of stability* (180) being a quadratic, there must, by the rules of algebra, always be two "roots" or values of x that will satisfy it; if one root is positive and the other negative, we take the positive root as the value of x. If one root is zero, and the other is negative, it shows that a buttress is not required, the wall itself being strong enough to resist the thrust. In the first example (fig. 107) suppose $T = 11\cdot25 \cdot \delta$; then the equation (180) becomes

$$11\cdot25 = 15x + 11\cdot25 + x^2,$$

or,
$$x^2 + 15x = 0,$$

the "roots" of which are $x = 0$, and $x = -15$; so that in this case no buttress is required. Again, let $T = 9\cdot25\,\delta$, then the equation (180) is

$$9\cdot25 = 15x + 11\cdot25 + x^2,$$

or,
$$x^2 + 15x + 2 = 0,$$

in which case both the "roots" are negative, being $x = -\cdot135$ and $x = -29\cdot73$; showing that the strength of the wall is more than sufficient to resist the thrust. In order that the equation may have a real positive "root," the third term which is independent of x, must be negative.

105. FLYING BUTTRESS. — A "flying-buttress" is an arched rib of masonry, the base of which rests on the top of one wall, while the vertex presses against another wall at a distance from the former. This form of buttress is adopted where a horizontal thrust has to be sustained by the higher wall, the weight of which is greatly reduced by openings, so that it is of insufficient strength to bear the thrust, which is conveyed by the flying buttress to a lower and outer wall against which an ordinary buttress can be built. This arrangement is shown by fig. 111 where BCE is the flying - buttress resting on the wall CEXY, the arch form being adopted to prevent it from breaking up by its own weight. The thrust T is received at the top, AB, of the higher wall, whose weight however we will neglect in the

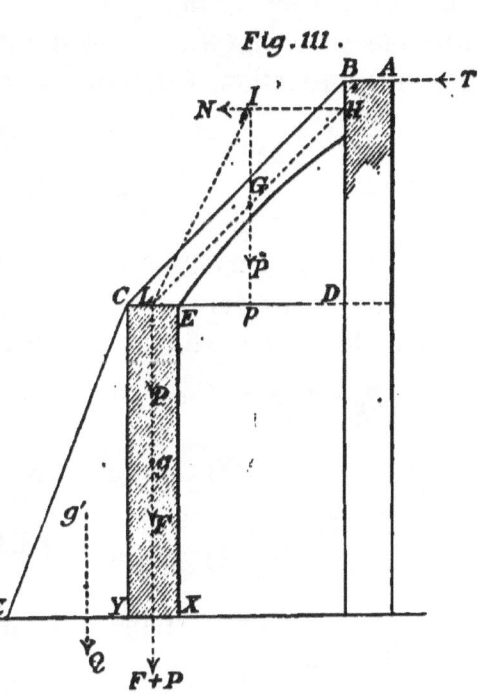

Fig. 111.

consideration of the problem. We will suppose that the thrust on the top of the outer or lower wall is counteracted by the weight, Q, of a triangular buttress, and the weight, F, of the wall itself. The flying buttress acts as an inclined beam, and produces a thrust N upon the top of the wall independently of the thrust T. Draw the line HGL passing through G the centre of gravity of the flying buttress and meeting the top of the wall at its middle point L. Let N be the reaction against the wall at H acting horizontally in HI; draw the vertical IGp, meeting the horizontal line CED in p; and let P be the weight of the flying-buttress acting vertically at G. Then, as we have previously shown (4), the resultant of N and P must act in the direction of the line IL, and we have

$$N : P = Lp : Ip$$

or,

$$N = P \frac{Lp}{Ip}.$$

The forces N and P can be considered as acting horizontally and vertically at L. Let a be the angle which HL makes with DC; and let DE = s, EC = t, CY = h, YZ = x; then DL = $s + \frac{t}{2} = \frac{2s+t}{2}$; DC = $s + t$; BD = $(s + t)$ tan. a; Ip = HD = $\frac{2s+t}{2}$ tan. a,

HL = $\frac{DL}{\cos. a} = \frac{2s+t}{2 \cos. a}$; L$p$ = $\frac{1}{2}$ DL = $\frac{2s+t}{4}$.

Let b be the breadth and d the average depth of the flying buttress; then

$$P = b \cdot d \times HL \times \delta = b \cdot d \cdot \frac{2s+t}{2 \cos. a} \cdot \delta,$$

$$N = P \frac{Lp}{Ip} = b \cdot d \cdot \frac{2s+t}{2 \cos. a} \times \frac{\frac{2s+t}{4}}{\frac{2s+t}{2} \tan. a} \delta$$

$$= b \cdot d \cdot \frac{2s+t}{4 \sin. a} \cdot \delta.$$

If F is the weight of the lower wall from centre to

centre of the buttresses, l the length, h the height, and t the thickness; we have

$$F = h \cdot t \cdot l \cdot \delta;$$

and the *moment* of F about Z is—

$$F\left(\frac{t}{2} + x\right).$$

The *moment* of P, acting at L, about Z, is—

$$P\left(\frac{t}{2} + x\right).$$

The *moment* of the buttress Q about Z is—

$$Q \times \tfrac{2}{3} x = \tfrac{1}{2} h \cdot b \cdot x \times \tfrac{2}{3} x \cdot \delta$$
$$= \tfrac{1}{3} b \cdot h \cdot x^2 \cdot \delta.$$

The *equation of stability* by which to determine x is,

$$2\left(N \times h + T(h + BD)\right)$$
$$= (P + F)\left(\frac{t}{2} + x\right) + \tfrac{2}{3} Q \cdot x \qquad . \quad (181)$$

For example, let $a = 30°$, $h = 20$, $t = 1·5$, $b = 2$, $d = 2$, $s = 10$, $l = 10$; then sin. $a = ·5$, cos. $a = ·866$, tan. $a = ·5774$; BD $= 11·5 \times ·5774 = 6·64$.

$$F = 20 \times 1·5 \times 10 \cdot \delta = 300\,\delta$$

$$P = 2 \times 2 \, \frac{21·5}{2 \times ·866} \, \delta = 50 \cdot \delta$$

$$N = 2 \times 2 \, \frac{21·5}{2} \, \delta = 43 \cdot \delta,$$

$$Q = \tfrac{1}{2} \times 20 \times 2\,x \cdot \delta = 20\,x \cdot \delta.$$

Suppose T to be the thrust from the ribs of a vaulted roof (92), and to be equal to 44·3 . δ, as in the example given. Then the *equation of stability* (181) becomes—

$$2 \, (43 \times 20 + 44 \cdot 3 \times 26 \cdot 64) = 350 \, (\tfrac{3}{4} + x) + 13 \cdot 3 \, x^2,$$

which reduces to

$$x^2 + 26 \, x - 287 = 0 \, ;$$

or,

$$x = 8 \cdot 35 \text{ ft.}$$

Now, suppose that $a = 45°$, all the other dimensions remaining as above. Then cos. a = sin. a = ·707, tan. a = 1, BD = $s + t$ = 11·5; P = 61 . δ, N = 30·4 . δ; and the equation (181) is

$$2 \, (30 \cdot 4 \times 20 + 44 \cdot 3 \times 31 \cdot 5) = 361 \, (\tfrac{3}{4} + x) + 13 \cdot 3 \, x^2,$$

which reduces to—

$$x^2 + 27 \, x - 281 = 0 \, ;$$

$$x = 8 \text{ ft.}$$

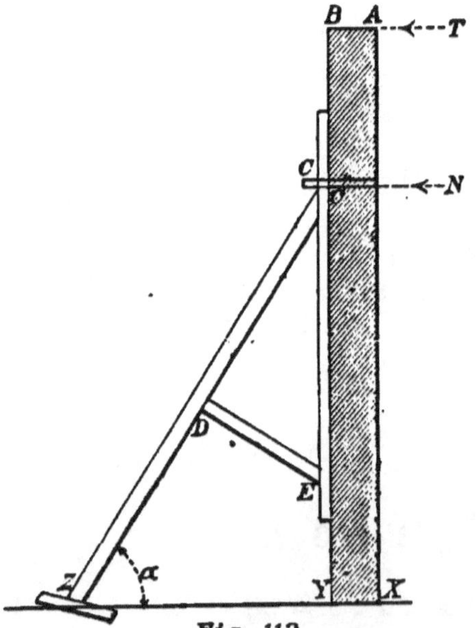

106. SHORING.—When the wall of a building has a tendency to fall outwards from any thrust that may act upon it from within, a slanting piece of timber called a "raking-shore" is placed against it to act as a kind of temporary buttress until it can be secured by a more permanent structure. The shore CZ (fig. 112) rests firmly on a template laid on the ground at Z and the

Fig. 112

on a template laid on the ground at Z and the

upper end C is secured to a *walling-piece* CE and prevented from moving upwards by a *needle* driven through the wall at C. In this case the weight of the shore itself has very little to do with resisting the thrust at C, being kept in its place by the weight of the part above C which presses upon the head. Suppose that a horizontal thrust, T, acting at AB tends to push the wall over about its base Y, and that when the shore is fixed at C the wall is just upon the point of falling. Then the pressure of the head of the shore produces a reaction N at C; and when the forces N and T are in equilibrium about Y, we have

$$T \times BY = N \times CY.$$

Let $CY = h$, $BC = y$; W the weight of wall below C, P the weight of wall above C, t its thickness. Then the above equation becomes

$$T(h + y) = N \times h.$$

Taking *moments* of T, W, and P, about Y, we have in equilibrium

$$T(h + y) = (W + P)\frac{t}{2}$$

$$= N \cdot h$$

Therefore, $N = (W + P)\dfrac{t}{2h}$.   .   (182)

Now in order that the shore may not slip upwards from the pressure N, we must have the weight, P of the part above C, sufficient to keep it down; and we have to determine the value of y which will suffice for this purpose, as this will give the highest point at which the head of the shore can be fixed against the wall that

x

will render it effective. Let s be the length of wall to be supported by the shore, δ its weight per cubic foot. Then we have

$$W = h \times t \times s \times \delta$$
$$P = y \times t \times s \times \delta.$$

Let a be the angle which the shore makes with the horizontal, then taking *moments* of P and N about Z, the shore foot, we have—

$$N \times h = P \times h \,.\, \cot. a,$$
or, $$P = N \times \tan. a$$

$$= (W + P)\frac{t \,.\, \tan. a}{2\,h}, \text{ from equation (182)},$$

from which we obtain

$$P (2h - t \,.\, \tan. a) = W \,.\, t \,.\, \tan. a \qquad . \qquad . \quad (183)$$

Substituting for P and W, their values as given above, in this equation, we have an equation from which to determine the minimum value of y that will preserve equilibrium; the angle a being given, as well as the dimensions t, h, and s.

For example, suppose $a = 60^\circ$, or, $\tan. a = 1{\cdot}732$: let $h = 20$ feet, $t = 1\frac{1}{2}$ feet, $s = 10$ feet; then W $= 300\,\delta$, P $= 15\,y\,\delta$; and the equation (183) becomes

$$15\,y\,(40 - 2{\cdot}6) = 300 \times 2{\cdot}6$$

$$y = \frac{300 \times 2{\cdot}6}{15 \times 37{\cdot}4} = 1{\cdot}39 \text{ ft.}$$

which is the least height of wall above C, supposing the whole length of the wall to act as a solid mass; but if this is not the case, or the wall is liable to fracture from the upward pressure of the shore, a much greater height will be necessary to secure the stability of the shore.

As another example, let $a = 75°$, or tan. $a = 3·732$; $h = 30$ feet, $t = 1·5$ feet, $s = 10$ feet. Then t . tan. $a = 5·6$, $W = 450$ δ, $P = 15y$. δ; and equation (183) becomes

$$15 y \times 54·4 = 450 \times 5·6$$
$$y = 3·1 \text{ feet.}$$

If there is a roof on the top of the wall, its weight must be taken into consideration, as it will have an important effect in keeping down the head of the shore. Let R be the weight of roof distributed over the length, s, of the wall, then we must add R to P in equation (182), so that we have

$$N = (W + P + R) \frac{t}{2h}$$

and taking *moments* about Z of N and P + R, we get

$$N \times h = (P + R) h \text{ . cot. } a,$$

or, $$P + R = N \text{ . tan. } a$$

$$= (W + P + R) \frac{t \text{ . tan. } a}{2 h},$$

therefore,

$$P (2h - t \text{ . tan.} a)$$
$$= (W + R) t \text{ . tan.} a - R \times 2h \quad . \quad . \quad (184)$$

Let us apply this to the examples given above, and suppose R to represent the weight of one side of a roof, 10 feet long, whose rafters are 20 feet long, and the load is 20 lbs. per square foot of roofing. Then in the first example

$$R = 20 \times 10 \times 20 = 4,000 \text{ lbs.}$$

Let the weight of the wall be 120 lbs. per cubic foot; then

$$W = 300 \times 120 = 36000, \quad P = 15\,y \times 120 = 1800\,y.$$

Equation (184) becomes therefore

$$1800\,y \times 37 \cdot 4 = 40000 \times 2 \cdot 6 - 4000 \times 40$$

$$y = -\frac{280}{336 \cdot 6} = -\cdot 83 \text{ ft.}$$

The height BC or y being *negative* in this case, shows that the pressure, R, of the roof alone is more than sufficient to keep down the head of the shore, so that the point C may be placed at B, the top of the wall, or immediately below the wall-plate of the roof.

In the second example we have

$$R = 4000, \quad W = 450 \times 120 = 54000,$$
$$P = 15\,y \times 120 = 1800\,y.$$

Then we have from equation (184)

$$y \times 54 \cdot 4 \times 1800 = 58000 \times 5 \cdot 6 - 4000 \times 60$$

$$y = \frac{424}{490} = \cdot 865 \text{ feet.}$$

Here we find y is positive, and the head of the shore must be at least $10\frac{1}{2}$ inches below the top of the wall.

We have now to determine the strength of the shore necessary to resist the pressure placed upon its head. The forces acting down the shore are the resolved parts of P, R, and N; and if we put F for the compression down the shore, we have

$$F = (P + R) \sin. a + N. \cos. a \qquad . \quad (185)$$

Apply this to the examples given above, and first

suppose that $R = 0$; then in the first example, we have, cot. $a = \cdot 5774$, sin. $a = \cdot 866$, cos. $a = \cdot 5$, $P = 21 \cdot \delta$, $N = P \times$ cot. $a = 12 \cdot \delta$; then equation (185) becomes

$$F = (21 \times \cdot 866 + 12 \times \cdot 5) \, \delta = 2{,}904 \text{ lbs.}$$

Let l be the length of the shore, then

$$l = \frac{h}{\sin. \, a} = 24 \text{ ft.}$$

From equation (86) for long square pillars of fir we have, for safe load—

$$F = 2240 \frac{d^i}{l^2}$$

or,

$$d^4 = \frac{2904}{2240} \times 576 = 808 \cdot 5 ;$$

therefore

$$d = 5\tfrac{1}{3} \text{ inches,}$$

which is the minimum scantling for a square shore.

If a strut, as DE, is fixed to the middle of the shore, it prevents the shore from bending in the middle by the compression, and it becomes virtually divided into two pillars each of which is half the length of the shore; and we have to divide l by 2 in the above equation, which gives $d = 4$ inches, as the minimum scantling.

In the second example, we have, sin. $a = \cdot 966$, cos. $a = \cdot 259$, cot. $a = \cdot 268$; $P = 15 \times 3 \cdot 1 \, \delta = 46 \cdot 3 \, \delta$,

$$N = P \cdot \text{cot.} \, a = 46 \cdot 3 \times \cdot 268 \, \delta = 12 \cdot 41 \, \delta.$$

Then we have from equation (185)

$$F = (46 \cdot 3 \times \cdot 966 + 12 \cdot 41 \times \cdot 259) \times 120$$
$$= 5{,}753 \text{ lbs.}$$

Then for the safe load of a pillar we have from equation (86)

$$d^4 = \frac{5753}{2240} \times 31^2 = 2470;$$

therefore, $d = 7\frac{1}{8}$ inches,

which is the minimum scantling for a square shore.

With a strut, DE, at the middle we can reduce the minimum scantling to $d = 5$ inches.

Now, let R = 4,000 lbs. in these examples, and we have in the first example, P = − 1494, P + R = 2506, N = (P + R) cot. a = 2,506 × ·5774 = 1446.
Then we have by equation (185)

F = 2506 × ·866 + 1446 × ·5 = 2,893 lbs.

Putting $l = 24$ feet in equation (86), we have for safe load in a long square pillar of fir,

$$d^4 = \frac{2893}{2240} \times 576$$

$d = 5\frac{1}{4}$ inches.

In the second example, putting R = 4000 lbs., we have

P = ·865 × 1·5 × 10 × 120 = 1,557 lbs.,
P + R = 5,557 lbs.

N = (P + R) cot. a = 5557 × ·268 = 1,489 lbs.

From equation (185) we have

F = 5557 × ·966 + 1489 × ·259 = 5,754 lbs.

which is the same as before, and $d = 7\frac{1}{8}$ inches is the minimum scantling when there is no strut at DE.

107. RETAINING WALLS. PRESSURE OF WATER.—

When a solid vertical wall, as ABCD
(fig. 113), is required to resist the
pressure of water, as in a reservoir
or tank, the normal pressure, P,
upon AD, is shown in treatises on
hydrostatics to be quite independent
of the horizontal area of the tank,
or on the quantity of water it contains, but depends wholly upon the

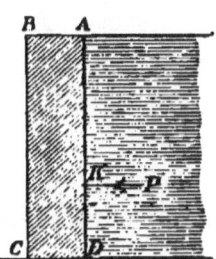

Fig. 113

depth of water in the tank. The pressure of water, or
of any other liquid, upon a surface immersed in it, is
equal to the weight of a column of liquid, whose base is
the area of the surface immersed, and whose height is the
depth of the centre of gravity of that surface below the
surface of the liquid. Consequently, the pressure P
on the rectangular side AD of the wall whose length is
1 foot and depth equals d, (p being the weight of a
cubic foot of the liquid, and $\frac{1}{2} d$ the depth of the centre
of gravity of AD) will be found to be

$$P = \tfrac{1}{2} p \cdot d^2 \qquad . \qquad . \qquad (186)$$

For example, let $d = 20$ feet, p
$= 62{\cdot}5$ lbs. for water, then $P = \frac{1}{2}$
$\times 62{\cdot}5 \times 20^2 = 12{,}500$ lbs. $= 5{\cdot}6$
tons, for the pressure on 1 foot
length of wall, whatever the area or
extent of the tank may be; supposing the surface of the water to be
level with the top of the wall.

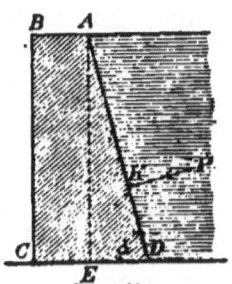

Fig. 114.

If the side AD (fig. 114) of the
wall which is next to the water is sloped at an angle
with the horizontal, the same rule holds good, but

in this case AD does not equal d the depth of the water, which is AE, but we have

$$AD = \frac{d}{\sin. a}.$$

Then the pressure P on 1 foot length of wall, is

$$P = \tfrac{1}{2} p . d \times AD$$

$$= p \frac{d^2}{2 \sin. a} \qquad . \qquad . \qquad (187)$$

Suppose for example, that $a = 60°$, or sin. $a = ·866$, $d = 20$ feet, then we find

$$P = 36 \times d^2 = 14,400 \text{ lbs.} = 6·43 \text{ tons.}$$

The resultant pressure of the liquid acts perpendicularly to the plane AD, and at a point on its surface called the "centre of pressure," the position of which is determined by mathematical analysis, and varies according to the shape of the surface immersed. When the surface is rectangular and wholly immersed in the liquid, the centre of pressure K is found to be at a distance of $\frac{1}{3}$ AD from the bottom edge D; or KD = $\frac{1}{3}$ AD, in both figures.

Having obtained the pressure P upon the surface of the wall and also the point of application of its resultant, we can easily determine the thickness that must be given to the wall in order that it may be strong enough to resist the pressure of the liquid. We will suppose the wall to be composed of solid concrete whose weight is w lbs. per cubic foot; then for a wall of rectangular section, as fig. 113, of which t is the thickness and d the height in feet, W being the weight of 1 foot length, we have

$$W = w \cdot d \cdot t.$$

And in order to secure *stability* we must have the *moment* of W about the outer edge C, equal to *twice* the *moment* of P, acting at K, taken about C, or

$$\tfrac{1}{2}W \cdot t = \tfrac{2}{3}P \cdot d,$$

or, from equation (186)—

$$\tfrac{1}{2}w \cdot d \cdot t^2 = \tfrac{1}{3}p \cdot d^3;$$

therefore, $\qquad t = \cdot 82 \sqrt{\dfrac{p}{w}} \cdot d \quad . \qquad . \quad (188)$

For example, let $p = 62\cdot 5$, $w = 125$, $d = 20$, $\dfrac{p}{w} = \tfrac{1}{2}$; then from equation (188) we find

$$t = \cdot 82 \times \cdot 707 \times 20 = 11\cdot 6 \text{ feet,}$$

which is the thickness of the wall sufficient to ensure *stability* when the tank is full of water.

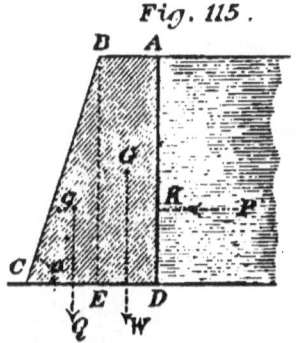

Fig. 115.

Next, suppose that the outside face of the wall is made to " batter," as BC (fig. 115) making a given angle, a, with the horizontal CD. Let W, as before, represent the weight of the rectangular part ABED, acting at G its centre of gravity, AB $= t =$ ED, AD $= d$; let Q be the weight of the triangular part BCE, acting at g; then

$$EC = d \cdot \cot. a.$$

The *moment* of W about C, is

$$W \times \left(CE + \frac{t}{2}\right) = w . d . t \left(d . \cot. a + \frac{t}{2}\right).$$

The *moment* of Q about C is—

$$Q \times \tfrac{2}{3} EC = \frac{w}{3} d . EC^2 = \frac{w}{3} d^3 . \cot.^2 a \; ;$$

consequently, the *equation of stability* becomes—

$$w . d^2 . t . \cot. a + \frac{w}{2} d . t^2 + \frac{w}{3} d^3 . \cot.^2 a = \frac{p}{3} d^3,$$

or, $3 . t^2 + 6 . d . \cot. a . t + 2 . d^2 . \cot.^2 a -$

$$2 \frac{p}{w} . d^2 = 0 \qquad . \qquad . \quad (189)$$

a quadratic equation from which to obtain the value of *t*, when *a* and *d* are known.

For example, suppose $a = 75°$, or cot. $a = ·268$, cot.^{2} $a = ·072$, $d = 20$, $\frac{p}{w} = \frac{1}{2}$; then equation (189) becomes

$$3 t^2 + 32 t + 58 - 400 = 0,$$

or, $$t^2 + 11 t - 114 = 0;$$

$$t = 6·5 \text{ feet.}$$

DC $= t + d . \cot. a = 6·5 + 5·36 = 11·86$ feet.

Comparing this result with that obtained for a rectangular wall (fig. 113), we find the area of section of the wall (fig. 115), when $a = 75°$, is to that of the rectangular wall in the proportion of

$$184 : 232$$

so that one-fifth of the material is saved by battering the wall.

We will now take the case of the wall whose section is shown by fig. 114, where the "batter" is towards the water, the outer face of the wall being vertical.

In this case we have from equation (187)

$$P = p \cdot \frac{d^2}{2 \sin. a}.$$

Since P acts at K perpendicularly to AD, we must resolve it vertically and horizontally, P sin. a being its horizontal and P . cos. a its vertical component; the *moment* about C of P . sin. a, is, P . sin. a . $\frac{d}{3}$; and the *moment* of P . cos. a, is,

$$P \left(t + \tfrac{2}{3} d \cdot \cot. a\right) \cdot \cos. a.$$

The *moment* of W about C, is—

$$W \times \frac{t}{2}.$$

The *moment* of Q about C, is—

$$Q \left(t + \tfrac{1}{3} ED\right) = Q \left(t + \tfrac{1}{3} d \cdot \cot a\right)$$

We have then for the *equation of stability*, putting 2P for P—

$$W \frac{t}{2} + Q \left(t + \tfrac{1}{3} d \cdot \cot. a\right)$$

$$+ 2P \left(t + \tfrac{2}{3} d \cdot \cot. a\right) \cos. a = 2P \cdot \sin. a \cdot \frac{d}{3},$$

or, $3 t^2 + 3 d \cdot \cot. a \left(1 + 2 \frac{p}{w}\right) t + 4 \frac{p}{w} d^2 \cdot \cot.^2 a -$

$2 \frac{p}{w} \cdot d^2 = 0$.   .   .   . (190)

a quadratic equation from which to find the value of t when a and d are known.

Suppose, as before, that $a = 75°$, $d = 20$ ft., $\frac{p}{w} = \frac{1}{2}$, then equation (190) becomes

$$t^2 + 11\,t - 114 = 0;$$

$$t = 6·5 \text{ ft.}$$

which is the same as was obtained by equation (189) with the battering face turned in the opposite direction; so that the stability of the wall is the same whichever way it is battered.

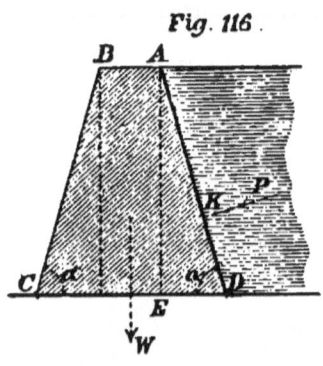

Fig. 116.

The usual section of a tank wall is that shown by fig. 116, where both sides are "battered," and we will here suppose the angle, a, of batter to be the same on both sides, then the weight, W, will act at the centre of the wall. Putting $AB = t$, and $AE = d$, we have—

$$CD = t + 2\,d \cdot \cot a$$

$$W = \frac{w}{2}\,d\,(AB + CD) = w \cdot d\,(t + d \cdot \cot a)$$

$$P = \frac{pd^2}{2 \sin a}$$

The horizontal component of P is, as above, $P \cdot \sin a$ and its vertical component is $P \cdot \cos a$; the *moment* of $P \cdot \sin a$, about C, is

$$P \cdot \sin a \cdot \frac{d}{3} = \frac{p \cdot d^2}{6}$$

The *moment* of P . cos. a, about C, is

$$P \cos. a \times \frac{5}{3} CD = \frac{5}{6} p . d^2 . \cot. a \ (t + 2 d . \cot. a).$$

The moment of W, about C, is

$$W \times \frac{CD}{2} = \frac{w}{2} d \ (t + d . \cot. a) \ (t + 2 d . \cot. a)$$

$$= \tfrac{1}{2} w . d \ (t^2 + 3 d . \cot. a . t + 2 d^2 . \cot.^2 a)$$

The *equation of stability* becomes

$$\tfrac{1}{2} W \times CD = 2 P . \sin. a . \frac{d}{3} - 2 P . \cos. a . \frac{5}{3} CD,$$

which reduces to the quadratic equation

$$3 t^2 + d . \cot. a \left(9 + 20 \frac{p}{w}\right) t$$

$$- d^2 . \cot.^2 a \left(40 \frac{p}{w} - 6\right) = 0 \quad . \quad (191)$$

Applying this to the case where $a = 75°$, cot. $a = \cdot268$, cot.$^2 a = \cdot072$, $d = 20$, $\frac{p}{w} = \frac{1}{2}$, we have from equation (191), $\quad 3 t^2 + 5\cdot36 \times 19 t - 28\cdot8 \times 14 = 0$,

or, $\qquad\qquad t^2 + 34 t - 134 = 0$;

$$t = 3\cdot6 \text{ feet,}$$

CD $= t + 2 d . \cot. a = 3\cdot6 + 10\cdot72 = 14\cdot32$ feet.

The area of section in this case is $\frac{1}{2}$ (14·32 + 3·6) $d = 8\cdot96 \ d$, while that of a rectangular wall (fig. 113) was found to be $11\cdot6 . d$, so that there is a saving of nearly one-fourth of the material by using this form of section in preference to the rectangular form; and a greater saving still is effected by increasing the slope or diminishing the angle a.

**108. RETAINING WALLS. EARTH PRESSURE. SUR-
FACE LEVEL WITH TOP OF WALL.**—When a mass of
loose earth is piled up against a vertical wall, it will
press against it with a force which varies inversely as
the cohesion of the particles of which it is composed.
All kinds of earth which are not solid rock, have a
tendency to form into a slope if left unsupported, and
the angle of inclination to the horizontal which the
earth finally takes, is called the "angle of repose" for
that particular material, and varies from 0° for an
absolute liquid up to about 55° for very compact earth.
The "angle of repose" for different kinds of earth, for
which we use the symbol ϕ, has been carefully
measured and its value is given in the following table,
together with the weight per cubic foot of the material.

	ϕ'.	p = Weight per Cubic Foot.	Δ in Lbs.
Compact earth . . .	55°	126 lbs.	12·53
Dry earth	45°	120	20·60
Dry clay	40°	120	26·10
Dry shingle	40°	112	24·35
Dry fine sand . . .	40°	100	21·75
Gravel	27°—37°	110	41·31 to 27·34
Common sand . . .	22°	118	54·94
Wet clay	20°	130	63·74
Water	0°	62½	62·50

In the case of sand which has become saturated with
water, the pressure will be similar to that produced by
water, as such material takes a very low angle of
repose; but as the weight per cubic foot will be greater
than that of water, its actual pressure will be increased
in like proportion.

Suppose ABCD (fig. 117) to represent the section of a retaining-wall with earth filled up against it level with its summit, as AE; and let DE be the "natural slope" that the earth would take if left to itself, the angle EDX being equal to ϕ the "angle of repose." Then the only portion which presses against the wall is the wedge AED of earth filled in between the wall and the natural slope DE. Now supposing the wall was

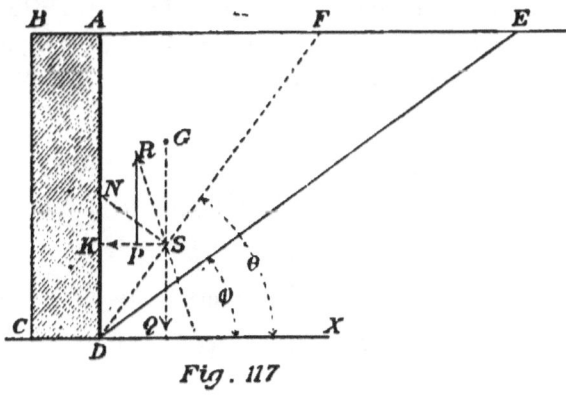

Fig. 117

just on the point of giving way under the pressure P of the earth, the whole mass AED would not yield at once, but a portion as ADF would be the first to fall, breaking away from the general mass in some line DF between DE and DA, making an angle θ with the horizontal. This wedge of earth is supported by P, the resistance of the wall, and by the resistance of the surface DF on which it tends to slide. Putting Q for the weight of this wedge 1 foot wide, p the weight of a cubic foot of earth, d the depth AD; then we have—

$$Q = \tfrac{1}{2}\, p \,.\, d^2 \,.\, \cot \theta \qquad . \qquad . \ (192)$$

The weight Q may be considered as acting vertically at

G the centre of gravity of the wedge, and will meet the horizontal force P at the point S on the inclined plane DF. If SN is the *normal* to the plane at S, then R the resultant of the resistances of the different points on the plane DF is inclined to SN at the angle ϕ, which is the "limiting angle of resistance" between any two contiguous surfaces of earth.* We have then the forces P, Q, and R in equilibrium with each other at the point S; consequently, by the principle of the "inclined plane" (77) they will be proportional respectively to the sides of the triangle SP, PR, RS; and we have—

$$P : Q = SP : PR$$
$$= \sin. PRS : \sin. RSP$$
$$= \sin. GSR : \cos. GSR$$
$$= \sin. (\theta - \phi) : \cos. (\theta - \phi)$$

since GSN $= \theta$, RSN $= \phi$, GSR $= \theta - \phi$.

Therefore

$$P = Q \times \frac{\sin. (\theta - \phi)}{\cos. (\theta - \phi)}$$
$$= Q \times \tan. (\theta - \phi)$$

or putting for Q its value from equation (192),

$$P = \tfrac{1}{2} p . d^2 . \cot. \theta . \tan. (\theta - \phi) \qquad . \qquad . (193)$$

We have now to determine the value of θ which makes P a maximum, since the pressure on the wall will vary according to the values given to the angle θ; and if the wall is made strong enough to support the wedge of earth whose inclination θ corresponds with the

* Moseley's *Engineering.*

maximum value of P, and which thus requires the greatest resistance to support it; then will the earth be prevented from slipping at any inclination whatever. If the wall supplies a resistance which is equal to the maximum value of P in respect to the variable angle θ, it will not be pushed over by the pressure of the earth against it; but if its resistance is less than the maximum value of P, it will be overthrown. We have then to determine when the quantity involving the angle θ, (equation 193)

$$\cot. \theta \times \tan. (\theta - \phi)$$

is greatest, with the variation of θ.

Now, by the rules of Trigonometry,

$$\cot. \theta . \tan. (\theta - \phi) = \frac{\cos. \theta . \sin. (\theta - \phi)}{\sin. \theta . \cos. (\theta - \phi)}$$

$$\sin. (\theta + \theta - \phi) = \sin. (2\theta - \phi) = \sin. \theta . \cos. (\theta - \phi)$$
$$+ \cos. \theta . \sin. (\theta - \phi)$$

$$\sin. (\theta - \overline{\theta - \phi}) = \sin. \phi = \sin. \theta . \cos. (\theta - \phi)$$
$$- \cos. \theta . \sin. (\theta - \phi).$$

By subtraction

$$\sin. (2\theta - \phi) - \sin. \phi = 2 \cos. \theta . \sin. \overline{\theta - \phi}.$$

By addition

$$\sin. (2\theta - \phi) + \sin. \phi = 2 \sin. \theta . \cos. \overline{\theta - \phi}.$$

Therefore we have

$$\cot. \theta . \tan. (\theta - \phi) = \frac{\sin. (2\theta - \phi) - \sin. \phi}{\sin. (2\theta - \phi) + \sin. \phi}$$

$$= 1 - \frac{2 \sin. \phi}{\sin. (2\theta - \phi) + \sin. \phi.}$$

Y

It will be evident that this quantity is greatest when the fractional part is least, or when

$$\sin. (2\,\theta - \phi) = 1 = \sin. 90°,$$

or, when $\theta = 45° + \dfrac{\phi}{2}$, or $\theta - \phi = 45° - \dfrac{\phi}{2}$.

Hence the equation (193) becomes

$$P = \tfrac{1}{2}\,p\,.\,d^2\,.\,\text{cotan.}\left(45° + \frac{\phi}{2}\right).\tan.\left(45° - \frac{\phi}{2}\right)$$

$$= \tfrac{1}{2}\,p\,.\,d^2 \tan.^2\left(45° - \frac{\phi}{2}\right) \qquad . \qquad . \quad (194)$$

$$= \tfrac{1}{2}\,p\,.\,d^2\,\frac{1 - \sin.\,\phi}{1 + \sin.\,\phi} = \tfrac{1}{2}\,d^2 \times \Delta.$$

Comparing this with equation (186) we see that the pressure of earth against a wall is the same as that of a fluid whose weight per cubic foot is—

$$\Delta = p\,.\,\tan.^2\left(45° - \frac{\phi}{2}\right) = p\,.\,\frac{1 - \sin.\,\phi}{1 + \sin.\,\phi}.$$

Also since SD $= \tfrac{1}{3}$ DF, the point K where P acts is given by DK $= \tfrac{1}{3}$ DA, since the pressure against the wall increases from the top downwards, as in the case of the pressure of water.

The value of Δ is given in the foregoing Table for each kind of earth, and has only to be multiplied by half the square of the height of the wall in feet to obtain the pressure P on the wall at K.

Suppose, as before, that w is the weight per cubic foot of the wall, t its thickness, when the wall is of rectangular section; the *equation of stability* is—

$$\tfrac{1}{2} w . d . t^2 = 2 P \times \frac{d}{3}$$

$$= \tfrac{1}{3} d^3 \times \Delta$$

$$t = \cdot 82 \sqrt{\frac{\Delta}{w}} . d \qquad . \quad (195)$$

Putting, $w = 125$ lbs., we have, when $\Delta = 20 \cdot 6$, or

$\frac{\Delta}{w} = \cdot 165$, $t = \cdot 82 \times \cdot 406 . d = \tfrac{1}{3} d$,

When $\Delta = 26 \cdot 1$, as for dry clay, $\frac{\Delta}{w} = \cdot 21$,

$$t = \cdot 82 \times \cdot 458 d = \cdot 38 d;$$

when $\Delta = 63 \cdot 74$, for wet clay, $\frac{\Delta}{w} = \cdot 51$,

$$t = \cdot 82 \times \cdot 714 d = \cdot 586 d.$$

Putting $d = 20$ in each of these cases, we have for the thickness t, $6\tfrac{3}{4}$ feet, $7\tfrac{1}{2}$ feet, and $11\tfrac{3}{4}$ feet, respectively.

Let us now suppose that the thickness of the wall at the base is double that at the top, the outer face of the wall being battered as in fig. 115, then if. a is the angle of slope which BC makes with the horizontal, and t is the thickness of the wall at the summit, we have—

$$\cot. a = \frac{t}{d}, \text{ or, } d . \cot. a = t.$$

Then from (**107**) we have for the *moment* of resistance of the weight of the wall, taken about C,

$$w . d . t \left(d . \cot. a + \frac{t}{2} \right) + \tfrac{1}{3} w . d^3 . \cot.^2 a$$

$$= \tfrac{3}{2} w . d . t^2 + \tfrac{1}{3} w . d . t^2$$

$$= \frac{11}{6} . x . d . t^2 \qquad . \qquad . \qquad . \qquad . \qquad . (196)$$

So that the *equation of stability* in this case is

$$\frac{11}{6} w . d . t^2 = \frac{1}{3} . d^3 . \Delta$$

$$t = \cdot 427 \sqrt{\frac{\Delta}{w}} . d \qquad (197)$$

If $\frac{\Delta}{w} = \cdot 165$, we have—

$$t = \cdot 427 \times \cdot 406 \times d = \cdot 173 \, d.$$

If $d = 20$ feet, then $t = 3\frac{1}{2}$ feet, and the base is 7 feet; so that the quantity of material in the wall is the same as that in a rectangular wall $5\frac{1}{4}$ feet thick, or nearly one-fourth of the material is saved by using this form of section instead of the rectangular one, where the thickness was found above to be $6\frac{2}{3}$ feet.

If $\frac{\Delta}{w} = \cdot 21$, then $t = \cdot 427 \times \cdot 458 . d = \cdot 196 \, d$;
and when $d = 20$ ft., $t = 3 \cdot 92$ ft.

If $\frac{\Delta}{w} = \cdot 51$, then $t = \cdot 427 \times \cdot 714 \, d = \cdot 305 \, d$;
and when $d = 20$ ft., $t = 7 \cdot 1$ ft.

109. RETAINING WALLS. PRESSURE OF EARTH: SURFACE SLOPING.—Instead of having the surface of the earth level with the top of the retaining wall, as in the last article, we now suppose it to be filled up to a height h or AL (fig. 118) above the top of the wall, and sloped back therefrom at an angle ϕ with the horizontal, the summit of the earth being level. Suppose that when the wall is about to yield to the pressure, the tendency of the earth is to break off in the line DF, making the angle θ with the horizontal DX. Let Q be the weight of the mass AKFD ; then if AD = d, AL = h, p = the weight of a cubic foot of earth ; we have—

$$Q = \frac{p \cdot h^2}{2} \left\{ \left(\frac{h+d}{h} \right)^2 \cot. \theta - \cot. \phi \right\}.$$

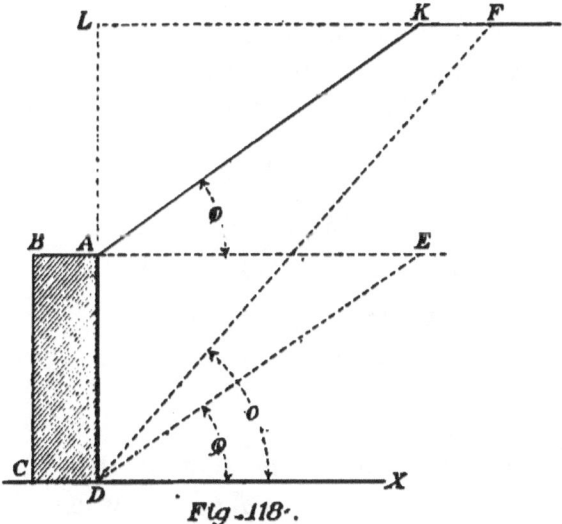

Fig. 118.

If P is the horizontal pressure against the wall, we have, as before,

$$P = Q \cdot \tan. (\theta - \phi)$$
$$= \frac{p h^2}{2} \left\{ \left(\frac{h+d}{h} \right)^2 \cot. \theta - \cot. \phi \right\} \tan. (\theta - \phi).$$

In order to find the pressure against the wall, we have to determine the value of θ which makes P a maximum. This however involves a too complicated process to give here, but the reader will find it worked out in Moseley's "Mechanical Principles of Engineering." We shall here proceed to ascertain by calculation the value of θ for different values of ϕ the angle of repose, by taking a particular case: and the same method can easily be applied to any other case.

Suppose that $h = d$, so that $\dfrac{h+d}{h} = 2$; then we have to take different values of θ and find which one makes P greatest for a given value of ϕ. Putting P $= \frac{1}{2} p h^2 \times$ P', we have to calculate the values of

$$P' = (4 \cot. \theta - \cot. \phi) \tan. (\theta - \phi)$$

Let $\phi = 45°$, cot. $\phi = 1$;
for $\theta = 55°$, P' $= \cdot 3173$,
for $\theta = 57\frac{1}{2}°$, P' $= \cdot 3432$,
for $\theta = 60°$, P' $= \cdot 3508$,
for $\theta = 62\frac{1}{2}°$, P' $= \cdot 3417$,

Hence we may conclude that in this case the maximum value of P is obtained when $\theta = 60°$, or $\theta - \phi = 15°$; then since cot. $\theta = \cdot 5774$,

tan. $(\theta - \phi) = \cdot 268$; we have—

$$P = \tfrac{1}{2} p \cdot d^2 \times \cdot 3508.$$

Equating twice the *moment* of P, acting at $\frac{1}{3} d$ above D, with that of the weight of the wall, taken about C, we have—

$$\tfrac{1}{2} w \cdot d \cdot t^2 = \tfrac{1}{3} p \cdot d^3 \times \cdot 3508,$$

$$\text{or,} \quad t = \cdot 484 \sqrt{\frac{p}{w}} \cdot d.$$

If we put $p = w$, then $t = \cdot484\,d$; and when $d = 20$, $t = 9\cdot68$.

Now take $\phi = 20°$, cot. $\phi = 2\cdot7475$.

$$\text{for } \theta = 30°, \quad \text{P}' = \cdot7358,$$
$$\text{for } \theta = 32\tfrac{1}{2}°, \quad \text{P}' = \cdot7820,$$
$$\text{for } \theta = 35°, \quad \text{P}' = \cdot7943,$$
$$\text{for } \theta = 37\tfrac{1}{2}°, \quad \text{P}' = \cdot7772.$$

We may therefore conclude that when $\phi = 20°$, $\theta = 35°$ gives the maximum value to P; so that in both cases we find $\theta - \phi = 15°$, or $\theta = \phi + 15°$; which we may, without material error, consider as the value of θ which makes P a maximum for any value of ϕ; so that the general equation for P will be—

$$P = \tfrac{1}{2}\,p\,.\,h^2\left\{\left(\frac{h+d}{h}\right)^2 \text{cot.}\,(\phi + 15°) - \text{cot.}\,\phi\right\}$$
$$\times \cdot268 \quad . \qquad . \qquad . \quad (198)$$

Taking $\phi = 20°$, we have, when $h = d$—
$$P = \tfrac{1}{2}\,p\,.\,d^2 \times \cdot7943.$$

Equating twice the *moment* of P with the *moment* of the weight of the wall, we have—

$$\tfrac{1}{2}\,w\,.\,d\,.\,t^2 = \tfrac{1}{3}\,p\,.\,d^3 \times \cdot7943$$
$$t = \cdot728\,\sqrt{\frac{p}{w}}\,.\,d.$$

If $p = w$, then $t = \cdot728\,.\,d$; and when $d = 20$, $t = 14\cdot56$.

When the face of the wall is made to *batter*, and the thickness at the base is twice that at the summit, or $2\,t$, we have, as before, by equation (196), when $\phi = 45°$—

$$\frac{11}{6} \; w \cdot d \cdot t^2 = \frac{1}{3} \; p \cdot d^3 \times \cdot 3508$$

$$t = \cdot 253 \sqrt{\frac{p}{w}} \cdot d.$$

If $p = w$, then we have $t = \cdot 253 \; d$; and when $d = 20$ feet, $t = 5 \cdot 06$ feet, the width of the base being $10 \cdot 12$ feet.

When $\phi = 20°$, we have—

$$\frac{11}{6} \; w \cdot d \cdot t^2 = \frac{1}{3} \; p \cdot d^3 \times \cdot 7943$$

$$t = \cdot 38 \sqrt{\frac{p}{w}} \cdot d.$$

If $p = w$, then $t = \cdot 38 \; d$; and when $d = 20$ feet, $t = 7 \cdot 6$ feet, the width at base being $15 \cdot 2$ feet.

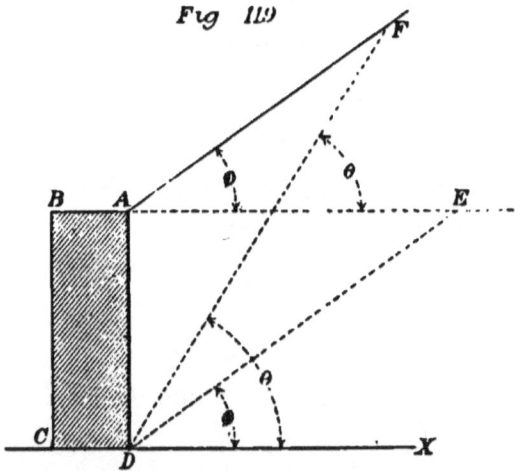

Fig 119

We will now suppose that the earth slopes away to an indefinite distance, as AF (fig. 119), or that the height h is indefinite. Supposing that when the wall is about to give way, the tendency of the earth is to break off in the line DF, making an unknown angle θ with the horizontal; we have—

$$Q = \tfrac{1}{2} p \cdot d \cdot AF \cdot \cos. \phi,$$

and, $\qquad AF : d = \cos. \theta : \sin. (\theta - \phi).$

Therefore, $\qquad Q = \tfrac{1}{2} p \cdot d^2 \frac{\cos. \theta \cdot \cos. \phi}{\sin. (\theta - \phi)}$

$$P = Q \cdot \tan. (\theta - \phi) = \tfrac{1}{2} p \cdot d^2 \cdot \frac{\cos. \theta}{\cos. (\theta - \phi)} \cdot \cos. \phi.$$

Now since θ can never be less than ϕ, the greatest value of P must be when $\theta = \phi$, and cos. $(\theta - \phi) = 1$, or DF coincides with the natural slope DE. Hence it appears that in this case the maximum value of P is —

$$P = \tfrac{1}{2} p \cdot d^2 \cdot \cos.^2 \phi \qquad . \qquad . \quad (199)$$

For a rectangular wall, the *equation of stability* is

$$\tfrac{1}{2} w \cdot d \cdot t^2 = \tfrac{1}{3} p \cdot d^3 \cdot \cos.^2 \phi$$

$$t = \cdot 816 \sqrt{\frac{p}{w}} \cdot d \cdot \cos. \phi.$$

If $p = w$, $t = \cdot 816 \cdot \cos. \phi \cdot d.$

When $\phi = 45°$, cos. $\phi = \cdot 707$, $t = \cdot 577 d$; if $d = 20$, $t = 11\cdot 54.$

When $\phi = 20°$, cos. $\phi = \cdot 9397$, $t = \cdot 767 d$; if $d = 20$, $t = 15\tfrac{1}{3}.$

When the thickness of the wall at the base is double that at the top, or is equal to $2t$, we have from equation (196)—

$$\frac{11}{6} w \cdot d \cdot t^2 = \frac{1}{3} p \cdot d^3 \cdot \cos.^2 \phi$$

$$t = \cdot 426 \sqrt{\frac{p}{w}} \; \cos. \; \phi \; . \; d.$$

If $p = w$, we have, $t = \cdot 426 \; . \; \cos. \; \phi \; . \; d.$

When $\phi = 45°$, cos. $\phi = \cdot 707$, $t = \cdot 301 \; . \; d$; $d = 20$ feet, $t = 6 \cdot 02$ feet, for the thickness of the wall at top; the thickness at base being $12 \cdot 04$ feet.

When $\phi = 20°$, cos. $\phi = \cdot 9397$; $t = \cdot 4 \; d$; $d = 20$ feet, $t = 8$ feet, for the thickness at top of wall, the base being 16 feet.

110. RETAINING WALLS. EARTH BUTTRESS. — We now suppose that the earth is filled in against a wall and left to form its natural slope, as AED (fig. 120), and

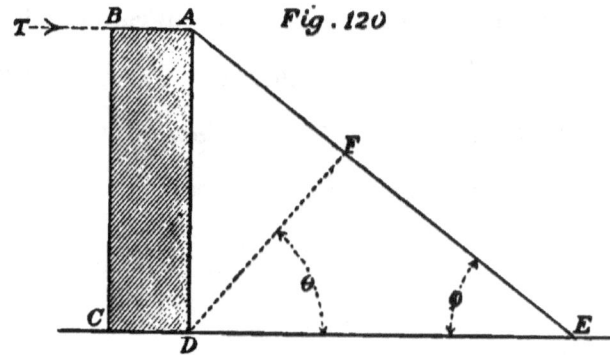

Fig. 120

we have to determine the pressure which it exerts against the wall. We will suppose, as before, that when the wall is about to give way under the pressure, the earth has a tendency to yield in the line DF making some angle θ with the horizontal. Then if Q is the weight of the wedge AFD, p that of a cubic foot of earth, we have—

$$Q = \tfrac{1}{2} \, p \; . \; AD \times AF \times \cos. \; \phi.$$

Now, $\qquad$ AF : AD $=$ cos. θ : sin. $(\phi + \theta)$

or, $\qquad$ AF $= \dfrac{d \cdot \cos. \theta}{\sin. (\theta + \phi)}$,

therefore, $\qquad$ Q $= \frac{1}{2} p \cdot d^2 \cdot \dfrac{\cos. \theta \cdot \cos. \phi}{\sin. (\theta + \phi)}$.

If P is the horizontal pressure against the wall, we have from equation (193)—

$$P = Q \cdot \tan. (\theta - \phi) = \frac{1}{2} p \cdot d^2 \cdot \frac{\cos. \theta \cdot \cos. \phi}{\sin. (\theta - \phi)} \cdot \tan. (\theta - \phi)$$

$$= \frac{1}{2} p \cdot d^2 \cdot \frac{\tan. \overline{\theta - \phi}}{\tan. \theta + \tan. \phi},$$

and we have to determine when P is a maximum for the variation of θ. Since

$$\tan. (\theta - \phi) = \frac{\tan. \theta - \tan. \phi}{1 + \tan. \theta \cdot \tan. \phi}, \text{ we have—}$$

$$P = \frac{1}{2} p \cdot d^2 \cdot \frac{\tan. \theta - \tan. \phi}{1 + \tan. \theta \cdot \tan. \phi} \times \frac{1}{\tan. \theta + \tan. \phi}.$$

If we put $\phi = 45°$, or tan. $\phi = 1$, then we find P to be a maximum when $\theta = 72°$, or $\theta - \phi = 27°$; and tan. $27°$ is very nearly equal to ·5 or $\frac{1}{2}$; tan. $72° = 3$; so that generally we can put $\theta = 27° + \phi$; therefore we have—

$$P = \frac{1}{4} \frac{p \cdot d^2.}{\tan. (27° + \phi) + \tan. \phi} \qquad . \qquad . \qquad (200)$$

When $\phi = 45°$, equation (200) becomes—

$$P = \frac{p \cdot d^2}{16}$$

Equating the *moment* of the wall about C with that

of *twice* P, which is supposed, as before, to act at a point one-third of d above D; we have for stability—

$$\tfrac{1}{2}\, w \, . \, d \, . \, t^2 = \frac{pd^2}{8} \times \frac{d}{3} = \frac{pd^3}{24}.$$

$$t = \cdot 3 \sqrt{\frac{p}{w}} \, . \, d.$$

If $p = w$, then $t = \cdot 3\, d$; and when $d = 20$ feet, $t = 6$ feet; the section of the wall being rectangular.

If there is a horizontal force, T, acting at the top of the wall and tending to push the wall over about its base D, then the embankment of earth will act as a *buttress*; and by equating the *moment* of T, about D, with the sum of the *moments* of W and P about D, we can ascertain what is the amount of the force, T, that would just be on the point of overturning the wall. We have then—

$$T \, . \, d = \frac{1}{2}\, w \, . \, d \, . \, t^2 + \frac{1}{48}\, p \, . \, d^3.$$

Let $w = p = 120$; then

$$T = 60 \left(t^2 + \frac{d^2}{24} \right) = 60 \,(t^2 + \cdot 0417 \, d^2).$$

Putting $t = 6$, and $d = 20$, we have—

$$T = 60 \,(36 + 16 \cdot 68) = 3161 \text{ lbs.,}$$

which is the pressure upon one foot length of wall.

Next, let $\phi = 20°$, then $\theta - \phi = 27°$, $\theta = 47°$; tan. $\phi = \cdot 364$, tan. $(27° + \phi) = \tan. 47° = 1 \cdot 0724$; therefore

$$P = p \, . \, d^2 \times \cdot 174.$$

Then the *equation of stability* for the wall becomes—

$$\tfrac{1}{2} w . d . t^2 = p . d^3 \times \cdot 116$$

$$t = \cdot 48 \sqrt{\tfrac{p}{w}} . d.$$

If $p = w$, then $t = \cdot 48 \, d$; and when $d = 20$ ft., $t = 9 \cdot 6$

$$T . d = \tfrac{1}{2} w . d . t^2 + \tfrac{1}{3} p . d^3 \times \cdot 174.$$

Putting $p = w = 120$; we have—

$$T = 60 \, (t^2 + \cdot 116 \, d^2).$$

If $t = 9 \cdot 6$, $d = 20$, we have—

$$T = 60 \, (92 \cdot 2 + 46 \cdot 4) = 8,316 \text{ lbs.},$$

and this is the pressure upon one foot length of wall that will just balance the resistances of the weight of wall and weight of the earth buttress.

111. Foundations.—When a heavy building is erected upon a yielding soil, the load must be distributed over a sufficiently wide area in order that the footings may not sink into the ground. It is also essential that the earth should be excavated to a considerable depth, so as to reach solid ground, and the trenches made perfectly level at the bottom before commencing to lay the foundations. [Speaking of foundations, Vitruvius (Book 3) says, " Fundationes eorum operum fodiantur si queat inveniri ad solidum et in solido quantum ex amplitudine operis pro ratione videbitur, extruaturque structura totum solum quam solidissima."] The amount of pressure per square foot that the earth will bear must depend on the nature of the soil, as it will be less with loose soils or those in which ϕ (the *angle of repose*) is small, than with those in which ϕ is large; so that the resistance of the soil must evidently be some *function* of the angle ϕ. The following Theorem is given by Rankine as the basis of his investigations upon this subject; " It is necessary to the stability of a granular mass, that the direction of the pressure between the portions into which it is divided by any plane should not at any point make with the normal to that plane an angle exceeding the angle of repose." If then p_1 is the greatest and p_2 the least pressure, we must have—

$$\frac{p_1 - p_2}{p_1 + p_2} \text{ not greater than sin. } \phi,$$

or putting $\quad \dfrac{p_1 - p_2}{p_1 + p_2} = \text{sin. } \phi,$

we have $\quad 1 + \text{sin. } \phi = \dfrac{p_1 + p_2 + p_1 - p_2}{p_1 + p_2} = \dfrac{2 p_1}{p_1 + p_2}$

$$1 - \text{sin. } \phi = \frac{p_1 + p_2 - p_1 + p_2}{p_1 + p_2} = \frac{2 p_2}{p_1 + p_2}.$$

Therefore, $\qquad \dfrac{p_1}{p_2} = \dfrac{1 + \text{sin. } \phi}{1 - \text{sin. } \phi}.$

If we put p' for the greatest horizontal pressure at a depth d, consistent with stability, w the weight of a cubic foot of earth, we have—

$$p' = w \cdot d \frac{1 + \text{sin. } \phi}{1 - \text{sin. } \phi}.$$

And if P is the greatest vertical pressure per square foot—

$$P = p' \left(\frac{1 + \text{sin. } \phi}{1 - \text{sin. } \phi} \right) = w \cdot d \left(\frac{1 + \text{sin. } \phi}{1 - \text{sin. } \phi} \right)^2 \quad (201)$$

Putting A for the area of the footings or foundation

of the wall, in feet; and Q for the weight of earth displaced by it, we have—

$$Q = w \cdot d \cdot A.$$

If W is the weight of the wall, then

$$W = P \times A.$$

Then, the limit of the ratio in which the weight of the building exceeds the weight of the earth displaced by it, when the pressure is uniformly distributed, is

$$\frac{W}{Q} = \frac{P \cdot A}{w \cdot d \cdot A} = \frac{P}{w \cdot d}$$

$$= \left(\frac{1 + \sin. \phi}{1 - \sin. \phi}\right)^2, \text{ from equation (201)}$$

When $\phi = 45°$, we find, $\dfrac{W}{Q} = 34$;

When $\phi = 20°$, we find, $\dfrac{W}{Q} = 4\cdot2$.

Suppose, for example, that $w = 120$ lbs. $d = 5$ feet; then for $\phi = 45°$,

P = 120 × 5 × 34 = 20,400 lbs. per square foot.

When $\phi = 20°$,

P = 120 × 5 × 4·2 = 2,520 lbs. per square foot.

Multiplying the value of P thus obtained by the area A of the foundations in feet, we obtain the utmost load that the earth will bear. To ensure stability we must either increase the area A at least threefold, or else divide the weight W by three or more.

112. PILING.—When the soil is too soft to allow of ordinary foundations being laid upon it, it becomes necessary to reach the firmer substrata by means of long baulks of timber, called "piles," shod with iron having a point or knife-edge, which are driven through the soft earth by means of a heavy "ram," or "monkey," let fall from a height upon the head of the pile. [The use of piling for the foundations of buildings to be erected on very loose soil, has been known from the earliest period, as is shown by the allusion made by Vitruvius (Book 3), in the following words, to this method of construction: "sin autem solidum non invenietur, sed locus erit congesticius ad imum aut paluster, tunc is locus fodiatur exinamiaturque et palis alneis aut oleagineis aut robusteis ustilatis configatur, sublicaeque machinis adigantur quam creberrimae, carbonibusque expleantur intervalla palorum et tunc structures solidissimis fundamenta impleantur."]

On the top of these piles the foundations of the walls are laid, and it is essential that we should be able to determine beforehand what load the piles will safely bear after they have been driven in as far as they will go. The investigation of a formula which approximately gives the maximum pressure which a pile will bear when the weight of the ram, the weight of the pile, the height of fall, and the distance through which it is driven at the last blow, are known, will be found in Weisbach's "Mechanics of Engineering," and is based on the following laws which govern the impact of bodies :—

First Law.—The *momentum*, or *mass × velocity*, of two bodies which impinge, at the *end* of the impact, is equal to the *momentum* at the *beginning* of the impact.

Second Law.—The *vis-viva*, or *mass × velocity-*

squared, lost by the inelastic impact is equal to the sum of the products of the masses and the squares of their gain or loss of velocity.

Third Law.—If a body falling from a height h, acquires the velocity v in feet per second, h is called the height *due* to the velocity v; and if g represents the force of gravity, or the velocity that a body acquires in falling from rest in one second of time, we have

$$v^2 = 2\,g\,.\,h,$$

the value of g being generally taken as 32·2 feet.

In this investigation we suppose the pile to be inelastic and not to be perceptibly compressed by the blows, so as not to complicate the formula. If then the ram falls through a height of h feet, v being the velocity with which it strikes the pile, we have by the 3rd law,

$$v^2 = 2\,g\,h.$$

Let W be the weight of the ram, P that of the pile; the pile will have moved through a certain space during the time of impact, and will have a velocity V at the end of that time; and if H is the height *due* to the velocity V, we have by the 3rd law,

$$V^2 = 2\,g\,H.$$

Now by the first law the *momentum* at the end of the time of impact equals the *momentum* at the beginning of the impact, or

$$V\,(W + P) = v\,.\,W;$$

therefore, $$V = \frac{W}{W + P}\,v$$

$$H = \frac{V^2}{2\,g} = \left(\frac{W}{W + P}\right)^2 \cdot \frac{v^2}{2\,g}$$
$$= \left(\frac{W}{W + P}\right)^2 \cdot h.$$

Now suppose that at the last blow the pile sinks a distance of s feet, v_1 being the velocity at the end of the last blow; then by the 3rd law,

$$v_1^2 = 2\,g\,s.$$

Let L be the resistance of the earth, or the load which the pile can just support without sinking further into the ground, then by the 2nd law, we have—

$$L \times v_1^2 = (W + P)\,V^2,$$

$$\text{or,}\quad L = \frac{V^2}{v_1^2}\,(W + P) = \frac{W^2}{(W + P)^2}\,(W + P)\,\frac{2\,gh}{2\,gs}$$

$$= \frac{W^2}{W + P} \cdot \frac{h}{s} \qquad \cdot \qquad \cdot \qquad \cdot \qquad (202)$$

neglecting altogether the action of W and P in opposition to the resistance of the earth.

In using the equation (202) to find the greatest load a pile can bear, we must express *both* h and s in feet or *both* in inches, and each of the quantities, W, P, and L, either in tons, cwts. or lbs.

For example, let W = 1 ton, P = $\frac{3}{4}$ ton, h = 10 feet, and suppose s is found by measurement to be ·36 inch or ·03 feet; then by equation (202) we have—

$$L = \frac{1}{1·75} \times \frac{10}{·03} = 190·5 \text{ tons.}$$

If W and P are expressed in cwts. or lbs., L will also be in cwts. or lbs.

According to Rankine, piles are generally driven till L amounts to between 2000 and 3000 lbs. per square inch of the area of section of the pile, and are loaded permanently with from 200 to 1000 lbs. per square inch, so that the *factor of safety* is from 10 to 3.

Thus, if we suppose in the foregoing example that the pile is 13 inches square, its area of section will be 169 square inches ; turning 190·5 tons into lbs. and dividing by 169 we get 2525 lbs. per square inch for the maximum load ; taking 5 for the factor of safety when the pile has been driven down to firm ground, we find that it can be safely loaded with 505 lbs. per square inch ; if however it has failed to reach a firm bottom, we must take 10 as the factor of safety and not load it with more than 253 lbs. per square inch.

The *momentum* M, with which the ram strikes the pile, is, $M = \text{mass} \times \text{velocity} = W \times \sqrt{64\cdot4 \times h}$; which is the force in tons with which a ram weighing W tons strikes the pile. Thus when $h = 1$ foot, and $W = 1$ ton, then $M = 8$ tons; or if the ram weighs 12 cwt., then

$$M = 8 \times 12 = 96 \text{ cwts.}$$

If $h = 5$ feet, $M = 18$ tons, when the ram is 1 ton ; or $M = 219$ cwts. when the ram is 12 cwts.

If $h = 10$ feet, $M = 25\cdot4$ tons, when the ram is 1 ton, or $M = 305$ cwts. when the ram is 12 cwts.

The pressure M, which the ram of weight W produces on the head of the pile at the moment of striking, with a fall of h feet, can therefore be approximately calculated by the formula

$$M = W \sqrt{64\cdot4 \cdot h}$$

z 2

CHAPTER XI.

113. PRESSURE OF WIND ON A PLANE.—The action of wind upon lofty buildings or those erected in exposed situations is of too powerful a character to allow of its being neglected by the architect. Tredgold estimated that the pressure of wind upon roofs amounted to more than all the other forces combined to which they were subjected; and since the wind acts directly upon one side only of the roof at a time, it produces a far greater strain on its timbers than an equal pressure placed uniformly over both sides would do.

It has been found as the result of experiment, that the *normal* pressure of wind against a plane surface is very nearly proportional to the *square* of the velocity with which the wind is moving; and it has been ascertained that wind moving with a velocity of 35 miles an hour, or 51 feet a second, produces a normal pressure of 6 lbs. on every square foot of a plane surface; and by assuming that the pressure varies as the square of the velocity, we can calculate the pressure for any given velocity. When the velocity is 2 × 35 or 70 miles an hour, the pressure will be 2^2 × 6 or 24 lbs. per square foot, and when it is 3 × 35 or 105 miles an hour, the pressure will be 3^2 × 6 or 54 lbs. If p is the pressure on a square foot due to a velocity v, A the given area, in feet, we have for the pressure P on the area A,

$$P = p \times A.$$

And since the pressure p is 6 lbs. when v is 35 miles, we have—

$$\frac{p}{6} = \left(\frac{v}{35}\right)^2,$$

or, $$p = \frac{6}{1225} v^2 = \frac{v^2}{204} = \cdot 0049\, v^2 \quad . \quad (203)$$

Also we can find v when p is given, from the equation—

$$v = \sqrt{204\, p} \quad . \quad . \quad (204)$$

The following Table gives the value of p for different values of v.

v, in Miles per Hour.	p, in lbs. per Foot.	v, in Miles per Hour.	p, in lbs. per Foot.
20	1·96	70	24·00
30	4·41	80	31·36
35	6·00	90	39·69
40	7·84	105	54·00
50	12·25	128	80·00
60	17·64		

The highest pressure that has been observed is about 80 lbs. per square foot, but this is a very exceptional amount and is rarely attained in the British Isles.

In considering the pressure of the wind upon a vertical wall against which it is blowing horizontally, it may be assumed to be uniform from top to bottom, so that the resultant pressure will act at the middle of the wall. If h is the height of the wall, the pressure, P, upon 1 foot length of wall will be,

$$P = p.\, h,$$

and the *moment* of this pressure about the outer edge of the wall, will be

$$P \frac{h}{2} = \frac{1}{2} p \cdot h^2.$$

Putting W for the weight of a wall one foot long and h feet in height, t for its thickness supposed uniform throughout; w the weight of a cubic foot of walling; we have—

$$W = w \cdot h \cdot t$$

and its *moment* about the outer edge, is

$$W \cdot \frac{t}{2} = \frac{1}{2} w \cdot h \cdot t^2.$$

Equating this to *twice* the *moment* of P, we have for stability,

$$\frac{1}{2} w \cdot h \cdot t^2 = 2 \frac{p}{2} h^2$$

$$t = \sqrt{\left(\frac{2p}{w} h\right)} \qquad . \qquad . \quad (205)$$

For example, let $p = 24$, $w = 144$, $h = 36$, then by equation (205)

$$t = \frac{6}{\sqrt{3}} = 2\sqrt{3} = 3\cdot464 \text{ ft.}$$

If the thickness at the base is double that at the summit, we have from equation (196), putting t for the least thickness,

$$\frac{11}{6} w \cdot h \cdot t^2 = 2 \frac{p}{2} \cdot h^2,$$

$$t = \sqrt{\frac{6}{11} \cdot \frac{p}{w} \cdot h} . \qquad . \quad (206)$$

from which we find,

$$t = \frac{6}{\sqrt{11}} = \frac{6\sqrt{11}}{11} = 1\cdot81 \text{ ft.},$$

and the thickness at base is 3·62 feet.

Example 2.—Let the pressure p be 80 lbs.; then for a wall of uniform thickness and 36 feet high, we have from equation (205)—

$$t = \sqrt{\frac{160}{144}} \times 6 = 6\cdot324 \text{ ft.}$$

If t is thickness at top and $2\,t$ that at base, we have by equation (206)—

$$t = \sqrt{\frac{6}{11}\ \frac{80}{144}} \times 6 = 3\cdot3 \text{ ft.}$$

and the base of the wall is $2\,t$ or 6·6 feet.

114. PRESSURE OF WIND ON A CYLINDER.—To determine the pressure of wind on the surface of a cylindrical tower, we must assume that the direction of the pressure is everywhere parallel to one axis of the cylinder. Let ABC (fig.121) represent a quarter plan of the tower, C the centre and AB the

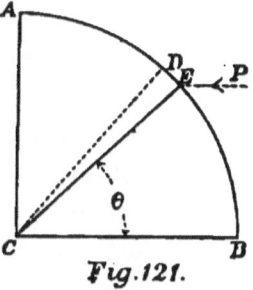

Fig.121.

surface pressed upon by the wind which we will suppose to be acting parallel to BC. Take any two points, D and E, very near together on the arc AB, and let the radius EC = r, making the angle θ with BC; and call the angle ECD the *differential* of θ, or $d\theta$; then the area $d\mathrm{A}$, of a small part of the surface of the tower whose height is h, is

$$d\mathrm{A} = h\,.\,r\,.\,d\theta.$$

If v is the velocity of wind parallel to BC, then the velocity normal to the surface at E is

$$v \cdot \cos \theta,$$

and the pressure on dA is proportional to

$$v^2 \cdot \cos^2 \theta.$$

Resolving this pressure parallel to BC, we have the pressure in that direction proportional to

$$v^2 \cdot \cos^3 \theta.$$

Consequently, the pressure on the curved surface AB is to that on a flat surface equal to AB, as

$$\frac{\int r \cdot \cos^3 \theta \cdot d\theta}{r} = \int \cos^3 \theta \cdot d\theta$$

$$= \frac{\sin \theta}{3} (\cos^2 \theta + 2).$$

Taking the limits of integration from $\theta = 0°$ to $\theta = 90°$, this ratio becomes equal to $2 : 3$, or the pressure on the cylindrical tower is two-thirds of that on a square one of same diameter.* If we put 2 R for the diameter, h for the height, P for the pressure on the surface, P' for that on a square tower, we have—

$$P' = p \cdot 2 \, R \cdot h$$
$$P = \tfrac{2}{3} P' = \tfrac{4}{3} p \, R \cdot h. \quad . \quad . \quad . \quad . \quad (207)$$

Let it be required to find the necessary thickness for *stability* of the walls of a cylindrical tower for a given

* Rankine says, "the total pressure of the wind against the side of a cylinder is *about one-half* of the total pressure against a diametral plane of that cylinder."

value of p, t being the thickness, and r the internal radius ; then $t = R - r$, or, $r = R - t$; W being the weight of the tower, and w the weight of 1 cubic foot of its material.   Then, $W = w \cdot \pi \cdot (R^2 - r^2) \cdot h = w \cdot \pi (2R - t) t \cdot h,$

and the *moment* of W about the outer edge is—

$$W \cdot R = w \cdot \pi (2R - t) R \cdot t \cdot h.$$

The *moment* of the pressure is, from equation (207)—

$$P \frac{h}{2} = \frac{2}{3} p \cdot R \cdot h^2.$$

Equating the *moment* of W with *twice* that of P, we have for *stability*—

$$W \cdot R = 2 P \frac{h}{2},$$

or,
$$\frac{4}{3} p \cdot h = w \cdot \pi (2 Rt - t^2),$$

or,
$$t^2 - 2 R \cdot t + \frac{4}{3 \pi} \cdot \frac{ph}{w} = 0;$$

or,
$$t^2 - 2 R \cdot t + \cdot 424 \frac{p}{w} \cdot h = 0 \qquad (208)$$

For example, let $2 R = 10$ ft., $\frac{p}{w} = \frac{24}{144} = \frac{1}{6}$, $h = 100$ ft.

Then equation (208) becomes—

$$t^2 - 10 t + 7 \cdot 1 = 0;$$
$$t = \cdot 769 \text{ ft.} = 9\frac{1}{4} \text{ ins.}$$

If $\frac{p}{w} = \frac{54}{144} = \frac{3}{8}$, the equation (208) becomes—

$$t^2 - 10\,t + 16 = 0\,;$$

$$t = 2\text{ ft.}$$

115. WIND PRESSURE ON ROOFS.—The experiments which have been made to determine the pressure on surfaces inclined at different angles to the direction of the wind have produced rather anomalous results, arising from the use of small planes moving rapidly through the air; but these offer a very different resistance to that offered by a large roof completely covered in; since with a plane moving through the air a partial vacuum is formed behind it which greatly affects the resistance; so that little reliance can be placed on such experiments as far as roofs are concerned.

In the case of a roof we have the wind impinging directly on a large surface enclosed all round, so that no wind can get behind it; and we may therefore consider its effect as following the ordinary laws of dynamics.

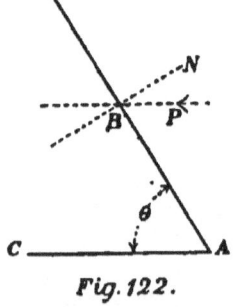

Fig. 122.

Let AB (fig. 122) represent the sloping side of a roof inclined at an angle θ with the horizontal AC, the angle θ being the "pitch" of the roof. We will suppose the wind to be blowing horizontally with a velocity v; then as before (**114**), we have the velocity *normal* to the plane at B represented by the quantity—

$$v \,.\, \sin\,\theta$$

and the pressure p on a square foot of inclined surface

is proportional to $v^2 . \sin.^2 \theta$; therefore from equation (203)

$$p = \frac{v^2 . \sin.^2 \theta}{204} \qquad . \qquad . \quad (209)$$

If $\theta = 30°$, $\sin.^2 \theta = \frac{1}{4}$, and the pressure per square foot on the sloping roof is one-fourth of that on a vertical plane; if $\theta = 60°$, $\sin.^2 \theta = \frac{3}{4}$, and the pressure per square foot is three-fourths of that on a vertical plane, or three times as much as when the angle is 30°. In the following Table the pressure per square foot has been calculated for roofs of different pitch, and for various velocities of wind.

PITCH OF ROOF.	20°.	30°.	40°.	50°.	60°.	70°.
Velocity of Wind. Miles per Hour.	p in lbs. per ft.	p in lbs. per ft.	p in lbs. per ft.	p in lbs. per ft.	p in lbs. per ft.	p in lbs. per ft.
20	·23	·49	·81	1·15	1·47	1·73
30	·52	1·10	1·82	2·59	3·31	3·89
35	·70	1·50	2·48	3·52	4·50	5·30
40	·92	1·96	3·24	4·60	5·88	6·92
50	1·43	3·06	5·06	7·19	9·19	10·82
60	2·06	4·41	7·29	10·35	13·23	15·58
70	2·81	6·00	9·88	14·08	18·00	21·19
80	3·67	7·84	12·96	18·40	23·52	27·70
90	4·64	9·92	16·40	23·29	29·76	35·05
105	6·32	13·50	22·31	31·69	40·50	47·68
128	9·36	20·00	33·06	46·94	60·00	70·64

We will now show how the stress on the timbers of a king-post roof can be calculated, which is caused by the wind blowing upon one side of the roof. In calculating the stress upon a roof (68) we included the pressure of wind as part of the dead load of 66 lbs. per foot superficial, as adopted by Tredgold. A more

accurate method, however, is to take the dead load
of covering, timbers and snow, separately from the
pressure of wind, and form the stress diagram as before
(**68**); and then to take the pressure of the wind as a
dead load on one side only of the roof.

Let ABC (fig. 123) represent the truss of a king-

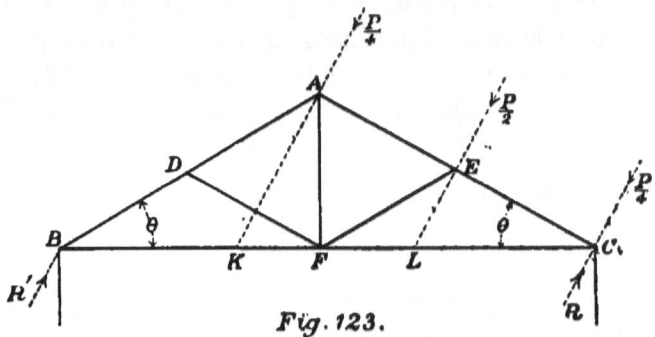

Fig. 123.

post roof in which the rafters AC and AB make the
angle θ with the tie-beam BC. Suppose the wind to
be blowing on the side AC, and P to be the total
pressure resolved perpendicularly to AC. Then if A is
the area of roof supported by AC, we have—

$$P = p \cdot A = p \cdot l \cdot b,$$

where l is the length of rafter, and b the breadth between
two trusses. Draw AK and EL perpendicular to AC,
and cutting BC in K and L. We will consider that
half the pressure, P, is borne at E and one-fourth at
each point A and C. The pressure $\frac{1}{4}$ P at A produces
a reaction R_1 at B, and a reaction R_2 at C; taking
moments of R_1 and $\frac{P}{4}$ about C, and of R_2 and $\frac{1}{4}$ P about
B, we have—

$$R_1 \times BC = \frac{P}{4} \times CK, \text{ or, } R_1 = \frac{P}{4} \times \frac{CK}{BC} = \frac{P}{8} \cdot \frac{1}{\cos^2 \theta},$$

$$R_2 \times BC = \frac{P}{4} \times BK, \text{ or, } R_2 = \frac{P}{4} \cdot \frac{BK}{BC} = \frac{P}{8} \cdot \frac{2\cos^2 \theta - 1}{\cos^2 \theta}.$$

The pressure $\frac{1}{2} P$ at E produces a reaction R_3 at B, and a reaction R_4 at C; taking *moments* of R_3 and $\frac{1}{2} P$ about C, and of R_4 and $\frac{P}{2}$ about B, we have—

$$R_3 \times BC = \frac{P}{2} \times LC, \text{ or, } R_3 = \frac{P}{2} \times \frac{CL}{BC} = \frac{P}{8} \cdot \frac{1}{\cos^2 \theta},$$

$$R_4 \times BC = \frac{P}{2} \times BL, \text{ or, } R_4 = \frac{P}{2} \times \frac{BL}{BC} = \frac{P}{8} \cdot \frac{4\cos^2 \theta - 1}{\cos^2 \theta}.$$

The total reaction R' at B is—

$$R' = R_1 + R_3 = \frac{P}{4} \cdot \frac{1}{\cos^2 \theta} = \frac{P}{4} (1 + \tan^2 \theta).$$

The total reaction R at C is—

$$R = \frac{P}{4} + R_2 + R_4 = \frac{P}{4} \left(\frac{3\cos^2 \theta - 1}{\cos^2 \theta} + 1 \right)$$

$$= \frac{P}{4} \cdot \frac{4\cos^2 \theta - 1}{\cos^2 \theta}.$$

For example, let $\theta = 30°$, then $\cos^2 \theta = \frac{3}{4}$, $R' = \frac{4}{3} \cdot \frac{P}{4}$

$$= \frac{1}{3} P, \quad R = \frac{8}{3} \times \frac{P}{4} = \frac{2}{3} P.$$

We now proceed to draw the stress diagram, as before (**68**), by first taking a line *bh* (fig. 124) parallel to the direction of the pressure, P; that is, perpendicular to AC (fig. 123). Take *ab* on this line to represent

¼ P, or the pressure at C, and *bc* to represent R the reaction at C. Draw *cd* horizontally at *c* meeting *ad* drawn at right angles to *bac*. Take *ac* to represent

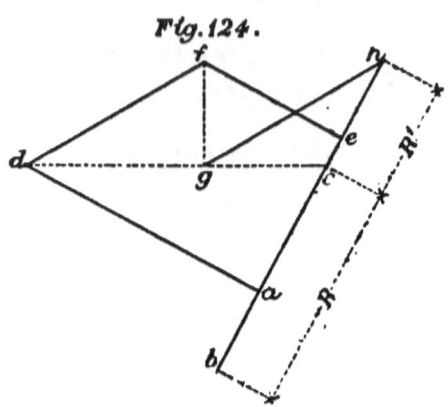

Fig. 124.

½ P, the pressure at E, and draw *ef* parallel to AC and at right angles to *ae*, meeting *df* drawn parallel to EF.

Draw a vertical line *fg* meeting *cd* in *g*, and draw *gh* parallel to AB. Then *bc* represents the reaction R at C, *hc* the reaction R' at B.

For the forces in equilibrium at C, we have *ab* representing ¼ P, *bc* representing R, *cd* acting *from* the centre C, representing the *tension* in CF, *da* acting towards C, and representing the compression in EC.

For the forces in equilibrium at E, we have *ea* representing ½ P, *ad* acting *towards* E, the *compression* in CE; *df* acting *towards* E, the *compression* in FE; *fe* acting *towards* E the *compression* in EA.

For the forces at A, we have *he* the force ¼ P, *ef* the pressure in EA, *fg* acting *from* A the tension in AF, *gh* acting *towards* A the *compression* in AD.

For the forces at B, *ch* is the reaction R'; *hg* acting *towards* B the *compression* in BA; *gc* acting *from* B, the *tension* in BF.

The strong lines in fig. 124 show the parts in *compression* and the dotted lines those in *tension*. We have, by calculation, when $\theta = 30°$, the following values of the different forces represented by the lines in the stress diagram (fig. 124) :—

$$ab = \frac{1}{4}\,P, \quad ae = 2\,\frac{P}{4}, \quad ac = \frac{5}{3} \cdot \frac{P}{4}, \quad he = \frac{P}{4}, \quad hc =$$

$$\frac{4}{3}\,\frac{P}{4}, \quad dc = 2\,ac = \frac{10}{3}\,\frac{P}{4}, \quad ad = dc \cdot \cos. \theta = \frac{5\,\sqrt{3}}{3} \cdot \frac{P}{4},$$

$$cg = ch = \frac{4}{3}\,\frac{P}{4}, \quad df = gh = \frac{4\,\sqrt{3}}{3} \cdot \frac{P}{4}, \quad fg = \frac{1}{2}\,df =$$

$$\frac{2\,\sqrt{3}}{3}\,\frac{P}{4}, \quad fe = \sqrt{3}\,\frac{P}{4}.$$

CHAPTER XII.

1. Find the dimensions of rolled iron joists placed 8 feet apart, to carry the floor of a public reading-room 26 feet wide, the depth to be as little as possible, the rest of the material being of wood.

We can only solve a problem of this kind by trying our previous formulæ upon two or three different "stock" sections of rolled joists. Take for example the section 12″ × 7¼″, the thickness of metal being ⅞ inch. Assuming the utmost load the floor has to bear to be 120 lbs. per square foot, we have for the load distributed over each iron joist—

$$W = 8 \times 120 \times 26 = 11 \text{ tons.}$$

Putting I for the "moment of inertia" of the section, we have—

$$I = \frac{7\frac{1}{4} \times 12^3 - 6\frac{3}{8}(10\frac{1}{4})^3}{12} = \frac{5668}{12} = 472.$$

By equation (57) the safe-load on the centre of such a beam is

$$W = 5 \times \frac{2}{3} \times \frac{5668}{12 \times 312} = 5\cdot05 \text{ tons.}$$

This is equivalent to a distributed load of 10·1 ton, so

that this section may be considered as very nearly *strong* enough for the purpose.

If however we take into consideration the resistance to deflexion or *stiffness* of the beams, we find that with a distributed load of 11 tons, by equation (64), the deflexion (δ) in inches is

$$\delta = \frac{5}{8} \times \frac{1}{48} \times \frac{11}{7600} \times \frac{(312)^3}{472} = 1 \cdot 2 \text{ ins.}$$

Now by Tredgold's rule (**50**) the deflexion of a floor joist should not exceed $\frac{1}{40}$ inch for every foot of its length; so that δ should not exceed $\frac{26}{40} = \cdot 65$ inch. The calculated deflexion is therefore nearly double that allowed by the rule, and in order to obtain the required *stiffness*, we must adopt a stronger section. As however the load of 120 lbs. per foot is much in excess of that which the beam has usually to carry, the section given above would probably suffice for the purpose; but if a greater depth can be obtained, the stiffness and rigidity of the floor would be much increased.

2. Find the distributed breaking-weight of a box girder (fig. 40) whose span is 33 feet, depth 18 inches, breadth 14 inches, thickness of top and bottom plates $\frac{1}{2}$ inch, thickness of web-plates $\frac{3}{8}$ inch.

By equation (57) we have for safe-load at centre of the beam—

$$W = 5 \times \frac{2}{3} \frac{14 \times 18^3 - 13\frac{1}{4} \times 17^3}{33 \times 12} = 6 \cdot 7 \text{ tons.}$$

Multiplying this by 2 for distributed load we have 13·4 tons for *safe* distributed load; and again multiplying by 4, we have 53·6 tons as the distributed *breaking*

A A

load. No account however has been taken of the resistance of the angle-irons, the dimensions of which we will suppose to be $3'' \times 3'' \times \frac{1}{4}''$. If A is the area of section of each angle-iron, then $A = \frac{11}{4}$, and we find by (40) and (41) that the centre of gravity of each angle-iron is $7\frac{1}{2}$ inches from the neutral axis Nn, or $Ga = 7\frac{1}{2}$ (fig. 40).

The *moment of inertia*, I′, of each angle-iron about an axis through its centre of gravity is (40)

$$ I' = \frac{3 \times \frac{1}{8}}{12} + 3 \times \frac{1}{2} \times \frac{9}{16} + \frac{\left(\frac{5}{2}\right)^3 \times \frac{1}{2}}{12} + \frac{5}{2} \times \frac{1}{2} \times \frac{9}{16} $$

$$ = \frac{107}{48} = 2\frac{1}{4}, \text{ very nearly.} $$

Then by (41) we have (referring to fig. 40)—

$$ I_4 = 4\,I' + 4\,A \times (Ga)^2 = 9 + 11 \times \left(\frac{15}{2}\right)^2 = 628. $$

The *moment of inertia*, I_1, of the plates is—

$$ I_1 = \frac{14 \times 18^3 - 13\frac{1}{4} \times 17^3}{12} = 1380. $$

If I is the *moment of inertia* of the whole section, including angle-irons, we have—

$$ I = I_1 + I_4 = 2008. $$

If W is the load at the centre, we have—

$$ \frac{1}{4}\,W \cdot l = S\frac{I}{z} = S \times \frac{2008}{9}. $$

For breaking-weight we may put $S = 20$ tons; then

$$W = \frac{4 \times 20 \times 2008}{38 \times 12 \times 9} = 39.1 \text{ tons.}$$

Therefore the *distributed* breaking load is 78.2 tons, when the resistance of the angle-irons is taken into consideration, or about one-third more than when they were omitted.

3. Required to find the safe-load at the centre of a composite wrought-iron beam of 38 feet bearing, formed of two rolled joists 16″ × 6″ × ¾″, riveted to a top and bottom plate 14″ × ½″.

Here the total depth of the beam is 17 inches.

If I_1 is the *moment of inertia* of one of the plates, I_2 that of one of the joists, taken about a line through the centre of the section; we have (**37**)—

$$I_1 = \frac{14 \times (\frac{1}{2})^3}{12} + \frac{1}{2} \times 14 \times (8\frac{1}{4})^2 = 476.$$

By (**39**) we have—

$$I_2 = \frac{6 \times 16^3 = 5\frac{1}{4}\,(14\frac{1}{2})^3}{12} = 714.$$

Putting I for the *moment of inertia* of the whole section, we have—

$$I = 2\,(I_1 + I_2) = 2380.$$

Let M be the *moment of stress* at the middle of the beam, and we have—

$$M = \frac{1}{4}\,W \cdot l = S\,\frac{I}{\frac{17}{2}} = \frac{5 \times 2380 \times 2}{17} = \frac{23800}{17};$$

$$W = \frac{4 \times 23800}{17 \times 38 \times 12} = 12.23 \text{ tons,}$$

which is the safe-load on the centre of the beam, allowing 5 tons per square inch of section as the value of S.

4. A shed 90 feet wide is covered by a tie-beam roof in two spans, resting in the middle on cast-iron columns 30 feet high, and on solid brick walls on the outside. Each roof has a span of 45 feet, and length of rafter 25 feet, the load on the roof being half a hundred-weight per foot superficial, and the principals ten feet apart. Find the diameter of the columns in the centre when solid or hollow.

Let W be the load on each side of each roof-truss, then $W = 25 \times 10 \times \frac{1}{2} = 125$ cwts. $= 6\frac{1}{4}$ tons. The load W is borne by each 10 feet length of walling, and $2W$ by each centre pillar; so that $12\frac{1}{2}$ tons is the load which the pillar has to carry with safety, and $6 \times 12\frac{1}{2}$ or 75 tons may therefore be taken as the breaking-weight. Then by equation (82) we have

$$75 = 42 \frac{d^{3\cdot5}}{30^{1\cdot68}} = 42 \frac{d^{3\cdot5}}{4^{1\cdot68} \times 7\cdot5^{1\cdot68}} = 42 \frac{d^{3\cdot5}}{9\cdot6 \times 26\cdot7}$$

$$= 42 \frac{d^{3\cdot5}}{256} ;$$

therefore, $d^{3\cdot5} = \frac{75}{42} \times 256 = 457.$

By the Table (56) we find that $(5\cdot75)^{3\cdot5} = 455\cdot9$, so that we may take $5\frac{3}{4}$ inches as the required diameter of a *solid* iron column sufficiently strong to carry this load in safety.

Now suppose the column to be *hollow* and 10 inches in external diameter, required to find d_1 its internal diameter, or the necessary thickness of metal. Using the equation (83) we have

$$75 = 42 \frac{10^{3\cdot5} - d_1^{3\cdot5}}{256} = 42 \frac{3162 - d_1^{3\cdot5}}{256},$$

or,

$$d_1^{3\cdot5} = 3162 - \frac{75}{42} \times 256 = 2705.$$

Referring to the Table we find that $(9\cdot5)^{3\cdot5} = 2643$, so that we may take $9\frac{1}{2}$ inches as the internal diameter, or the thickness of metal $\frac{1}{4}$ inch.

5. Find the pressure of wind per square foot that would overturn a 14″ brick wall 10 feet high, weighing 108 lbs. per cubic foot. The adhesion of the mortar being neglected.

Taking 1 foot length of the wall, its weight will be

$$W = 108 \times 10 \times 1\frac{1}{6} = 1260 \text{ lbs.}$$

This load acts vertically down the middle of the wall, and its *moment* about the bottom edge is

$$W \times \frac{7}{12} = 735.$$

The *moment* of the wind pressure (p) is equal to this, or

$$p \times 5 \times 10 = 735$$

$$p = \frac{735}{50} = 14\cdot7 \text{ lbs.}$$

6. A collar roof 38 feet span, has a pitch of 45°, the collar placed half-way up the rafters, and the weight on each truss 14 tons, the trusses 9 feet 8 apart. The supporting walls are 14″ thick and eighteen feet high; required the projection of a buttress 14″ thick and of uniform projection its whole height, sufficient to resist the thrust of the truss.

Putting W (= 7 tons) for the weight on one side of the roof, we have by equation (102) for the horizontal thrust

$$T = \tfrac{1}{4} W \cdot \text{cotan.} \; \theta.$$

And in this case, cotan. $\theta = 1$; and the *moment* of T, about the base of the wall, is

$$T \times 18 = \tfrac{7}{4} \times 2240 \times 18.$$

Putting x for the projection of the buttress from the face of the wall, we have for the *moments* of the wall and buttress taken about *its* outer edge, allowing 108 lbs. to the foot cube for the weight of the wall,

$$108 \times 9\tfrac{2}{3} \times \tfrac{7}{6} \times 18 \, (x + \tfrac{7}{12}) + 108 \times 18 \times \tfrac{7}{12} x^2$$

which must equal the *moment* of the thrust when the forces are in equilibrium. Hence the *equation of equilibrium* is

$$9x^2 + 174x - 458 = 0,$$

or,
$$x^2 + 19\tfrac{1}{3}x - 51 = 0;$$

from which we find $x = 2\tfrac{1}{3}$ feet.

This is the minimum thickness, or projection, of the buttress. For *stability* we may put 2 T for T, and the value of x will then be found to be $x = 4\tfrac{1}{2}$ feet.

7. What weight can be supported on a cast-iron stanchion of H section 11 feet high, with flanges 5″ wide and depth 9 inches, the thickness of the metal being 1 inch?

From (**59**) we have for the breaking load (f) per square inch, where r is the ratio of length to diameter,

$$f = \frac{a}{1 + b \cdot r^2} = \frac{36}{1 + \tfrac{1}{500} \times 15^2} = 25.$$

Multiplying this by the area of section which is 17 inches, we have W, the breaking-weight,

$$W = 17 \times 25 = 425 \text{ tons.}$$

$$\text{Safe-load} = \tfrac{1}{10} W = 42 \cdot 5 \text{ tons.}$$

This is however too high a value for W, since the section is not square but only 5″ wide; if then we take

$$r = \frac{132}{5} = 26 \text{ (nearly), we have for breaking weight,}$$

$$W = 17 \times \frac{36}{1 + \tfrac{1}{500} \times 26^2} = 17 \times 15 \cdot 36 = 261 \text{ tons,}$$

or, for safe-load $= \dfrac{W}{10} = 26 \cdot 1$ ton.

This will be too little, and to get the true result we must take the *mean* of the two, which is 343 tons breaking weight, and 34·3 tons safe-load.

8. A beam AB with 12 feet bearing carries a distributed load of 15 cwt. If a pillar is placed under it at a point C which is 4 feet from A, what weight can now be laid upon each part of the beam, so that the stress shall be the same as before?

Let M_1 be the *moment of stress* at the middle of the beam without the pillar, having a distributed load of $w_1 = \tfrac{15}{12}$ cwts. per foot run. Let M be the same for the middle of the length AC, w the load per foot; M′ that for the length BC, w' the load per foot. Then in order that the stresses may be the same in both cases, we must have—

$$M_1 = M = M'$$

or, $w \times 4^2 = w' \times 8^2$; or, $w' = \tfrac{1}{4} w$; and $w_1 = \tfrac{5}{4}$ cwts.

$$w_1 \times 12^2 = w \times 4^2;$$

or,
$$w = 9 \times w_1 = \frac{45}{4} \text{ cwts.}$$

$$w' = \tfrac{1}{4} w = \frac{45}{16} \text{ cwts.}$$

Therefore the load upon AC is $w \times 4$ or 45 cwts., and the load on BC is $w' \times 8$ or $22\frac{1}{2}$ cwts.

9. What weight at the middle will break a beam of Riga fir, 20 feet bearing, 12″ deep and 9″ wide?

From equation (42) we have—

$$W = \tfrac{2}{3} S \frac{b \cdot d^2}{l};$$

From the Table (32) we have S = 6 × 775 lbs. for breaking stress; therefore,

$$W = 3100 \times \frac{9 \times 12^2}{12 \times 20} = 16,740 \text{ lbs.} = 7\cdot47 \text{ tons.}$$

The safe-load will be one-sixth of this.

10. A beam AB having a span of 32 feet, is supported at A and B, and loaded uniformly through-out a part of its length, AC, equal to 24 feet, with $1\frac{1}{4}$ cwt. per foot run. It is required to find the reactions, W_1 and W_2, of the supports at A and B.

Let W be the load on AC, equal to $24 \times 1\frac{1}{4}$ or 30 cwts. which acts at a point halfway between A and C or 12 feet from A and 20 feet from B. Taking *moments* of the forces W_1 and W about B, we have—

$$W_1 \times 32 = W \times 20 = 30 \times 20;$$

therefore,
$$W_1 = \frac{30 \times 20}{32} = 18\frac{3}{4} \text{ cwts.}$$

Again, taking *moments* of W_2 and W about A, we have

$$W_2 \times 32 = W \times 12 = 30 \times 12;$$

therefore, $\quad W_2 = \dfrac{30 \times 12}{32} = 11\frac{1}{4}$ cwts.

11. A girder AB having a bearing of 50 feet 6 inches is supported at A and B, its weight, w, being $\frac{1}{5}$th ton per foot run. It has also a load, W, of 64·8 tons distributed over a length, BC, of 23 feet. Required to find the point in the beam where the stress is greatest.

The greatest stress is at the point where the Resultant (2) of all the forces acts on the beam. Now the resultant of the load $W = 64·8$ tons distributed over BC acts at a point half-way between B and C, or at 11 feet 6 inches from B. The resultant of the weight $w = 10·1$ tons acts at the centre of the beam. We have then to find the position of the resultant of the two forces W and w, whose distance from B we call x.

Taking *moments* of the forces about B, we have—

$$(W + w)\, x = W\, \frac{BC}{2} + w\, \frac{AB}{2}$$

$$= W \times 11·5 + w \times 25·25;$$

therefore,

$$x = \frac{64·8 \times 11·5 + 10·1 \times 25·25}{74·9} = 13·32 \text{ feet from B.}$$

12. A beam is supported at A and B and has a span of 18 feet; it is loaded at C, 5 feet from A, with 3 tons, and at D, 7 feet from B, with 2 tons. It is required to

find the reactions P and Q at A and B, or the proportion of load which is carried at each point.

Taking *moments* about B of all the forces, we have

$$P \times AB = 3 \times BC + 2 \times BD$$
$$= 3 \times 13 + 2 \times 7$$
$$= 53$$

Therefore, $P = \dfrac{53}{18}$ tons.

Taking *moments* about A, we have—

$$Q \times BA = 2 \times AD + 3 \times AC$$
$$= 2 \times 11 + 3 \times 5$$
$$= 37$$

Therefore, $Q = \dfrac{37}{18}$ tons.

$$P + Q = \dfrac{90}{18} = 5 \text{ tons.}$$

13. A horizontal beam, AB (fig. 125) of which C is the centre, is supported at each end and loaded uniformly over its middle portion EF only, the parts AE and BF being

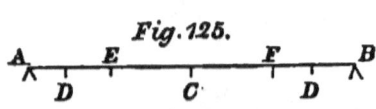
Fig. 125.

unloaded; to find the *moment* of stress at the middle point C.

Let w be the weight per foot of load on EF, AB = l, BF = AE = x, M the *moment of stress* at C.

First, suppose the beam to be loaded uniformly throughout its length with w per foot, and M' the *moment of stress* at C. Let M₁ represent the *moment of stress* at C produced by the load on AE or BF, the resultant of which acts at D, where AD or BD $= \frac{x}{2}$.

By equation (25) we have—

$$M' = \tfrac{1}{8} w \cdot l^2.$$

The load on AE or BF is $w \cdot x$, and by equation (19)

$$M_1 = w \cdot x \frac{\frac{x}{2} \times \frac{l}{2}}{l} = 2\,w \frac{x^2}{l} \times \frac{l}{8} = \frac{w}{4} \cdot x^2$$

Therefore, $M = M' - 2\,M_1 = wl \times \frac{l}{8} - 4\,w \cdot x \frac{x}{l} \times \frac{l}{8}$

$$= w \frac{l}{8}\left(l - 4\,x \frac{x}{l}\right)$$

For example, let $x = \frac{l}{4}$; $w = 1$ ton, $l = 20$ feet, $x = 5$ ft., $\frac{x}{l} = \frac{1}{4}$. Then we have—

$$M = \frac{5}{2}\left(20 - 20 \times \frac{1}{4}\right) = \frac{75}{2}$$

Also, $\qquad M' = 20 \times 2\tfrac{1}{2} = 50 = \frac{100}{2}$

So that, $\qquad M : M' = 3 : 4$

Therefore $M = \frac{3}{4} M'$, or the *moment of stress* at C with the load w per foot over the middle half only, is three-fourths of the *moment of stress* produced by w per foot over the whole length of the beam.

It will be seen that the value of M_1 is independent of the length l of the beam, and depends only on the length x of AE and BF. In the above example, $M_1 =$

6·25, as the value of the moment of stress about C of the load on AE or BF.

Let us apply the above equations to determine the safe-load per foot in the beam of Riga fir described at **(33)**, where $l = 15$ ft. $= 180$ inches, $b = 6''$, $d = 12''$. Here the moment of resistance is—

$$\tfrac{4}{3} \, S \, . \, b \, . \, d^2 = 6200 \times 144 = 892800,$$

when $S = 775$ for safe-load.

Let w' be the load per foot when the beam is loaded uniformly throughout its entire length; then we have

$$M' = \tfrac{1}{8} w' \, . \, l^2 = 892800,$$

or,
$$w' = \frac{892800 \times 8}{180^2} = 221 \text{ lbs.}$$

Let w_1 be the load per foot when the beam is loaded only on AE and BF; then if $x = \tfrac{1}{4} l$, we have

$$2 \, M_1 = \tfrac{3}{4} w_1 \, . \, x^2 = 892800,$$

or,
$$w_1 = \frac{892800 \times 2}{45^2} = 882 \text{ lbs.}$$

Let w be the load per foot when the beam is loaded only on EF; then

$$M = \tfrac{1}{8} w \, . \, l \left(l - 4 \, x \, \frac{x}{l} \right) = 892800.$$

or,
$$w = \frac{892800 \times 8}{\tfrac{3}{4} \times 180^2} = 294 \text{ lbs.}$$

INDEX.

LIST OF USEFUL BOOKS FOR
BUILDERS, CONTRACTORS, DECORATORS, &c.

2

LOCKWOOD'S BUILDER'S PRICE-BOOK : a Comprehensive Handbook of the Latest Prices of every kind of Material and Labour in Trades connected with Building; including a great variety of useful Information in all matters concerning these Trades. With many useful Memoranda and Tables. New Edition, Re-written, Re-modelled, and greatly Enlarged. Edited by FRANCIS T. W. MILLER, A.R.I.B.A. 700 pages, Crown 8vo, 4s. cloth.

ARCHITECTURAL PERSPECTIVE : The Whole Course and Operations of the Draughtsman in Drawing a Large House in Linear Perspective. Illustrated by numerous Diagrams. By F. O. FERGUSON. Demy 8vo, 3s. 6d. boards.

THE HANDBOOK OF SPECIFICATIONS. By Professor T. L. DONALDSON. 8vo, £1 11s. 6d. cloth.

SPECIFICATIONS FOR PRACTICAL ARCHITECTURE. By F. ROGERS, Architect. 8vo, 15s. cloth.

THE HOUSE-OWNER'S ESTIMATOR; or, What will it Cost to Build, Alter, or Repair? By JAMES D. SIMON and FRANCIS T. W. MILLER. Crown 8vo, 3s. 6d. cloth.

THE POCKET TECHNICAL GUIDE, MEASURER, AND ESTIMATOR FOR BUILDERS AND SURVEYORS. By A. C. BEATON. 1s. 6d.

QUANTITIES AND MEASUREMENTS, How to Calculate and Take. By A. C. BEATON. 1s. 6d. cloth.

PLUMBING : a Text-book to the Practice of the Art or Craft of the Plumber. By W. P. BUCHAN. Sixth Edition, Enlarged. With 380 Illustrations. 4s. cloth.

VENTILATION : a Text-book to the Practice of the Art of Ventilating Buildings. By W. P. BUCHAN. With 170 Illustrations. 4s. cloth.

SANITARY WORK IN THE SMALLER TOWNS AND IN VILLAGES. By CHARLES SLAGG, A.M. Inst. C.E. 12mo, 3s. 6d. cloth.

HOUSE-PAINTING, GRAINING, MARBLING, AND SIGN-WRITING. By ELLIS A. DAVIDSON. Fifth Edition. 6s. cloth.

A GRAMMAR OF COLOURING, applied to DECORATIVE PAINTING AND THE ARTS. By GEORGE FIELD. Revised by E. A. DAVIDSON. 3s. 6d. cloth.

SCHOOL OF PAINTING FOR THE IMITATION OF WOODS AND MARBLES. By A. R. VAN DER BURG and P. VAN DER BURG. Royal folio, 18½ by 12½ in. Illustrated with 24 full-size Coloured Plates. Also 12 Plain Plates, comprising 154 Figures. £1 11s. 6d.

ELEMENTARY DECORATION, as applied to the Interior and Exterior Decoration of Dwelling Houses, &c. By JAMES WILLIAM FACEY. 2s. cloth.

PRACTICAL HOUSE DECORATION. By JAMES WILLIAM FACEY. 2s. 6d. cloth.

LONDON : CROSBY LOCKWOOD & SON, 7, Stationers' Hall Court, Ludgate Hill.

7, STATIONERS' HALL COURT, LONDON, E.C.

October, 1891.

A

CATALOGUE OF BOOKS

INCLUDING NEW AND STANDARD WORKS IN

ENGINEERING: CIVIL, MECHANICAL, AND MARINE, MINING AND METALLURGY, ELECTRICITY AND ELECTRICAL ENGINEERING, ARCHITECTURE AND BUILDING, INDUSTRIAL AND DECORATIVE ARTS, SCIENCE, TRADE, AGRICULTURE, GARDENING, LAND AND ESTATE MANAGEMENT, LAW, &c.

PUBLISHED BY

CROSBY LOCKWOOD & SON.

MECHANICAL ENGINEERING, etc.

New Pocket-Book for Mechanical Engineers.
THE MECHANICAL ENGINEER'S POCKET-BOOK OF TABLES, FORMULÆ, RULES AND DATA. A Handy Book of Reference for Daily Use in Engineering Practice. By D. KINNEAR CLARK, M.Inst.C.E., Author of "Railway Machinery," "Tramways," &c. &c. Small 8vo, nearly 700 pages. With Illustrations. Rounded edges, cloth limp, 7s. 6d.; or leather, gilt edges, 9s. [*Just Published.*]

New Manual for Practical Engineers.
THE PRACTICAL ENGINEER'S HAND-BOOK. Comprising a Treatise on Modern Engines and Boilers: Marine, Locomotive and Stationary. And containing a large collection of Rules and Practical Data relating to recent Practice in Designing and Constructing all kinds of Engines, Boilers, and other Engineering work. The whole constituting a comprehensive Key to the Board of Trade and other Examinations for Certificates of Competency in Modern Mechanical Engineering. By WALTER S. HUTTON, Civil and Mechanical Engineer, Author of "The Works' Manager's Handbook for Engineers," &c. With upwards of 370 Illustrations. Third Edition, Revised, with Additions. Medium 8vo, nearly 500 pp., price 18s. Strongly bound.

☞ *This work is designed as a companion to the Author's* "WORKS' MANAGER'S HAND-BOOK." *It possesses many new and original features, and contains, like its predecessor, a quantity of matter not originally intended for publication, but collected by the author for his own use in the construction of a great variety of modern engineering work.*

*** OPINIONS OF THE PRESS.

" A thoroughly good practical handbook, which no engineer can go through without learning something that will be of service to him."—*Marine Engineer.*

" An excellent book of reference for engineers, and a valuable text-book for students of engineering."—*Scotsman.*

" This valuable manual embodies the results and experience of the leading authorities on mechanical engineering."—*Building News.*

" The author has collected together a surprising quantity of rules and practical data, and has shown much judgment in the selections he has made. . . . There is no doubt that this book is one of the most useful of its kind published, and will be a very popular compendium."—*Engineer.*

" A mass of information, set down in simple language, and in such a form that it can be easily referred to at any time. The matter is uniformly good and well chosen, and is greatly elucidated by the illustrations. The book will find its way on to most engineers' shelves, where it will rank as one of the most useful books of reference."—*Practical Engineer.*

" Should be found on the office shelf of all practical engineers."—*English Mechanic.*

B

Handbook for Works' Managers.

THE WORKS' MANAGER'S HANDBOOK OF MODERN RULES, TABLES, AND DATA. For Engineers, Millwrights, and Boiler Makers; Tool Makers, Machinists, and Metal Workers; Iron and Brass Founders, &c. By W. S. HUTTON, C.E., Author of "The Practical Engineer's Handbook." Fourth Edition, carefully Revised, and partly Re-written. In One handsome Volume, medium 8vo, 15s. strongly bound. [Just published.

☞ The Author having compiled Rules and Data for his own use in a great variety of modern engineering work, and having found his notes extremely useful, decided to publish them—revised to date—believing that a practical work, suited to the DAILY REQUIREMENTS OF MODERN ENGINEERS, would be favourably received.

In the Third Edition, the following among other additions have been made, viz.: Rules for the Proportions of Riveted Joints in Soft Steel Plates, the Results of Experiments by PROFESSOR KENNEDY for the Institution of Mechanical Engineers—Rules for the Proportions of Turbines—Rules for the Strength of Hollow Shafts of Whitworth's Compressed Steel, &c.

₊ OPINIONS OF THE PRESS.

"The author treats every subject from the point of view of one who has collected workshop notes for application in workshop practice, rather than from the theoretical or literary aspect. The volume contains a great deal of that kind of information which is gained only by practical experience, and is seldom written in books."—Engineer.

"The volume is an exceedingly useful one, brimful with engineers' notes, memoranda, and rules, and well worthy of being on every mechanical engineer's bookshelf."—Mechanical World.

"The information is precisely that likely to be required in practice. . . . The work forms a desirable addition to the library not only of the works manager, but of anyone connected with general engineering."—Mining Journal.

"A formidable mass of facts and figures, readily accessible through an elaborate index . . . . Such a volume will be found absolutely necessary as a book of reference in all sorts of 'works' connected with the metal trades."—Ryland's Iron Trades Circular.

"Brimful of useful information, stated in a concise form, Mr. Hutton's books have met a pressing want among engineers. The book must prove extremely useful to every practical man possessing a copy."—Practical Engineer.

Practical Treatise on Modern Steam-Boilers.

STEAM-BOILER CONSTRUCTION. A Practical Handbook for Engineers, Boiler-Makers, and Steam Users. Containing a large Collection of Rules and Data relating to the Design, Construction, and Working of Modern Stationary, Locomotive, and Marine Steam-Boilers. By WALTER S. HUTTON, C.E., Author of "The Works' Manager's Handbook," &c. With upwards of 300 Illustrations. Medium 8vo, 18s. cloth. [Just published.

"Every detail, both in boiler design and management, is clearly laid before the reader. The volume shows that boiler construction has been reduced to the condition of one of the most exact sciences; and such a book is of the utmost value to the fin de siecle Engineer and Works' Manager."—Marine Engineer.

"There has long been room for a modern handbook on steam boilers; there is not that room now, because Mr. Hutton has filled it. It is a thoroughly practical book for those who are occupied in the construction, design, selection, or use of boilers."—Engineer.

"The Modernised Templeton."

THE PRACTICAL MECHANIC'S WORKSHOP COMPANION. Comprising a great variety of the most useful Rules and Formulæ in Mechanical Science, with numerous Tables of Practical Data and Calculated Results for Facilitating Mechanical Operations. By WILLIAM TEMPLETON, Author of "The Engineer's Practical Assistant," &c. &c. Sixteenth Edition, Revised, Modernised, and considerably Enlarged by WALTER S. HUTTON, C.E., Author of "The Works' Manager's Handbook," "The Practical Engineer's Handbook," &c. Fcap. 8vo, nearly 500 pp., with Eight Plates and upwards of 250 Illustrative Diagrams, 6s., strongly bound for workshop or pocket wear and tear. [Just published.

₊ OPINIONS OF THE PRESS.

"In its modernised form Hutton's 'Templeton' should have a wide sale, for it contains much valuable information which the mechanic will often find of use, and not a few tables and notes which he might look for in vain in other works. This modernised edition will be appreciated by all who have learned to value the original editions of 'Templeton.'"—English Mechanic.

"It has met with great success in the engineering workshop, as we can testify; and there are a great many men who, in a great measure, owe their rise in life to this little book."—Building News.

"This familiar text-book—well known to all mechanics and engineers—is of essential service to the every-day requirements of engineers, millwrights, and the various trades connected with engineering and building. The new modernised edition is worth its weight in gold."—Building News. (Second Notice.)

"This well-known and largely used book contains information, brought up to date, of the sort so useful to the foreman and draughtsman. So much fresh information has been introduced as to constitute it practically a new book. It will be largely used in the office and workshop."—Mechanical World.

Stone-working Machinery.

STONE-WORKING MACHINERY, and the Rapid and Economical Conversion of Stone. With Hints on the Arrangement and Management of Stone Works. By M. POWIS BALE, M.I.M.E. With Illusts. Crown 8vo, 9s.
"Should be in the hands of every mason or student of stone-work."—*Colliery Guardian.*
"A capital handbook for all who manipulate stone for building or ornamental purposes."—*Machinery Market.*

Pump Construction and Management.

PUMPS AND PUMPING : A Handbook for Pump Users. Being Notes on Selection, Construction and Management. By M. POWIS BALE, M.I.M.E., Author of "Woodworking Machinery," &c. Crown 8vo, 2s. 6d.
"The matter is set forth as concisely as possible. In fact, condensation rather than diffuseness has been the author's aim throughout; yet he does not seem to have omitted anything likely to be of use."—*Journal of Gas Lighting.*

Milling Machines, etc.

MILLING : A Treatise on Machines, Appliances, and Processes employed in the Shaping of Metals by Rotary Cutters, including Information on Making and Grinding the Cutters. By PAUL N. HASLUCK, Author of "Lathe-work." With upwards of 300 Engravings. Large crown 8vo, 12s. 6d. cloth.
[Just published.

Turning.

LATHE-WORK : A Practical Treatise on the Tools, Appliances, and Processes employed in the Art of Turning. By PAUL N. HASLUCK. Fourth Edition, Revised and Enlarged. Cr. 8vo, 5s. cloth.
"Written by a man who knows, not only how work ought to be done, but who also knows how to do it, and how to convey his knowledge to others. To all turners this book would be valuable."—*Engineering.*
"We can safely recommend the work to young engineers. To the amateur it will simply be invaluable. To the student it will convey a great deal of useful information."—*Engineer.*

Screw-Cutting.

SCREW THREADS : And Methods of Producing Them. With Numerous Tables, and complete directions for using Screw-Cutting Lathes. By PAUL N. HASLUCK, Author of "Lathe-Work," &c. With Fifty Illustrations. Third Edition, Enlarged. Waistcoat-pocket size, 1s. 6d. cloth.
"Full of useful information, hints and practical criticism. Taps, dies and screwing-tools generally are illustrated and their action described."—*Mechanical World.*
"It is a complete compendium of all the details of the screw cutting lathe; in fact a *multum-in-parvo* on all the subjects it treats upon."—*Carpenter and Builder.*

Smith's Tables for Mechanics, etc.

TABLES, MEMORANDA, AND CALCULATED RESULTS, FOR MECHANICS, ENGINEERS, ARCHITECTS, BUILDERS, etc. Selected and Arranged by FRANCIS SMITH. Fifth Edition, thoroughly Revised and Enlarged, with a New Section of ELECTRICAL TABLES, FORMULÆ, and MEMORANDA. Waistcoat-pocket size, 1s. 6d. limp leather. *[Just published.*
"It would, perhaps, be as difficult to make a small pocket-book selection of notes and formulæ to suit ALL engineers as it would be to make a universal medicine; but Mr. Smith's waistcoat-pocket collection may be looked upon as a successful attempt."—*Engineer.*
"The best example we have ever seen of 250 pages of useful matter packed into the dimensions of a card-case."—*Building News.* "A veritable pocket treasury of knowledge."—*Iron.*

Engineer's and Machinist's Assistant.

THE ENGINEER'S, MILLWRIGHT'S, and MACHINIST'S PRACTICAL ASSISTANT. A collection of Useful Tables, Rules and Data. By WILLIAM TEMPLETON. 7th Edition, with Additions. 18mo, 2s. 6d. cloth.
"Occupies a foremost place among books of this kind. A more suitable present to an apprentice to any of the mechanical trades could not possibly be made."—*Building News.*
"A deservedly popular work, it should be in the 'drawer' of every mechanic."—*English Mechanic.*

Iron and Steel.

"IRON AND STEEL" : A Work for the Forge, Foundry, Factory, and Office. Containing ready, useful, and trustworthy Information for Iron-masters; Managers of Bar, Rail, Plate, and Sheet Rolling Mills; Iron and Metal Founders; Iron Ship and Bridge Builders; Mechanical, Mining, and Consulting Engineers; Contractors, Builders, &c. By CHARLES HOARE. Eighth Edition, Revised and considerably Enlarged. 32mo, 6s. leather.
"One of the best of the pocket books."—*English Mechanic.*
"We cordially recommend this book to those engaged in considering the details of all kinds of iron and steel works."—*Naval Science.*

Engineering Construction.

PATTERN-MAKING : A Practical Treatise, embracing the Main Types of Engineering Construction, and including Gearing, both Hand and Machine made, Engine Work, Sheaves and Pulleys, Pipes and Columns, Screws, Machine Parts, Pumps and Cocks, the Moulding of Patterns in Loam and Greensand, &c., together with the methods of Estimating the weight of Castings; to which is added an Appendix of Tables for Workshop Reference. By a FOREMAN PATTERN MAKER. With upwards of Three Hundred and Seventy Illustrations. Crown 8vo, 7s. 6d. cloth.

"A well-written technical guide, evidently written by a man who understands and has practised what he has written about. . . . We cordially recommend it to engineering students, young journeymen, and others desirous of being initiated into the mysteries of pattern-making."—*Builder.*
"We can confidently recommend this comprehensive treatise.'—*Building News.*
"Likely to prove a welcome guide to many workmen, especially to draughtsmen who have lacked a training in the shops, pupils pursuing their practical studies in our factories, and to employers and managers in engineering works."—*Hardware Trade Journal.*
"More than 370 illustrations help to explain the text, which is, however, always clear and explicit, thus rendering the work an excellent *vade mecum* for the apprentice who desires to become master of his trade."—*English Mechanic.*

Dictionary of Mechanical Engineering Terms.

LOCKWOOD'S DICTIONARY OF TERMS USED IN THE PRACTICE OF MECHANICAL ENGINEERING, embracing those current in the Drawing Office, Pattern Shop, Foundry, Fitting, Turning, Smith's and Boiler Shops, &c. &c. Comprising upwards of 6,000 Definitions. Edited by A FOREMAN PATTERN-MAKER, Author of "Pattern Making." Crown 8vo, 7s. 6d. cloth.

"Just the sort of handy dictionary required by the various trades engaged in mechanical engineering. The practical engineering pupil will find the book of great value in his studies, and every foreman engineer and mechanic should have a copy."—*Building News.*
"After a careful examination of the book, and trying all manner of words, we think that the engineer will here find all he is likely to require. It will be largely used."—*Practical Engineer.*
"One of the most useful books which can be presented to a mechanic or student."—*English Mechanic.*
"Not merely a dictionary, but, to a certain extent, also a most valuable guide. It strikes us as a happy idea to combine with a definition of the phrase useful information on the subject of which it treats."—*Machinery Market.*
"No word having connection with any branch of constructive engineering seems to be omitted. No more comprehensive work has been, so far, issued."—*Knowledge.*
"We strongly commend this useful and reliable adviser to our friends in the workshop, and to students everywhere."—*Colliery Guardian.*

Steam Boilers.

A TREATISE ON STEAM BOILERS : Their Strength, Construction, and Economical Working. By ROBERT WILSON, C.E. Fifth Edition. 12mo, 6s. cloth.

"The best treatise that has ever been published on steam boilers."—*Engineer.*
"The author shows himself perfect master of his subject, and we heartily recommend all employing steam power to possess themselves of the work."—*Ryland's Iron Trade Circular.*

Boiler Chimneys.

BOILER AND FACTORY CHIMNEYS ; Their Draught-Power and Stability. With a Chapter on *Lightning Conductors.* By ROBERT WILSON, A.I.C.E., Author of "A Treatise on Steam Boilers," &c. Second Edition. Crown 8vo, 3s. 6d. cloth.

"Full of useful information, definite in statement, and thoroughly practical in treatment.—*The Local Government Chronicle.*
"A valuable contribution to the literature of scientific building."—*The Builder.*

Boiler Making.

THE BOILER-MAKER'S READY RECKONER & ASSISTANT. With Examples of Practical Geometry and Templating, for the Use of Platers, Smiths and Riveters. By JOHN COURTNEY, Edited by D. K. CLARK, M.I.C.E. Third Edition, 480 pp., with 140 Illusts. Fcap. 8vo, 7s. half-bound.

"No workman or apprentice should be without this book."—*Iron Trade Circular.*
"Boiler-makers will readily recognise the value of this volume. . . . The tables are clearly printed, and so arranged that they can be referred to with the greatest facility, so that it cannot be doubted that they will be generally appreciated and much used."—*Mining Journal.*

Warming.

HEATING BY HOT WATER ; with Information and Suggestions on the best Methods of Heating Public, Private and Horticultural Buildings. By WALTER JONES. With Illustrations, crown 8vo, 2s. cloth.

"We confidently recommend all interested in heating by hot water to secure a copy of this valuable little treatise."—*The Plumber and Decorator.*

Steam Engine.

TEXT-BOOK ON THE STEAM ENGINE. With a Supplement on Gas Engines, and PART II. ON HEAT ENGINES. By T. M. GOODEVE, M.A., Barrister-at-Law, Professor of Mechanics at the Normal School of Science and the Royal School of Mines; Author of "The Principles of Mechanics," "The Elements of Mechanism," &c. Eleventh Edition, Enlarged. With numerous Illustrations. Crown 8vo, 6s. cloth.

"Professor Goodeve has given us a treatise on the steam engine which will bear comparison with anything written by Huxley or Maxwell, and we can award it no higher praise."—*Engineer.*

"Mr. Goodeve's text-book is a work of which every young engineer should possess himself." —*Mining Journal.*

Gas Engines.

ON GAS-ENGINES. Being a Reprint, with some Additions, of the Supplement to the *Text-book on the Steam Engine,* by T. M. GOODEVE, M.A. Crown 8vo, 2s. 6d. cloth.

"Like all Mr. Goodeve's writings, the present is no exception in point of general excellence. It is a valuable little volume."—*Mechanical World.*

Steam.

THE SAFE USE OF STEAM. Containing Rules for Unprofessional Steam-users. By an ENGINEER. Sixth Edition. Sewed, 6d.

"If steam-users would but learn this little book by heart boiler explosions would become sensations by their rarity."—*English Mechanic.*

Reference Book for Mechanical Engineers.

THE MECHANICAL ENGINEER'S REFERENCE BOOK, for Machine and Boiler Construction. In Two Parts. Part I. GENERAL ENGINEERING DATA. Part II. BOILER CONSTRUCTION. With 51 Plates and numerous Illustrations. By NELSON FOLEY, M.I.N.A. Folio, £5 5s. halfbound. [*Just published.*

Coal and Speed Tables.

A POCKET BOOK OF COAL AND SPEED TABLES, for Engineers and Steam-users. By NELSON FOLEY, Author of "Boiler Construction." Pocket-size, 3s. 6d. cloth; 4s. leather.

"These tables are designed to meet the requirements of every-day use; and may be commended to engineers and users of steam."—*Iron.*

"This pocket-book well merits the attention of the practical engineer. Mr. Foley has compiled a very useful set of tables, the information contained in which is frequently required by engineers, coal consumers and users of steam."—*Iron and Coal Trades Review.*

Fire Engineering.

FIRES, FIRE-ENGINES, AND FIRE-BRIGADES. With a History of Fire-Engines, their Construction, Use, and Management; Remarks on Fire-Proof Buildings, and the Preservation of Life from Fire; Foreign Fire Systems, &c. By C. F. T. YOUNG, C E. With numerous Illustrations, 544 pp., demy 8vo, £1 4s. cloth.

"To such of our readers as are interested in the subject of fires and fire apparatus, we can most heartily commend this book."—*Engineering.*

"It displays much evidence of careful research; and Mr. Young has put his facts neatly together. It is evident enough that his acquaintance with the practical details of the construction of steam fire engines is accurate and full."—*Engineer.*

Estimating for Engineering Work, &c.

ENGINEERING ESTIMATES, COSTS AND ACCOUNTS: A Guide to Commercial Engineering. With numerous Examples of Estimates and Costs of Millwright Work, Miscellaneous Productions, Steam Engines and Steam Boilers; and a Section on the Preparation of Costs Accounts. By A GENERAL MANAGER. Demy 8vo, 12s. cloth.

"This is an excellent and very useful book, covering subject-matter in constant requisition in every factory and workshop. . . . The book is invaluable, not only to the young engineer, but also to the estimate department of every works."—*Builder.*

"We accord the work unqualified praise. The information is given in a plain, straightforward manner, and bears throughout evidence of the intimate practical acquaintance of the author with every phrase of commercial engineering."—*Mechanical World.*

Elementary Mechanics.

CONDENSED MECHANICS. A Selection of Formulæ, Rules, Tables, and Data for the Use of Engineering Students, Science Classes, &c. In Accordance with the Requirements of the Science and Art Department. By W. G. CRAWFORD HUGHES, A.M.I.C.E. Crown 8vo, 2s. 6d. cloth.

[*Just published.*

THE POPULAR WORKS OF MICHAEL REYNOLDS
("THE ENGINE DRIVER'S FRIEND").

Locomotive-Engine Driving.

LOCOMOTIVE-ENGINE DRIVING : A Practical Manual for *Engineers in charge of Locomotive Engines.* By MICHAEL REYNOLDS, Member of the Society of Engineers, formerly Locomotive Inspector L. B. and S. C. R. Eighth Edition. Including a KEY TO THE LOCOMOTIVE ENGINE. With Illustrations and Portrait of Author. Crown 8vo. 4s. 6d. cloth.

"Mr. Reynolds has supplied a want, and has supplied it well. We can confidently recommend the book, not only to the practical driver, but to everyone who takes an interest in the performance of locomotive engines."—*The Engineer.*

"Mr. Reynolds has opened a new chapter in the literature of the day. This admirable practical treatise, of the practical utility of which we have to speak in terms of warm commendation."—*Athenaeum.*

"Evidently the work of one who knows his subject thoroughly."—*Railway Service Gazette.*

"Were the cautions and rules given in the book to become part of the every-day working of our engine-drivers, we might have fewer distressing accidents to deplore."—*Scotsman.*

Stationary Engine Driving.

STATIONARY ENGINE DRIVING : A Practical Manual for *Engineers in charge of Stationary Engines.* By MICHAEL REYNOLDS. Fourth Edition, Enlarged. With Plates and Woodcuts. Crown 8vo, 4s. 6d. cloth.

"The author is thoroughly acquainted with his subjects, and his advice on the various points treated is clear and practical. . . . He has produced a manual which is an exceedingly useful one for the class for whom it is specially intended."—*Engineering.*

"Our author leaves no stone unturned. He is determined that his readers shall not only know something about the stationary engine, but all about it."—*Engineer.*

"An engineman who has mastered the contents of Mr. Reynolds's book will require but little actual experience with boilers and engines before he can be trusted to look after them."—*English Mechanic.*

The Engineer, Fireman, and Engine-Boy.

THE MODEL LOCOMOTIVE ENGINEER, FIREMAN, and ENGINE-BOY. Comprising a Historical Notice of the Pioneer Locomotive Engines and their Inventors. By MICHAEL REYNOLDS. With numerous Illustrations and a fine Portrait of George Stephenson. Crown 8vo, 4s. 6d. cloth.

"From the technical knowledge of the author it will appeal to the railway man of to-day more forcibly than anything written by Dr. Smiles. . . . The volume contains information of a technical kind, and facts that every driver should be familiar with."—*English Mechanic.*

"We should be glad to see this book in the possession of everyone in the kingdom who has ever laid, or is to lay, hands on a locomotive engine."—*Iron.*

Continuous Railway Brakes.

CONTINUOUS RAILWAY BRAKES : A Practical Treatise on *the several Systems in Use in the United Kingdom ; their Construction and Performance.* With copious Illustrations and numerous Tables. By MICHAEL REYNOLDS. Large crown 8vo, 9s. cloth.

"A popular explanation of the different brakes. It will be of great assistance in forming public opinion, and will be studied with benefit by those who take an interest in the brake."—*English Mechanic.*

"Written with sufficient technical detail to enable the principle and relative connection of the various parts of each particular brake to be readily grasped."—*Mechanical World.*

Engine-Driving Life.

ENGINE-DRIVING LIFE : *Stirring Adventures and Incidents in the Lives of Locomotive-Engine Drivers.* By MICHAEL REYNOLDS. Second Edition, with Additional Chapters. Crown 8vo. 2s. cloth.

"From first to last perfectly fascinating. Wilkie Collins's most thrilling conceptions are thrown into the shade by true incidents, endless in their variety, related in every page."—*North British Mail.*

"Anyone who wishes to get a real insight into railway life cannot do better than read 'Engine-Driving Life' for himself ; and if he once take it up he will find that the author's enthusiasm and real love of the engine-driving profession will carry him on till he has read every page."—*Saturday Review.*

Pocket Companion for Enginemen.

THE ENGINEMAN'S POCKET COMPANION AND PRAC-TICAL EDUCATOR FOR ENGINEMEN, BOILER ATTENDANTS, AND MECHANICS. By MICHAEL REYNOLDS. With Forty-five Illustrations and numerous Diagrams. Second Edition, Revised. Royal 18mo, 3s. 6d., strongly bound for pocket wear.

"This admirable work is well suited to accomplish its object, being the honest workmanship of a competent engineer."—*Glasgow Herald.*

"A most meritorious work, giving in a succinct and practical form all the information an engine-minder desirous of mastering the scientific principles of his daily calling would require."—*Miller.*

"A boon to those who are striving to become efficient mechanics."—*Daily Chronicle.*

French-English Glossary for Engineers, etc.

A POCKET GLOSSARY of TECHNICAL TERMS: ENGLISH-FRENCH, FRENCH-ENGLISH; with Tables suitable for the Architectural, Engineering, Manufacturing and Nautical Professions. By John James Fletcher, Engineer and Surveyor. 200 pp. Waistcoat-pocket size, 1s. 6d., limp leather.

"It ought certainly to be in the waistcoat-pocket of every professional man."—*Iron.*

"It is a very great advantage for readers and correspondents in France and England to have so large a number of the words relating to engineering and manufacturers collected in a lilliputian volume. The little book will be useful both to students and travellers.'—*Architect.*

"The glossary of terms is very complete, and many of the tables are new and well arranged. We cordially commend the book."—*Mechanical World*

Portable Engines.

THE PORTABLE ENGINE; ITS CONSTRUCTION AND MANAGEMENT. A Practical Manual for Owners and Users of Steam Engines generally. By William Dyson Wansbrough. With 90 Illustrations. Crown 8vo, 3s. 6d. cloth.

"This is a work of value to those who use steam machinery. . . . Should be read by every-one who has a steam engine, on a farm or elsewhere."—*Mark Lane Express.*

"We cordially commend this work to buyers and owners of steam engines, and to those who have to do with their construction or use."—*Timber Trades Journal.*

"Such a general knowledge of the steam engine as Mr. Wansbrough furnishes to the reader should be acquired by all intelligent owners and others who use the steam engine."—*Building News.*

"An excellent text-book of this useful form of engine, which describes with all necessary minuteness the details of the various devices. . . ' The Hints to Purchasers contain a good deal of commonsense and practical wisdom."—*English Mechanic.*

CIVIL ENGINEERING, SURVEYING, etc.

MR. HUMBER'S IMPORTANT ENGINEERING BOOKS.

The Water Supply of Cities and Towns.

A COMPREHENSIVE TREATISE on the WATER-SUPPLY OF CITIES AND TOWNS. By William Humber, A-M.Inst.C.E., and M. Inst. M.E., Author of "Cast and Wrought Iron Bridge Construction," &c. &c. Illustrated with 50 Double Plates, 1 Single Plate, Coloured Frontispiece, and upwards of 250 Woodcuts, and containing 400 pages of Text. Imp. 4to, £6 6s. elegantly and substantially half-bound in morocco.

List of Contents.

I. Historical Sketch of some of the means that have been adopted for the Supply of Water to Cities and Towns.—II. Water and the Foreign Matter usually associated with it.—III. Rainfall and Evaporation.—IV. Springs and the water-bearing formations of various districts.—V. Measurement and Estimation of the flow of Water.—VI. On the Selection of the Source of Supply.—VII. Wells.—VIII. Reservoirs.—IX. The Purification of Water.—X. Pumps.—XI. Pumping Machinery — XII. Conduits.—XIII. Distribution of Water.—XIV. Meters, Service Pipes, and House Fittings.—XV. The Law and Economy of Water Works.—XVI. Constant and Intermittent Supply.—XVII. Description of Plates. — Appendices, giving Tables of Rates of Supply, Velocities, &c. &c., together with Specifications of several Works illustrated, among which will be found: Aberdeen, Bideford, Canterbury, Dundee, Halifax, Lambeth, Rotherham, Dublin, and others.

"The most systematic and valuable work upon water supply hitherto produced in English, or in any other language. . . . Mr. Humber's work is characterised almost throughout by an exhaustiveness much more distinctive of French and German than of English technical treatises."—*Engineer.*

"We can congratulate Mr. Humber on having been able to give so large an amount of information on a subject so important as the water supply of cities and towns. The plates, fifty in number, are mostly drawings of executed works, and alone would have commanded the attention of every engineer whose practice may lie in this branch of the profession."—*Builder.*

Cast and Wrought Iron Bridge Construction.

A COMPLETE AND PRACTICAL TREATISE ON CAST AND WROUGHT IRON BRIDGE CONSTRUCTION, including Iron Foundations. In Three Parts—Theoretical, Practical, and Descriptive. By William Humber, A.M.Inst.C.E., and M.Inst.M.E. Third Edition, Revised and much improved, with 115 Double Plates (20 of which now first appear in this edition), and numerous Additions to the Text. In Two Vols., imp. 4to, £6 16s. 6d. half-bound in morocco.

"A very valuable contribution to the standard literature of civil engineering. In addition to elevations, plans and sections, large scale details are given which very much enhance the instructive worth of those illustrations."—*Civil Engineer and Architect's Journal.*

"Mr. Humber's stately volumes, lately issued—in which the most important bridges erected during the last five years, under the direction of the late Mr. Brunel, Sir W. Cubitt, Mr. Hawkshaw, Mr. Page, Mr. Fowler, Mr. Hemans, and others among our most eminent engineers, are drawn and specified in great detail."—*Engineer.*

MR. HUMBER'S GREAT WORK ON MODERN ENGINEERING.

Complete in Four Volumes, imperial 4to, price £12 12s., half-morocco. Each
Volume sold separately as follows:—

A RECORD OF THE PROGRESS OF MODERN ENGINEER-
ING. FIRST SERIES. Comprising Civil, Mechanical, Marine, Hydraulic,
Railway, Bridge, and other Engineering Works, &c. By WILLIAM HUMBER,
A-M.Inst.C.E., &c. Imp. 4to, with 36 Double Plates, drawn to a large scale,
Photographic Portrait of John Hawkshaw, C.E., F.R.S., &c., and copious
descriptive Letterpress, Specifications, &c., £3 3s. half-morocco.

List of the Plates and Diagrams.

Victoria Station and Roof, L. B. & S. C. R. (8 plates); Southport Pier (2 plates); Victoria Station and Roo', L. C. & D. and G. W. R. (6 plates); Roof of Cremorne Music Hall; Bridge over G. N. Railway; Roof of Station, Dutch Rhenish Rail (2 plates); Bridge over the | Thames, West London Extension Railway (5 plates); Armour Plates: Suspension Bridge, Thames (4 plates); The Allen Engine; Suspension Bridge, Avon (3 plates); Underground Railway (3 plates).

"Handsomely lithographed and printed. It will find favour with many who desire to preserve in a permanent form copies of the plans and specifications prepared for the guidance of the contractors for many important engineering works."—*Engineer*.

HUMBER'S RECORD OF MODERN ENGINEERING. SECOND
SERIES. Imp. 4to, with 36 Double Plates, Photographic Portrait of Robert
Stephenson, C.E., M.P., F.R.S., &c., and copious descriptive Letterpress,
Specifications, &c., £3 3s. half-morocco.

List of the Plates and Diagrams.

Birkenhead Docks, Low Water Basin (15 plates); Charing Cross Station Roof, C. C. Railway (3 plates); Digswell Viaduct, Great Northern Railway; Robbery Wood Viaduct, Great Northern Railway; Iron Permanent Way; Clydach Viaduct, Merthyr, Tredegar, | and Abergavenny Railway; Ebbw Viaduct, Merthyr, Tredegar, and Abergavenny Railway; College Wood Viaduct, Cornwall Railway; Dublin Winter Palace Roof (3 plates); Bridge over the Thames, L. C. & D. Railway (6 plates); Albert Harbour, Greenock (4 plates).

"Mr. Humber has done the profession good and true service, by the fine collection of examples he has here brought before the profession and the public."—*Practical Mechanic's Journal.*

HUMBER'S RECORD OF MODERN ENGINEERING. THIRD
SERIES. Imp. 4to, with 40 Double Plates, Photographic Portrait of J. R.
M'Clean, late Pres. Inst. C.E., and copious descriptive Letterpress, Speci-
fications, &c., £3 3s. half-morocco.

List of the Plates and Diagrams.

MAIN DRAINAGE, METROPOLIS.—*North Side.*—Map showing Interception of Sewers; Middle Level Sewer (2 plates); Outfall Sewer, Bridge over River Lea (3 plates); Outfall Sewer, Bridge over Marsh Lane, North Woolwich Railway, and Bow and Barking Railway Junction; Outfall Sewer, Bridge over Bow and Barking Railway (3 plates); Outfall Sewer, Bridge over East London Waterworks' Feeder (2 plates); Outfall Sewer, Reservoir (2 plates); Outfall Sewer, Tumbling Bay and Outlet; Outfall Sewer, Penstocks. *South Side.*—Outfall Sewer, Bermondsey Branch (2 plates); Outfall | Sewer, Reservoir and Outlet (4 plates); Outfall Sewer, Filth Hoist; Sections of Sewers (North and South Sides). THAMES EMBANKMENT.—Section of River Wall; Steamboat Pier, Westminster (2 plates); Landing Stairs between Charing Cross and Waterloo Bridges; York Gate (2 plates); Overflow and Outlet at Savoy Street Sewer (3 plates); Steamboat Pier, Waterloo Bridge (3 plates); Junction of Sewers, Plans and Sections; Gullies, Plans and Sections; Rolling Stock; Granite and Iron Forts.

"The drawings have a constantly increasing value, and whoever desires to possess clear representations of the two great works carried out by our Metropolitan Board will obtain Mr. Humber's volume."—*Engineer.*

HUMBER'S RECORD OF MODERN ENGINEERING. FOURTH
SERIES. Imp. 4to, with 36 Double Plates, Photographic Portrait of John
Fowler, late Pres. Inst. C.E., and copious descriptive Letterpress, Speci-
fications, &c., £3 3s. half-morocco.

List of the Plates and Diagrams.

Abbey Mills Pumping Station, Main Drainage, Metropolis (4 plates); Barrow Docks (5 plates); Manquis Viaduct, Santiago and Valparaiso Railway (2 plates); Adam's Locomotive, St. Helen's Canal Railway (2 plates); Cannon Street Station Roof, Charing Cross Railway (3 plates); Road Bridge over the River Moka (2 plates); Telegraphic Apparatus for | Mesopotamia; Viaduct over the River Wye, Midland Railway (3 plates); St. Germans Viaduct, Cornwall Railway (2 plates); Wrought-Iron Cylinder for Diving Bell; Millwall Docks (6 plates); Milroy's Patent Excavator; Metropolitan District Railway (6 plates); Harbours, Forts, and Breakwaters (3 plates).

"We gladly welcome another year's issue of this valuable publication from the able pen of Mr. Humber. The accuracy and general excellence of this work are well known, while its usefulness in giving the measurements and details of some of the latest examples of engineering, as carried out by the most eminent men in the profession, cannot be too highly prized."—*Artisan.*

MR. HUMBER'S ENGINEERING BOOKS—continued.

Strains, Calculation of.

A HANDY BOOK FOR THE CALCULATION OF STRAINS IN GIRDERS AND SIMILAR STRUCTURES, AND THEIR STRENGTH. Consisting of Formulæ and Corresponding Diagrams, with numerous details for Practical Application, &c. By WILLIAM HUMBER, A-M.Inst.C.E., &c. Fourth Edition. Crown 8vo, nearly 100 Woodcuts and 3 Plates, 7s. 6d. cloth.

" The formulæ are neatly expressed, and the diagrams good."—*Athenæum.*
" We heartily commend this really *handy* book to our engineer and architect readers."—*English Mechanic.*

Barlow's Strength of Materials, enlarged by Humber

A TREATISE ON THE STRENGTH OF MATERIALS; with Rules for Application in Architecture, the Construction of Suspension Bridges, Railways, &c. By PETER BARLOW, F.R.S. A New Edition, revised by his Sons, P. W. BARLOW, F.R.S., and W. H. BARLOW, F.R.S.; to which are added, Experiments by HODGKINSON, FAIRBAIRN, and KIRKALDY; and Formulæ for Calculating Girders, &c. Arranged and Edited by W. HUMBER, A-M.Inst.C.E. Demy 8vo, 400 pp., with 19 large Plates and numerous Woodcuts, 18s. cloth.

" Valuable alike to the student, tyro, and the experienced practitioner, it will always rank in future, as it has hitherto done, as the standard treatise on that particular subject."—*Engineer.*
" There is no greater authority than Barlow."—*Building News.*
" As a scientific work of the first class, it deserves a foremost place on the bookshelves of every civil engineer and practical mechanic."—*English Mechanic.*

Trigonometrical Surveying.

AN OUTLINE OF THE METHOD OF CONDUCTING A TRIGONOMETRICAL SURVEY, for the Formation of Geographical and Topographical Maps and Plans, Military Reconnaissance, Levelling, &c., with Useful Problems, Formulæ, and Tables. By Lieut.-General FROME, R.E. Fourth Edition, Revised and partly Re-written by Major General Sir CHARLES WARREN, G.C.M.G., R.E. With 19 Plates and 115 Woodcuts, royal 8vo, 16s. cloth.

" The simple fact that a fourth edition has been called for is the best testimony to its merits. No words of praise from us can strengthen the position so well and so steadily maintained by this work. Sir Charles Warren has revised the entire work, and made such additions as were necessary to bring every portion of the contents up to the present date."—*Broad Arrow.*

Field Fortification.

A TREATISE ON FIELD FORTIFICATION, THE ATTACK OF FORTRESSES, MILITARY MINING, AND RECONNOITRING. By Colonel I. S. MACAULAY, late Professor of Fortification in the R.M.A., Woolwich. Sixth Edition, crown 8vo, cloth, with separate Atlas of 12 Plates, 12s.

Oblique Bridges.

A PRACTICAL AND THEORETICAL ESSAY ON OBLIQUE BRIDGES. With 13 large Plates. By the late GEORGE WATSON BUCK, M.I.C.E. Third Edition, revised by his Son, J. H. WATSON BUCK, M.I.C.E.; and with the addition of Description to Diagrams for Facilitating the Construction of Oblique Bridges, by W. H. BARLOW, M.I.C.E. Royal 8vo, 12s. cloth.

" The standard text-book for all engineers regarding skew arches is Mr. Buck's treatise, and it would be impossible to consult a better."—*Engineer.*
" Mr. Buck's treatise is recognised as a standard text-book, and his treatment has divested the subject of many of the intricacies supposed to belong to it. As a guide to the engineer and architect, on a confessedly difficult subject, Mr. Buck's work is unsurpassed."—*Building News.*

Water Storage, Conveyance and Utilisation.

WATER ENGINEERING : A Practical Treatise on the Measurement, Storage, Conveyance and Utilisation of Water for the Supply of Towns, for Mill Power, and for other Purposes. By CHARLES SLAGG, Water and Drainage Engineer, A.M.Inst.C.E., Author of " Sanitary Work in the Smaller Towns, and in Villages," &c. With numerous Illusts. Cr. 8vo, 7s. 6d. cloth.

" As a small practical treatise on the water supply of towns, and on some applications of water-power, the work is in many respects excellent."—*Engineering.*
" The author has collated the results deduced from the experiments of the most eminent authorities, and has presented them in a compact and practical form, accompanied by very clear and detailed explanations. . . . The application of water as a motive power is treated very carefully and exhaustively."—*Builder.*
" For anyone who desires to begin the study of hydraulics with a consideration of the practical applications of the science there is no better guide."—*Architect.*

Statics, Graphic and Analytic.

GRAPHIC AND ANALYTIC STATICS, *in their Practical Application to the Treatment of Stresses in Roofs, Solid Girders, Lattice, Bowstring and Suspension Bridges, Braced Iron Arches and Piers, and other Frameworks.* By R. HUDSON GRAHAM, C.E. Containing Diagrams and Plates to Scale. With numerous Examples, many taken from existing Structures. Specially arranged for Class-work in Colleges and Universities. Second Edition, Revised and Enlarged. 8vo, 16s. cloth.

"Mr. Graham's book will find a place wherever graphic and analytic statics are used or studied." —*Engineer.*

"The work is excellent from a practical point of view, and has evidently been prepared with much care. The directions for working are ample, and are illustrated by an abundance of well-selected examples. It is an excellent text-book for the practical draughtsman."—*Athenæum.*

Student's Text-Book on Surveying.

PRACTICAL SURVEYING: A Text-Book for Students preparing for Examination or for Survey-work in the Colonies. By GEORGE W. USILL, A.M.I.C.E., Author of "The Statistics of the Water Supply of Great Britain." With Four Lithographic Plates and upwards of 330 Illustrations. Second Edition, Revised. Crown 8vo, 7s. 6d. cloth.

"The best forms of instruments are described as to their construction, uses and modes of employment, and there are innumerable hints on work and equipment such as the author, in his experience as surveyor, draughtsman and teacher, has found necessary, and which the student in his inexperience will find most serviceable."—*Engineer.*

"The latest treatise in the English language on surveying, and we have no hesitation in saying that the student will find it a better guide than any of its predecessors . . . . Deserves to be recognised as the first book which should be put in the hands of a pupil of Civil Engineering, and every gentleman of education who sets out for the Colonies would find it well to have a copy."—*Architect.*

"A very useful, practical handbook on field practice. Clear, accurate and not too condensed."—*Journal of Education.*

Survey Practice.

AID TO SURVEY PRACTICE, *for Reference in Surveying, Levelling, and Setting-out ; and in Route Surveys of Travellers by Land and Sea.* With Tables, Illustrations, and Records. By LEWIS D'A. JACKSON, A.M.I.C.E., Author of "Hydraulic Manual," "Modern Metrology," &c. Second Edition, Enlarged. Large crown 8vo, 12s. 6d. cloth.

"Mr. Jackson has produced a valuable *vade-mecum* for the surveyor. We can recommend this book as containing an admirable supplement to the teaching of the accomplished surveyor."—*Athenæum.*

"As a text-book we should advise all surveyors to place it in their libraries, and study well the matured instructions afforded in its pages."—*Colliery Guardian.*

"The author brings to his work a fortunate union of theory and practical experience which, aided by a clear and lucid style of writing, renders the book a very useful one."—*Builder.*

Surveying, Land and Marine.

LAND AND MARINE SURVEYING, in Reference to the Preparation of Plans for Roads and Railways; Canals, Rivers, Towns' Water Supplies; Docks and Harbours. With Description and Use of Surveying Instruments. By W. D. HASKOLL, C.E., Author of "Bridge and Viaduct Construction," &c. Second Edition, Revised, with Additions. Large cr. 8vo, 9s. cl.

"This book must prove of great value to the student. We have no hesitation in recommending it, feeling assured that it will more than repay a careful study."—*Mechanical World.*

"A most useful and well arranged book for the aid of a student. We can strongly recommend it as a carefully-written and valuable text-book. It enjoys a well-deserved repute among surveyors." —*Builder.*

"This volume cannot fail to prove of the utmost practical utility. It may be safely recommended to all students who aspire to become clean and expert surveyors."—*Mining Journal.*

Tunnelling.

PRACTICAL TUNNELLING. Explaining in detail the Setting-out of the works, Shaft-sinking and Heading-driving, Ranging the Lines and Levelling underground, Sub-Excavating, Timbering, and the Construction of the Brickwork of Tunnels, with the amount of Labour required for, and the Cost of, the various portions of the work. By FREDERICK W. SIMMS, F.G.S., M.Inst.C.E. Third Edition, Revised and Extended by D. KINNEAR CLARK, M.Inst.C.E. Imperial 8vo, with 21 Folding Plates and numerous Wood Engravings, 30s. cloth.

"The estimation in which Mr. Simms's book on tunnelling has been held for over thirty years cannot be more truly expressed than in the words of the late Prof. Rankine:—'The best source of information on the subject of tunnels is Mr. F.W. Simms's work on Practical Tunnelling.'"—*Architect.*

"It has been regarded from the first as a text book of the subject. . . . Mr. Clarke has added immensely to the value of the book."—*Engineer.*

Levelling.

A TREATISE ON THE PRINCIPLES AND PRACTICE OF LEVELLING. Showing its Application to purposes of Railway and Civil Engineering, in the Construction of Roads; with Mr. TELFORD's Rules for the same. By FREDERICK W. SIMMS, F.G.S., M.Inst.C.E. Seventh Edition, with the addition of LAW's Practical Examples for Setting-out Railway Curves, and TRAUTWINE's Field Practice of Laying-out Circular Curves. With 7 Plates and numerous Woodcuts, 8vo, 8s. 6d. cloth. *** TRAUTWINE on Curves may be had separate, 5s.

"The text-book on levelling in most of our engineering schools and colleges."—*Engineer.*
"The publishers have rendered a substantial service to the profession, especially to the younger members, by bringing out the present edition of Mr. Simms's useful work."—*Engineering.*

Heat, Expansion by.

EXPANSION OF STRUCTURES BY HEAT. By JOHN KEILY, C.E., late of the Indian Public Works and Victorian Railway Departments. Crown 8vo, 3s. 6d. cloth.

SUMMARY OF CONTENTS.

Section I. FORMULAS AND DATA.	Section VI. MECHANICAL FORCE OF HEAT.
Section II. METAL BARS.	
Section III. SIMPLE FRAMES.	Section VII. WORK OF EXPANSION AND CONTRACTION.
Section IV. COMPLEX FRAMES AND PLATES.	
	Section VIII. SUSPENSION BRIDGES.
Section V. THERMAL CONDUCTIVITY.	Section IX. MASONRY STRUCTURES.

"The aim the author has set before him, viz., to show the effects of heat upon metallic and other structures, is a laudable one, for this is a branch of physics upon which the engineer or architect can find but little reliable and comprehensive data in books."—*Builder.*
"Whoever is concerned to know the effect of changes of temperature on such structures as suspension bridges and the like, could not do better than consult Mr. Keily's valuable and handy exposition of the geometrical principles involved in these changes."—*Scotsman.*

Practical Mathematics.

MATHEMATICS FOR PRACTICAL MEN: Being a Commonplace Book of Pure and Mixed Mathematics. Designed chiefly for the use of Civil Engineers, Architects and Surveyors. By OLINTHUS GREGORY, LL.D., F.R.A.S., Enlarged by HENRY LAW, C.E. 4th Edition, carefully Revised by J. R. YOUNG, formerly Professor of Mathematics, Belfast College. With 13 Plates, 8vo, £1 1s. cloth.

"The engineer or architect will here find ready to his hand rules for solving nearly every mathematical difficulty that may arise in his practice The rules are in all cases explained by means of examples, in which every step of the process is clearly worked out."—*Builder.*
"One of the most serviceable books for practical mechanics. . . It is an instructive book for the student, and a text-book for him who, having once mastered the subjects it treats of, needs occasionally to refresh his memory upon them."—*Building News.*

Hydraulic Tables.

HYDRAULIC TABLES, CO-EFFICIENTS, and FORMULÆ for finding the Discharge of Water from Orifices, Notches, Weirs, Pipes, and Rivers. With New Formulæ, Tables, and General Information on Rainfall, Catchment-Basins, Drainage, Sewerage, Water Supply for Towns and Mill Power. By JOHN NEVILLE, Civil Engineer, M.R.I.A. Third Ed., carefully Revised, with considerable Additions. Numerous Illusts. Cr. 8vo, 14s. cloth.

"Alike valuable to students and engineers in practice, its study will prevent the annoyance of avoidable failures, and assist them to select the readiest means of successfully carrying out any given work connected with hydraulic engineering."—*Mining Journal.*
"It is, of all English books on the subject, the one nearest to completeness. . . . From the good arrangement of the matter, the clear explanations, and abundance of formulæ, the carefully calculated tables, and, above all, the thorough acquaintance with both theory and construction, which is displayed from first to last, the book will be found to be an acquisition."—*Architect.*

Hydraulics.

HYDRAULIC MANUAL. Consisting of Working Tables and Explanatory Text. Intended as a Guide in Hydraulic Calculations and Field Operations. By LOWIS D'A. JACKSON, Author of "Aid to Survey Practice," "Modern Metrology," &c. Fourth Edition, Enlarged. Large cr. 8vo, 16s. cl.

"The author has had a wide experience in hydraulic engineering and has been a careful observer of the facts which have come under his notice, and from the great mass of material at his command he has constructed a manual which may be accepted as a trustworthy guide to this branch of the engineer's profession. We can heartily recommend this volume to all who desire to be acquainted with the latest development of this important subject."—*Engineering.*
"The standard-work in this department of mechanics."—*Scotsman.*
"The most useful feature of this work is its freedom from what is superannuated, and its thorough adoption of recent experiments; the text is, in fact, in great part a short account of the great modern experiments."—*Nature.*

Drainage.

ON THE DRAINAGE OF LANDS, TOWNS AND BUILD-INGS. By G. D. DEMPSEY, C.E., Author of "The Practical Railway Engineer," &c. Revised, with large Additions on RECENT PRACTICE IN DRAINAGE ENGINEERING, by D. KINNEAR CLARK, M.Inst.C.E. Author of "Tramways: Their Construction and Working," "A Manual of Rules, Tables, and Data for Mechanical Engineers." &c. &c. Crown 8vo, 7s. 6d. cloth.

"The new matter added to Mr. Dempsey's excellent work is characterised by the comprehensive grasp and accuracy of detail for which the name of Mr. D. K. Clark is a sufficient voucher."—*Athenæum.*
"As a work on recent practice in drainage engineering, the book is to be commended to all who are making that branch of engineering science their special study."—*Iron.*
"A comprehensive manual on drainage engineering, and a useful introduction to the student." *Building News.*

Tramways and their Working.

TRAMWAYS: THEIR CONSTRUCTION AND WORKING. Embracing a Comprehensive History of the System; with an exhaustive Analysis of the various Modes of Traction, including Horse-Power, Steam, Heated Water, and Compressed Air; a Description of the Varieties of Rolling Stock; and ample Details of Cost and Working Expenses: the Progress recently made in Tramway Construction, &c. &c. By D. KINNEAR CLARK, M.Inst.C.E. With over 200 Wood Engravings, and 13 Folding Plates. Two Vols., large crown 8vo, 30s. cloth.

"All interested in tramways must refer to it, as all railway engineers have turned to the author's work 'Railway Machinery.'"—*Engineer.*
"An exhaustive and practical work on tramways, in which the history of this kind of locomotion, and a description and cost of the various modes of laying tramways, are to be found."—*Building News.*
"The best form of rails, the best mode of construction, and the best mechanical appliances are so fairly indicated in the work under review, that any engineer about to construct a tramway will be enabled at once to obtain the practical information which will be of most service to him."—*Athenæum.*

Oblique Arches.

A PRACTICAL TREATISE ON THE CONSTRUCTION OF OBLIQUE ARCHES. By JOHN HART. Third Edition, with Plates. Imperial 8vo, 8s. cloth.

Curves, Tables for Setting-out.

TABLES OF TANGENTIAL ANGLES AND MULTIPLES *for Setting-out Curves from* 5 *to* 200 *Radius.* By ALEXANDER BEAZELEY, M.Inst.C.E. Third Edition. Printed on 48 Cards, and sold in a cloth box, waistcoat-pocket size, 3s. 6d.

"Each table is printed on a small card, which, being placed on the theodolite, leaves the hands free to manipulate the instrument—no small advantage as regards the rapidity of work."—*Engineer.*
"Very handy; a man may know that all his day's work must fall on two of these cards, which he puts into his own card-case, and leaves the rest behind."—*Athenæum.*

Earthwork.

EARTHWORK TABLES. Showing the Contents in Cubic Yards of Embankments, Cuttings, &c., of Heights or Depths up to an average of 80 feet. By JOSEPH BROADBENT, C.E., and FRANCIS CAMPIN, C.E. Crown 8vo, 5s. cloth.

"The way in which accuracy is attained, by a simple division of each cross section into three elements, two in which are constant and one variable, is ingenious."—*Athenæum.*

Tunnel Shafts.

THE CONSTRUCTION OF LARGE TUNNEL SHAFTS: A *Practical and Theoretical Essay.* By J. H. WATSON BUCK, M.Inst.C.E., Resident Engineer, London and North-Western Railway. Illustrated with Folding Plates, royal 8vo, 12s. cloth.

"Many of the methods given are of extreme practical value to the mason; and the observations on the form of arch, the rules for ordering the stone, and the construction of the templates will be found of considerable use. We commend the book to the engineering profession."—*Building News.*
"Will be regarded by civil engineers as of the utmost value, and calculated to save much time and obviate many mistakes."—*Colliery Guardian.*

Girders, Strength of.

GRAPHIC TABLE FOR FACILITATING THE COMPUTATION OF THE WEIGHTS OF WROUGHT IRON AND STEEL GIRDERS, *&c.,* for Parliamentary and other Estimates. By J. H. WATSON BUCK, M.Inst.C.E. On a Sheet, 2s. 6d.

River Engineering.

RIVER BARS: *The Causes of their Formation, and their Treatment by " Induced Tidal Scour ; "* with a Description of the Successful Reduction by this Method of the Bar at Dublin. By I. J. MANN, Assist. Eng. to the Dublin Port and Docks Board. Royal 8vo, 7s. 6d. cloth.

" We recommend all interested in harbour works—and, indeed, those concerned in the improvements of rivers generally—to read Mr. Mann's interesting work on the treatment of river bars."—*Engineer.*

Trusses.

TRUSSES OF WOOD AND IRON. *Practical Applications of Science in Determining the Stresses, Breaking Weights, Safe Loads, Scantlings, and Details of Construction,* with Complete Working Drawings. By WILLIAM GRIFFITHS, Surveyor, Assistant Master, Tranmere School of Science and Art. Oblong 8vo, 4s. 6d. cloth.

" This handy little book enters so minutely into every detail connected with the construction of roof trusses, that no student need be ignorant of these matters."—*Practical Engineer.*

Railway Working.

SAFE RAILWAY WORKING. *A Treatise on Railway Accidents: Their Cause and Prevention; with a Description of Modern Appliances and Systems.* By CLEMENT E. STRETTON, C.E., Vice-President and Consulting Engineer, Amalgamated Society of Railway Servants. With Illustrations and Coloured Plates. Second Edition, Enlarged. Crown 8vo, 3s. 6d. cloth. [*Just published.*

" A book for the engineer, the directors, the managers; and, in short, all who wish for information on railway matters will find a perfect encyclopædia in ' Safe Railway Working.' "—*Railway Review.*

" We commend the remarks on railway signalling to all railway managers, especially where a uniform code and practice is advocated."—*Herepath's Railway Journal.*

" The author may be congratulated on having collected, in a very convenient form, much valuable information on the principal questions affecting the safe working of railways."—*Railway Engineer.*

Field-Book for Engineers.

THE ENGINEER'S, MINING SURVEYOR'S, AND CONTRACTOR'S FIELD-BOOK. Consisting of a Series of Tables, with Rules, Explanations of Systems, and use of Theodolite for Traverse Surveying and Plotting the Work with minute accuracy by means of Straight Edge and Set Square only ; Levelling with the Theodolite, Casting-out and Reducing Levels to Datum, and Plotting Sections in the ordinary manner; setting-out Curves with the Theodolite by Tangential Angles and Multiples, with Right and Left-hand Readings of the Instrument: Setting-out Curves without Theodolite, on the System of Tangential Angles by sets of Tangents and Offsets ; and Earthwork Tables to 80 feet deep, calculated for every 6 inches in depth. By W. DAVIS HASKOLL, C.E. With numerous Woodcuts. Fourth Edition, Enlarged. Crown 8vo, 12s. cloth.

" The book is very handy; the separate tables of sines and tangents to every minute will make it useful for many other purposes, the genuine traverse tables existing all the same."—*Athenæum.*

" Every person engaged in engineering field operations will estimate the importance of such a work and the amount of valuable time which will be saved by reference to a set of reliable tables prepared with the accuracy and fulness of those given in this volume."—*Railway News.*

Earthwork, Measurement of.

A MANUAL ON EARTHWORK. By ALEX. J. S. GRAHAM, C.E. With numerous Diagrams. Second Edition. 18mo, 2s. 6d. cloth.

" A great amount of practical information, very admirably arranged, and available for rough estimates, as well as for the more exact calculations required in the engineer's and contractor's offices."—*Artisan.*

Strains in Ironwork.

THE STRAINS ON STRUCTURES OF IRONWORK; with Practical Remarks on Iron Construction. By F. W. SHEILDS, M.Inst.C.E. Second Edition, with 5 Plates. Royal 8vo, 5s. cloth.

" The student cannot find a better little book on this subject."—*Engineer.*

Cast Iron and other Metals, Strength of.

A PRACTICAL ESSAY ON THE STRENGTH OF CAST IRON AND OTHER METALS. By THOMAS TREDGOLD, C.E. Fifth Edition, including HODGKINSON'S Experimental Researches. 8vo, 12s. cloth.

ARCHITECTURE, BUILDING, etc.

Construction.
THE SCIENCE OF BUILDING : An Elementary Treatise on
the Principles of Construction. By E. WYNDHAM TARN, M.A., Architect.
Third Edition, Enlarged, with 59 Engravings. Fcap. 8vo, 4s. cloth.
"A very valuable book, which we strongly recommend to all students."—*Builder.*
"No architectural student should be without this handbook."—*Architect.*

Villa Architecture.
A HANDY BOOK OF VILLA ARCHITECTURE : Being a
Series of Designs for Villa Residences in various Styles. With Outline
Specifications and Estimates. By C. WICKES, Author of "The Spires and
Towers of England," &c. 61 Plates, 4to, £1 11s. 6d. half-morocco, gilt edges.
"The whole of the designs bear evidence of their being the work of an artistic architect, and
they will prove very valuable and suggestive."—*Building News.*

Text-Book for Architects.
THE ARCHITECT'S GUIDE : Being a Text-Book of Useful
Information for Architects, Engineers, Surveyors, Contractors, Clerks of
Works, &c. &c. By FREDERICK ROGERS, Architect, Author of " Specifica-
tions for Practical Architecture," &c. Second Edition, Revised and Enlarged.
With numerous Illustrations. Crown 8vo, 6s. cloth.
"As a text-book of useful information for architects, engineers, surveyors, &c., it would be
hard to find a handier or more complete little volume."—*Standard.*
"A young architect could hardly have a better guide-book."—*Timber Trades Journal.*

Taylor and Cresy's Rome.
THE ARCHITECTURAL ANTIQUITIES OF ROME. By
the late G. L. TAYLOR, Esq., F.R.I.B.A., and EDWARD CRESY, Esq. New
Edition, thoroughly Revised by the Rev. ALEXANDER TAYLOR, M.A. (son of
the late G. L. Taylor, Esq.), Fellow of Queen's College, Oxford, and Chap-
lain of Gray's Inn. Large folio, with 130 Plates, half-bound, £3 3s.
"Taylor and Cresy's work has from its first publication been ranked among those professional
books which cannot be bettered. . . . It would be difficult to find examples of drawings, even
among those of the most painstaking students of Gothic, more thoroughly worked out than are the
one hundred and thirty plates in this volume."—*Architect.*

Linear Perspective.
ARCHITECTURAL PERSPECTIVE : The whole Course and
Operations of the Draughtsman in Drawing a Large House in Linear Perspective. Illustrated by 39 Folding Plates. By F. O. FERGUSON. Demy
8vo, 3s. 6d. boards.   [Just published.

Architectural Drawing.
PRACTICAL RULES ON DRAWING, for the Operative Builder
and Young Student in Architecture. By GEORGE PYNE. With 14 Plates, 4to,
7s. 6d. boards.

Sir Wm. Chambers on Civil Architecture.
THE DECORATIVE PART OF CIVIL ARCHITECTURE.
By Sir WILLIAM CHAMBERS, F.R.S. With Portrait, Illustrations, Notes, and
an Examination of Grecian Architecture, by JOSEPH GWILT, F.S.A. Revised
and Edited by W. H. LEEDS, with a Memoir of the Author. 66 Plates, 4to,
21s. cloth.

House Building and Repairing.
THE HOUSE-OWNER'S ESTIMATOR ; or, What will it Cost
to Build, Alter, or Repair? A Price Book adapted to the Use of Unpro-
fessional People, as well as for the Architectural Surveyor and Builder. By
JAMES D. SIMON, A.R.I.B.A. Edited and Revised by FRANCIS T. W MILLER,
A.R.I.B.A. With numerous Illustrations. Fourth Edition, Revised. Crown
8vo, 3s. 6d. cloth.
"In two years it will repay its cost a hundred times over."—*Field.*

Cottages and Villas.
COUNTRY AND SUBURBAN COTTAGES AND VILLAS :
How to Plan and Build Them. Containing 33 Plates, with Introduction,
General Explanations, and Description of each Plate. By JAMES W. BOGUE,
Architect, Author of " Domestic Architecture," &c. 4to, 10s. 6d. cloth.

The New Builder's Price Book, 1891.

LOCKWOOD'S BUILDER'S PRICE BOOK FOR 1891. A
Comprehensive Handbook of the Latest Prices and Data for Builders,
Architects, Engineers and Contractors. Re-constructed, Re-written and
Greatly Enlarged. By FRANCIS T. W. MILLER. 640 closely-printed pages,
crown 8vo, 4s. cloth. *[Just published.*
"This book is a very useful one, and should find a place in every English office connected with
the building and engineering professions."—*Industries.*
"This Price Book has been set up in new type. . . . Advantage has been taken of the
transformation to add much additional information, and the volume is now an excellent book of
reference."—*Architect.*
"In its new and revised form this Price Book is what a work of this kind should be—comprehensive, reliable, well arranged, legible and well bound."—*British Architect.*
"A work of established reputation."—*Athenæum.*
"This very useful handbook is well written, exceedingly clear in its explanations and great
care has evidently been taken to ensure accuracy."—*Morning Advertiser*

Designing, Measuring, and Valuing.

**THE STUDENT'S GUIDE to the PRACTICE of MEASUR-
ING AND VALUING ARTIFICERS' WORKS.** Containing Directions for
taking Dimensions, Abstracting the same, and bringing the Quantities into
Bill, with Tables of Constants for Valuation of Labour, and for the Calcula-
tion of Areas and Solidities. Originally edited by EDWARD DOBSON, Architect.
With Additions on Mensuration and Construction, and a New Chapter on
Dilapidations, Repairs, and Contracts, by E. WYNDHAM TARN, M.A. Sixth
Edition, including a Complete Form of a Bill of Quantities. With 8 Plates and
63 Woodcuts. Crown 8vo, 7s. 6d. cloth.
"Well fulfils the promise of its title-page, and we can thoroughly recommend it to the class
for whose use it has been compiled. Mr. Tarn's additions and revisions have much increased the
usefulness of the work, and have especially augmented its value to students."—*Engineering.*
"This edition will be found the most complete treatise on the principles of measuring and
valuing artificers' work that has yet been published."—*Building News.*

Pocket Estimator and Technical Guide.

**THE POCKET TECHNICAL GUIDE, MEASURER AND
ESTIMATOR FOR BUILDERS AND SURVEYORS.** Containing Tech-
nical Directions for Measuring Work in all the Building Trades, Complete
Specifications for Houses, Roads, and Drains, and an easy Method of Estimat-
ing the parts of a Building collectively. By A. C. BEATON, Author of
"Quantities and Measurements," &c. Fifth Edition. With 53 Woodcuts,
waistcoat-pocket size, 1s. 6d. gilt edges.
"No builder, architect, surveyor, or valuer should be without his 'Beaton.'"—*Building News.*
"Contains an extraordinary amount of information in daily requisition in measuring and
estimating. Its presence in the pocket will save valuable time and trouble."—*Building World.*

Donaldson on Specifications.

THE HANDBOOK OF SPECIFICATIONS; or, Practical
Guide to the Architect, Engineer, Surveyor, and Builder, in drawing up
Specifications and Contracts for Works and Constructions. Illustrated by
Precedents of Buildings actually executed by eminent Architects and En-
gineers. By Professor T. L. DONALDSON, P.R.I.B.A., &c. New Edition, in
One large Vol., 8vo, with upwards of 1,000 pages of Text, and 33 Plates,
£1 11s. 6d. cloth.
"In this work forty-four specifications of executed works are given, including the specifications for parts of the new Houses of Parliament, by Sir Charles Barry, and for the new Royal
Exchange, by Mr. Tite, M.P. The latter, in particular, is a very complete and remarkable
document. It embodies, to a great extent, as Mr. Donaldson mentions, 'the bill of quantities
with the description of the works.' . . . It is valuable as a record, and more valuable still as a
book of precedents. . . . Suffice it to say that Donaldson's 'Handbook of Specifications'
must be bought by all architects."—*Builder.*

Bartholomew and Rogers' Specifications.

SPECIFICATIONS FOR PRACTICAL ARCHITECTURE.
A Guide to the Architect, Engineer, Surveyor, and Builder. With an Essay
on the Structure and Science of Modern Buildings. Upon the Basis of the
Work by ALFRED BARTHOLOMEW, thoroughly Revised, Corrected, and greatly
added to by FREDERICK ROGERS, Architect. Second Edition, Revised, with
Additions. With numerous Illustrations, medium 8vo, 15s. cloth.
"The collection of specifications prepared by Mr. Rogers on the basis of Bartholomew's work
is too well known to need any recommendation from us. It is one of the books with which every
young architect must be equipped; for time has shown that the specifications cannot be set aside
through any defect in them."—*Architect.*

Building; Civil and Ecclesiastical.

A BOOK ON BUILDING, Civil and Ecclesiastical, including Church Restoration ; with the Theory of Domes and the Great Pyramid, &c. By Sir EDMUND BECKETT, Bart., LL.D., F.R.A.S., Author of "Clocks and Watches, and Bells," &c. Second Edition, Enlarged. Fcap. 8vo, 5s. cloth.

"A book which is always amusing and nearly always instructive. The style throughout is in the highest degree condensed and epigrammatic."—*Times.*

Ventilation of Buildings.

VENTILATION. A Text Book to the Practice of the Art of Ventilating Buildings. With a Chapter upon Air Testing. By W. P. BUCHAN, R.P., Sanitary and Ventilating Engineer, Author of "Plumbing," &c. With 170 Illustrations. 12mo, 4s. cloth boards. [*Just published.*

The Art of Plumbing.

PLUMBING. A Text Book to the Practice of the Art or Craft of the Plumber, with Supplementary Chapters on House Drainage, embodying the latest Improvements. By WILLIAM PATON BUCHAN, R.P., Sanitary Engineer and Practical Plumber. Fifth Edition, Enlarged to 370 pages, and 380 Illustrations. 12mo, 4s. cloth boards.

"A text book which may be safely put in the hands of every young plumber, and which will also be found useful by architects and medical professors."—*Builder.*

"A valuable text book, and the only treatise which can be regarded as a really reliable manual of the plumber's art."—*Building News.*

Geometry for the Architect, Engineer, etc.

PRACTICAL GEOMETRY, for the Architect, Engineer and Mechanic. Giving Rules for the Delineation and Application of various Geometrical Lines, Figures and Curves. By E. W. TARN, M.A., Architect, Author of "The Science of Building," &c. Second Edition. With 172 Illustrations, demy 8vo, 9s. cloth.

"No book with the same objects in view has ever been published in which the clearness of the rules laid down and the illustrative diagrams have been so satisfactory."—*Scotsman.*

The Science of Geometry.

THE GEOMETRY OF COMPASSES; or, Problems Resolved by the mere Description of Circles, and the use of Coloured Diagrams and Symbols. By OLIVER BYRNE. Coloured Plates. Crown 8vo, 3s. 6d. cloth.

"The treatise is a good one, and remarkable—like all Mr. Byrne's contributions to the science of geometry—for the lucid character of its teaching."—*Building News.*

DECORATIVE ARTS, etc.

Woods and Marbles (Imitation of).

SCHOOL OF PAINTING FOR THE IMITATION OF WOODS AND MARBLES, as Taught and Practised by A. R. VAN DER BURG and P. VAN DER BURG, Directors of the Rotterdam Painting Institution. Royal folio, 18¼ by 12¼ in., Illustrated with 24 full-size Coloured Plates; also 12 plain Plates, comprising 154 Figures. Second and Cheaper Edition. Price £1 11s. 6d.

List of Plates.

1. Various Tools required for Wood Painting—2, 3. Walnut: Preliminary Stages of Graining and Finished Specimen—4. Tools used for Marble Painting and Method of Manipulation—5, 6. St. Remi Marble: Earlier Operations and Finished Specimen—7. Methods of Sketching different Grains, Knots, &c.—8, 9. Ash: Preliminary Stages and Finished Specimen—10. Methods of Sketching Marble Grains—11, 12. Breche Marble: Preliminary Stages of Working and Finished Specimen—13. Maple: Methods of Producing the different Grains—14, 15. Bird's-eye Maple: Preliminary Stages and Finished Specimen—16. Methods of Sketching the different Species of White Marble—17, 18. White Marble: Preliminary Stages of Process and Finished Specimen—19. Mahogany: Specimens of various Grains and Methods of Manipulation—20, 21. Mahogany: Earlier Stages and Finished Specimen—22, 23, 24. Sienna Marble: Varieties of Grain, Preliminary Stages and Finished Specimen—25, 26, 27. Juniper Wood: Methods of producing Grain, &c.; Preliminary Stages and Finished Specimen—28, 29, 30. Vert de Mer Marble: Varieties of Grain and Methods of Working Unfinished and Finished Specimens—31, 32, 33. Oak: Varieties of Grain, Tools Employed, and Methods of Manipulation, Preliminary Stages and Finished Specimen—34, 35, 36. Waulsort Marble: Varieties of Grain, Unfinished and Finished Specimens.

. OPINIONS OF THE PRESS.

"Those who desire to attain skill in the art of painting woods and marbles will find advantage in consulting this book. . . . Some of the Working Men's Clubs should give their young men the opportunity to study it."—*Builder.*

"A comprehensive guide to the art. The explanations of the processes, the manipulation and management of the colours, and the beautifully executed plates will not be the least valuable to the student who aims at making his work a faithful transcript of nature."—*Building News.*

House Decoration.

ELEMENTARY DECORATION. A Guide to the Simpler Forms of Everyday Art, as applied to the Interior and Exterior Decoration of Dwelling Houses, &c. By JAMES W. FACEY, Jun. With 68 Cuts. 12mo, 2s. cloth limp.

PRACTICAL HOUSE DECORATION : A Guide to the Art of Ornamental Painting, the Arrangement of Colours in Apartments, and the principles of Decorative Design. With some Remarks upon the Nature and Properties of Pigments. By JAMES WILLIAM FACEY, Author of " Elementary Decoration," &c. With numerous Illustrations. 12mo, 2s. 6d. cloth limp.

N.B.—The above Two Works together in One Vol., strongly half-bound, 5s.

Colour.

A GRAMMAR OF COLOURING. Applied to Decorative Painting and the Arts. By GEORGE FIELD. New Edition, Revised, Enlarged, and adapted to the use of the Ornamental Painter and Designer. By ELLIS A. DAVIDSON. With New Coloured Diagrams and Engravings. 12mo, 3s. 6d. cloth boards.

" The book is a most useful *resume* of the properties of pigments."—*Builder.*

House Painting, Graining, etc.

HOUSE PAINTING, GRAINING, MARBLING, AND SIGN WRITING, A Practical Manual of. By ELLIS A. DAVIDSON. Sixth Edition. With Coloured Plates and Wood Engravings. 12mo, 6s. cloth boards.

" A mass of information, of use to the amateur and of value to the practical man."—*English Mechanic.*
" Simply invaluable to the youngster entering upon this particular calling, and highly serviceable to the man who is practising it."—*Furniture Gazette.*

Decorators, Receipts for.

THE DECORATOR'S ASSISTANT : A Modern Guide to Decorative Artists and Amateurs, Painters, Writers, Gilders, &c. Containing upwards of 600 Receipts, Rules and Instructions ; with a variety of Information for General Work connected with every Class of Interior and Exterior Decorations, &c. Fourth Edition, Revised. 152 pp., crown 8vo, 1s. in wrapper.

" Full of receipts of value to decorators, painters, gilders, &c. The book contains the gist of larger treatises on colour and technical processes. It would be difficult to meet with a work so full of varied information on the painter's art."—*Building News.*
" We recommend the work to all who, whether for pleasure or profit, require a guide to decoration."—*Plumber and Decorator.*

Moyr Smith on Interior Decoration.

ORNAMENTAL INTERIORS, ANCIENT AND MODERN. By J. MOYR SMITH. Super-royal 8vo, with 32 full-page Plates and numerous smaller Illustrations, handsomely bound in cloth, gilt top, price 18s.

" The book is well illustrated and handsomely got up, and contains some true criticism and a good many good examples of decorative treatment."—*The Builder.*
" This is the most elaborate and beautiful work on the artistic decoration of interiors that we have seen. . . . The scrolls, panels and other designs from the author's own pen are very beautiful and chaste ; but he takes care that the designs of other men shall figure even more than his own."—*Liverpool Albion.*
" To all who take an interest in elaborate domestic ornament this handsome volume will be welcome."—*Graphic.*

British and Foreign Marbles.

MARBLE DECORATION and the Terminology of British and Foreign Marbles. A Handbook for Students. By GEORGE H. BLAGROVE, Author of " Shoring and its Application," &c. With 28 Illustrations. Crown 8vo, 3s. 6d. cloth.

" This most useful and much wanted handbook should be in the hands of every architect and builder."—*Building World.*
" It is an excellent manual for students, and interesting to artistic readers generally."—*Saturday Review.*
" A carefully and usefully written treatise ; the work is essentially practical."—*Scotsman.*

Marble Working, etc.

MARBLE AND MARBLE WORKERS : A Handbook for Architects, Artists, Masons and Students. By ARTHUR LEE, Author of " A Visit to Carrara," " The Working of Marble," &c. Small crown 8vo, 2s. cloth.

" A really valuable addition to the technical literature of architects and masons."—*Building News.*

DELAMOTTE'S WORKS ON ILLUMINATION AND ALPHABETS.

A PRIMER OF THE ART OF ILLUMINATION, for the Use of Beginners: with a Rudimentary Treatise on the Art, Practical Directions for its exercise, and Examples taken from Illuminated MSS., printed in Gold and Colours. By F. DELAMOTTE. New and Cheaper Edition. Small 4to, 6s. ornamental boards.

"The examples of ancient MSS. recommended to the student, which, with much good sense, the author chooses from collections accessible to all, are selected with judgment and knowledge, as well as taste."—*Athenæum.*

ORNAMENTAL ALPHABETS, Ancient and Mediæval, from the Eighth Century, with Numerals; including Gothic, Church-Text, large and small, German, ·Italian, Arabesque, Initials for Illumination, Monograms, Crosses, &c. &c., for the use of Architectural and Engineering Draughtsmen, Missal Painters, Masons, Decorative Painters, Lithographers, Engravers, Carvers, &c. &c. Collected and Engraved by F. DELAMOTTE, and printed in Colours. New and Cheaper Edition. Royal 8vo, oblong, 2s. 6d. ornamental boards.

"For those who insert enamelled sentences round gilded chalices, who blazon shop legends over shop-doors, who letter church walls with pithy sentences from the Decalogue, this book will be useful."—*Athenæum.*

EXAMPLES OF MODERN ALPHABETS, Plain and Ornamental; including German, Old English, Saxon, Italic, Perspective, Greek, Hebrew, Court Hand, Engrossing, Tuscan, Riband, Gothic, Rustic, and Arabesque; with several Original Designs, and an Analysis of the Roman and Old English Alphabets, large and small, and Numerals, for the use of Draughtsmen, Surveyors, Masons, Decorative Painters, Lithographers, Engravers, Carvers, &c. Collected and Engraved by F. DELAMOTTE, and printed in Colours. New and Cheaper Edition. Royal 8vo, oblong, 2s. 6d. ornamental boards.

"There is comprised in it every possible shape into which the letters of the alphabet and numerals can be formed, and the talent which has been expended in the conception of the various plain and ornamental letters is wonderful."—*Standard.*

MEDIÆVAL ALPHABETS AND INITIALS FOR ILLUMI-NATORS. By F. G. DELAMOTTE. Containing 21 Plates and Illuminated Title, printed in Gold and Colours. With an Introduction by J. WILLIS BROOKS. Fourth and Cheaper Edition. Small 4to, 4s. ornamental boards.

" A volume in which the letters of the alphabet come forth glorified in gilding and all the colours of the prism interwoven and intertwined and intermingled."—*Sun.*

THE EMBROIDERER'S BOOK OF DESIGN. Containing Initials, Emblems, Cyphers, Monograms, Ornamental Borders, Ecclesiastical Devices, Mediæval and Modern Alphabets, and National Emblems. Collected by F. DELAMOTTE, and printed in Colours. Oblong royal 8vo, 1s. 6d. ornamental wrapper.

"The book will be of great assistance to ladies and young children who are endowed with the art of plying the needle in this most ornamental and useful pretty work."—*East Anglian Times.*

Wood Carving.

INSTRUCTIONS IN WOOD-CARVING, for Amateurs; with Hints on Design. By A LADY. With Ten Plates. New and Cheaper Edition. Crown 8vo, 2s. in emblematic wrapper.

"The handicraft of the wood-carver, so well as a book can impart it, may be learnt from ' A Lady's ' publication."—*Athenæum.*
"The directions given are plain and easily understood."—*English Mechanic.*

Glass Painting.

GLASS STAINING AND THE ART OF PAINTING ON GLASS. From the German of Dr. GESSERT and EMANUEL OTTO FROMBERG. With an Appendix on THE ART OF ENAMELLING. 12mo, 2s. 6d. cloth limp.

Letter Painting.

THE ART OF LETTER PAINTING MADE EASY. By JAMES GREIG BADENOCH. With 12 full-page Engravings of Examples, 1s. 6d. cloth limp.

"The system is a simple one, but quite original, and well worth the careful attention of letter painters. It can be easily mastered and remembered."—*Building News.*

CARPENTRY, TIMBER, etc.

Tredgold's Carpentry, Revised & Enlarged by Tarn.

THE ELEMENTARY PRINCIPLES OF CARPENTRY.
A Treatise on the Pressure and Equilibrium of Timber Framing, the Resistance of Timber, and the Construction of Floors, Arches, Bridges, Roofs,
Uniting Iron and Stone with Timber, &c. To which is added an Essay
on the Nature and Properties of Timber, &c., with Descriptions of the kinds
of Wood used in Building; also numerous Tables of the Scantlings of Timber for different purposes, the Specific Gravities of Materials, &c. By THOMAS
TREDGOLD, C.E. With an Appendix of Specimens of Various Roofs of Iron
and Stone, Illustrated. Seventh Edition, thoroughly revised and considerably
enlarged by E. WYNDHAM TARN, M.A., Author of "The Science of Building," &c. With 61 Plates, Portrait of the Author, and several Woodcuts. In
one large vol., 4to, price £1 5s. cloth.
"Ought to be in every architect's and every builder's library."—*Builder.*
"A work whose monumental excellence must commend it wherever skilful carpentry is concerned. The author's principles are rather confirmed than impaired by time. The additional
plates are of great intrinsic value."—*Building News.*

Woodworking Machinery.

WOODWORKING MACHINERY: Its Rise, Progress, and
Construction. With Hints on the Management of Saw Mills and the Economical Conversion of Timber. Illustrated, with Examples of Recent Designs by
leading English, French, and American Engineers. By M. POWIS BALE,
A.M.Inst.C.E., M.I.M.E. Large crown 8vo, 12s. 6d. cloth.
"Mr. Bale is evidently an expert on the subject and he has collected so much information that
his book is all-sufficient for builders and others engaged in the conversion of timber."—*Architect.*
"The most comprehensive compendium of wood-working machinery we have seen. The
author is a thorough master of his subject."—*Building News.*
"The appearance of this book at the present time will, we should think, give a considerable
impetus to the onward march of the machinist engaged in the designing and manufacture of
wood-working machines. It should be in the office of every wood-working factory."—*English
Mechanic.*

Saw Mills.

SAW MILLS: Their Arrangement and Management, and the
Economical Conversion of Timber. (A Companion Volume to "Woodworking Machinery.") By M. POWIS BALE. With numerous Illustrations. Crown
8vo, 10s. 6d. cloth.
"The *administration* of a large sawing establishment is discussed, and the subject examined
from a financial standpoint. We could not desire a more complete or practical treatise."—*Builder.*
"We highly recommend Mr. Bale's work to the attention and perusal of all those who are engaged in the art of wood conversion, or who are about building or remodelling saw-mills on improved principles."—*Building News.*

Carpentering.

THE CARPENTER'S NEW GUIDE; or, Book of Lines for Car-
penters; comprising all the Elementary Principles essential for acquiring a
knowledge of Carpentry. Founded on the late PETER NICHOLSON's Standard
Work. A New Edition, Revised by ARTHUR ASHPITEL, F.S.A. Together
with Practical Rules on Drawing, by GEORGE PYNE. With 74 Plates,
4to, £1 1s. cloth.

Handrailing and Stairbuilding.

A PRACTICAL TREATISE ON HANDRAILING: Showing
New and Simple Methods for Finding the Pitch of the Plank, Drawing the
Moulds, Bevelling, Jointing-up, and Squaring the Wreath. By GEORGE
COLLINGS. Second Edition, Revised and Enlarged, to which is added A
TREATISE ON STAIRBUILDING. With Plates and Diagrams. 12mo, 2s. 6d.
cloth limp.
"Will be found of practical utility in the execution of this difficult branch of joinery."—*Builder.*
"Almost every difficult phase of this somewhat intricate branch of joinery is elucidated by the
aid of plates and explanatory letterpress."—*Furniture Gazette.*

Circular Work.

CIRCULAR WORK IN CARPENTRY AND JOINERY: A
Practical Treatise on Circular Work of Single and Double Curvature. By
GEORGE COLLINGS, Author of "A Practical Treatise on Handrailing." Illus-
trated with numerous Diagrams. Second Edition. 12mo, 2s. 6d. cloth limp.
"An excellent example of what a book of this kind should be. Cheap in price, clear in defini-
tion and practical in the examples selected."—*Builder.*

Timber Merchant's Companion.

THE TIMBER MERCHANT'S AND BUILDER'S COM-
PANION. Containing New and Copious Tables of the Reduced Weight and
Measurement of Deals and Battens, of all sizes, from One to a Thousand
Pieces, and the relative Price that each size bears per Lineal Foot to any
given Price per Petersburg Standard Hundred; the Price per Cube Foot of
Square Timber to any given Price per Load of 50 Feet; the proportionate
Value of Deals and Battens by the Standard, to Square Timber by the Load
of 50 Feet; the readiest mode of ascertaining the Price of Scantling per
Lineal Foot of any size, to any given Figure per Cube Foot, &c. &c.   By
WILLIAM DOWSING. Fourth Edition, Revised and Corrected. Cr. 8vo, 3s. cl.
"We are glad to see a fourth edition of these admirable tables, which for correctness and
simplicity of arrangement leave nothing to be desired."—*Timber Trades Journal.*
"An exceedingly well-arranged, clear, and concise manual of tables for the use of all who buy
or sell timber."—*Journal of Forestry.*

Practical Timber Merchant.

THE PRACTICAL TIMBER MERCHANT. Being a Guide
for the use of Building Contractors, Surveyors, Builders, &c., comprising
useful Tables for all purposes connected with the Timber Trade, Marks of
Wood, Essay on the Strength of Timber, Remarks on the Growth of Timber,
&c. By W. RICHARDSON. Fcap. 8vo, 3s. 6d. cloth.
"This handy manual contains much valuable information for the use of timber merchants,
builders, foresters, and all others connected with the growth, sale, and manufacture of timber."—
Journal of Forestry.

Timber Freight Book.

THE TIMBER MERCHANT'S, SAW MILLER'S, AND
IMPORTER'S FREIGHT BOOK AND ASSISTANT. Comprising Rules,
Tables, and Memoranda relating to the Timber Trade. By WILLIAM
RICHARDSON, Timber Broker; together with a Chapter on "SPEEDS OF SAW
MILL MACHINERY," by M. POWIS BALE, M.I.M.E., &c. 12mo, 3s. 6d. cl. boards.
"A very useful manual of rules, tables, and memoranda relating to the timber trade. We re-
commend it as a compendium of calculation to all timber measurers and merchants, and as supply-
ing a real want in the trade."—*Building News.*

Packing-Case Makers, Tables for.

PACKING-CASE TABLES; showing the number of Super-
ficial Feet in Boxes or Packing-Cases, from six inches square and upwards.
By W. RICHARDSON, Timber Broker. Third Edition. Oblong 4to, 3s. 6d. cl.
"Invaluable labour-saving tables."—*Ironmonger.*
"Will save much labour and calculation."—*Grocer.*

Superficial Measurement.

THE TRADESMAN'S GUIDE TO SUPERFICIAL MEA-
SUREMENT. Tables calculated from 1 to 200 inches in length, by 1 to 108
inches in breadth. For the use of Architects, Surveyors, Engineers, Timber
Merchants, Builders, &c. By JAMES HAWKINGS. Third Edition. Fcap.,
3s. 6d. cloth.
"A useful collection of tables to facilitate rapid calculation of surfaces. The exact area of any
surface of which the limits have been ascertained can be instantly determined. The book will be
found of the greatest utility to all engaged in building operations."—*Scotsman.*
"These tables will be found of great assistance to all who require to make calculations in superficial measurement."—*English Mechanic.*

Forestry.

THE ELEMENTS OF FORESTRY. Designed to afford In-
formation concerning the Planting and Care of Forest Trees for Ornament or
Profit, with Suggestions upon the Creation and Care of Woodlands. By F. B.
HOUGH. Large crown 8vo, 10s. cloth.

Timber Importer's Guide.

THE TIMBER IMPORTER'S, TIMBER MERCHANT'S AND
BUILDER'S STANDARD GUIDE. By RICHARD E. GRANDY. Compris-
ing an Analysis of Deal Standards, Home and Foreign, with Comparative
Values and Tabular Arrangements for fixing Nett Landed Cost on Baltic
and North American Deals, including all intermediate Expenses, Freight,
Insurance, &c. &c. Together with copious Information for the Retailer and
Builder. Third Edition, Revised. 12mo, 2s. cloth limp.
"Everything it pretends to be: built up gradually, it leads one from a forest to a treenail, and
throws in, as a makeweight, a host of material concerning bricks, columns, cisterns, &c."—*English
Mechanic.*

MARINE ENGINEERING, NAVIGATION, etc.

Chain Cables.

CHAIN CABLES AND CHAINS. Comprising Sizes and Curves of Links, Studs, &c., Iron for Cables and Chains, Chain Cable and Chain Making, Forming and Welding Links, Strength of Cables and Chains, Certificates for Cables, Marking Cables, Prices of Chain Cables and Chains, Historical Notes, Acts of Parliament, Statutory Tests, Charges for Testing, List of Manufacturers of Cables, &c. &c. By THOMAS W. TRAILL, F.E.R.N., M. Inst. C.E., Engineer Surveyor in Chief, Board of Trade, Inspector of Chain Cable and Anchor Proving Establishments, and General Superintendent, Lloyd's Committee on Proving Establishments. With numerous Tables, Illustrations and Lithographic Drawings. Folio, £2 2s. cloth, bevelled boards.

"It contains a vast amount of valuable information. Nothing seems to be wanting to make it a complete and standard work of reference on the subject."—*Nautical Magazine.*

Marine Engineering.

MARINE ENGINES AND STEAM VESSELS (A Treatise on). By ROBERT MURRAY, C.E. Eighth Edition, thoroughly Revised, with considerable Additions by the Author and by GEORGE CARLISLE, C.E., Senior Surveyor to the Board of Trade at Liverpool. 12mo, 5s. cloth boards.

"Well adapted to give the young steamship engineer or marine engine and boiler maker a general introduction into his practical work."—*Mechanical World.*

"We feel sure that this thoroughly revised edition will continue to be as popular in the future as it has been in the past, as, for its size, it contains more useful information than any similar treatise."—*Industries.*

"The information given is both sound and sensible, and well qualified to direct young sea-going hands on the straight road to the extra chief's certificate. Most useful to surveyors, inspectors, draughtsmen, and all young engineers who take an interest in their profession."—*Glasgow Herald.*

"An indispensable manual for the student of marine engineering."—*Liverpool Mercury.*

Pocket-Book for Naval Architects and Shipbuilders.

THE NAVAL ARCHITECT'S AND SHIPBUILDER'S POCKET-BOOK of Formulæ, Rules, and Tables, and MARINE ENGINEER'S AND SURVEYOR'S Handy Book of Reference. By CLEMENT MACKROW, Member of the Institution of Naval Architects, Naval Draughtsman. Fourth Edition, Revised. With numerous Diagrams, &c. Fcap., 12s. 6d. strongly bound in leather.

"Will be found to contain the most useful tables and formulæ required by shipbuilders, carefully collected from the best authorities, and put together in a popular and simple form."—*Engineer.*

"The professional shipbuilder has now, in a convenient and accessible form, reliable data for solving many of the numerous problems that present themselves in the course of his work."—*Iron.*

"There is scarcely a subject on which a naval architect or shipbuilder can require to refresh his memory which will not be found within the covers of Mr. Mackrow's book."—*English Mechanic.*

Pocket-Book for Marine Engineers.

A POCKET-BOOK OF USEFUL TABLES AND FOR-MULÆ FOR MARINE ENGINEERS. By FRANK PROCTOR, A.I.N.A. Third Edition. Royal 32mo, leather, gilt edges, with strap, 4s.

"We recommend it to our readers as going far to supply a long-felt want."—*Naval Science.*

"A most useful companion to all marine engineers."—*United Service Gazette.*

Introduction to Marine Engineering.

ELEMENTARY ENGINEERING: A Manual for Young Marine Engineers and Apprentices. In the Form of Questions and Answers on Metals, Alloys, Strength of Materials, Construction and Management of Marine Engines and Boilers, Geometry, &c. &c. With an Appendix of Useful Tables. By JOHN SHERREN BREWER, Government Marine Surveyor, Hong-kong. Small crown 8vo, 2s. cloth.

"Contains much valuable information for the class for whom it is intended, especially in the chapters on the management of boilers and eng nes."—*Nautical Magazine.*

"A useful introduction to the more elaborate text books."—*Scotsman.*

"To a student who has the requisite desire and resolve to attain a thorough knowledge, Mr. Brewer offers decidedly useful help."—*Athenæum.*

Navigation.

PRACTICAL NAVIGATION. Consisting of THE SAILOR'S SEA-BOOK, by JAMES GREENWOOD and W. H. ROSSER: together with the requisite Mathematical and Nautical Tables for the Working of the Problems, by HENRY LAW, C.E., and Professor J. R. YOUNG. Illustrated. 12mo, 7s. strongly half-bound.

MINING AND METALLURGY.

Metalliferous Mining in the United Kingdom.

BRITISH MINING: A Treatise on the History, Discovery, Practical Development, and Future Prospects of Metalliferous Mines in the United Kingdom. By ROBERT HUNT, F.R.S., Keeper of Mining Records; Editor of "Ure's Dictionary of Arts, Manufactures, and Mines," &c. Upwards of 950 pp., with 230 Illustrations. Second Edition, Revised. Super-royal 8vo, £2 2s. cloth.

"One of the most valuable works of reference of modern times. Mr. Hunt, as keeper of mining records of the United Kingdom, has had opportunities for such a task not enjoyed by anyone else, and has evidently made the most of them. . . . The language and style adopted are good, and the treatment of the various subjects laborious, conscientious, and scientific."—*Engineering.*

"The book is, in fact, a treasure-house of statistical information on mining subjects, and we know of no other work embodying so great a mass of matter of this kind. Were this the only merit of Mr. Hunt s volume, it would be sufficient to render it indispensable in this library of everyone interested in the development of the mining and metallurgical industries of this country."—*Athenæum.*

"A mass of information not elsewhere available, and of the greatest value to those who may be interested in our great mineral industries."—*Engineer.*

"A sound, business-like collection of interesting facts. . . . The amount of information Mr. Hunt has brought together is enormous. . . . The volume appears likely to convey more instruction upon the subject than any work hitherto published."—*Mining Journal.*

Colliery Management.

THE COLLIERY MANAGER'S HANDBOOK: A Comprehensive Treatise on the Laying-out and Working of Collieries, Designed as a Book of Reference for Colliery Managers, and for the Use of Coal-Mining Students preparing for First-class Certificates. By CALEB PAMELY, Mining Engineer and Surveyor; Member of the North of England Institute of Mining and Mechanical Engineers; and Member of the South Wales Institute of Mining Engineers. With nearly 500 Plans, Diagrams, and other Illustrations. Medium 8vo, about 600 pages. Price £1 5s. strongly bound.

[*Just published.*

Coal and Iron.

THE COAL AND IRON INDUSTRIES OF THE UNITED KINGDOM. Comprising a Description of the Coal Fields, and of the Principal Seams of Coal, with Returns of their Produce and its Distribution, and Analyses of Special Varieties. Also an Account of the occurrence of Iron Ores in Veins or Seams; Analyses of each Variety; and a History of the Rise and Progress of Pig Iron Manufacture. By RICHARD MEADE, Assistant Keeper of Mining Records. With Maps. 8vo, £1 8s. cloth.

"The book is one which must find a place on the shelves of all interested in coal and iron production, and in the iron, steel, and other metallurgical industries."—*Engineer.*

"Of this book we may unreservedly say that it is the best of its class which we have ever met. . . . A book of reference which no one engaged in the iron or coal trades should omit from his library."—*Iron and Coal Trades Review.*

Prospecting for Gold and other Metals.

THE PROSPECTOR'S HANDBOOK: A Guide for the Prospector and Traveller in Search of Metal-Bearing or other Valuable Minerals. By J. W. ANDERSON, M.A. (Camb.), F.R.G.S., Author of "Fiji and New Caledonia." Fifth Edition, thoroughly Revised and Enlarged. Small crown 8vo, 3s. 6d. cloth.

"Will supply a much felt want, especially among Colonists, in whose way are so often thrown many mineralogical specimens the value of which it is difficult to determine."—*Engineer.*

"How to find commercial minerals, and how to identify them when they are found, are the leading points to which attention is directed. The author has managed to pack as much practical detail into his pages as would supply material for a book three times its size."—*Mining Journal.*

Mining Notes and Formulæ.

NOTES AND FORMULÆ FOR MINING STUDENTS. By JOHN HERMAN MERIVALE, M.A., Certificated Colliery Manager, Professor of Mining in the Durham College of Science, Newcastle-upon-Tyne. Third Edition, Revised and Enlarged. Small crown 8vo, 2s. 6d. cloth.

"Invaluable to anyone who is working up for an examination on mining subjects."—*Coal and Iron Trades Review.*

"The author has done his work in an exceedingly creditable manner, and has produced a book that will be of service to students, and those who are practically engaged in mining operations."—*Engineer.*

"A vast amount of technical matter of the utmost value to mining engineers, and of considerable interest to students."—*Schoolmaster.*

Explosives.

A HANDBOOK ON MODERN EXPLOSIVES. Being a Practical Treatise on the Manufacture and Application of Dynamite, Gun-Cotton, Nitro-Glycerine and other Explosive Compounds. Including the Manufacture of Collodion-Cotton. By M. EISSLER, Mining Engineer and Metallurgical Chemist, Author of "The Metallurgy of Gold," &c. With about 100 Illustrations. Crown 8vo, 10s. 6d. cloth.

"Useful not only to the miner, but also to officers of both services to whom blasting and the use of explosives generally may at any time become a necessary auxiliary."—*Nature.*

"A veritable mine of information on the subject of explosives employed for military, mining and blasting purposes."—*Army and Navy Gazette.*

"The book is clearly written. Taken as a whole, we consider it an excellent little book and one that should be found of great service to miners and others who are engaged in work requiring the use of explosives."—*Athenæum.*

Gold, Metallurgy of.

THE METALLURGY OF GOLD: A Practical Treatise on the Metallurgical Treatment of Gold-bearing Ores. Including the Processes of Concentration and Chlorination, and the Assaying, Melting and Refining of Gold. By M. EISSLER, Mining Engineer and Metallurgical Chemist, formerly Assistant Assayer of the U. S. Mint, San Francisco. Third Edition, Revised and greatly Enlarged. With 187 Illustrations. Crown 8vo, 12s. 6d. cloth.

"This book thoroughly deserves its title of a 'Practical Treatise.' The whole process of gold milling, from the breaking of the quartz to the assay of the bullion, is described in clear and orderly narrative and with much, but not too much, fulness of detail."—*Saturday Review.*

"The work is a storehouse of information and valuable data, and we strongly recommend it to all professional men engaged in the gold-mining industry."—*Mining Journal.*

Silver, Metallurgy of.

THE METALLURGY OF SILVER: A Practical Treatise on the Amalgamation, Roasting and Lixiviation of Silver Ores. Including the Assaying, Melting and Refining of Silver Bullion. By M. EISSLER, Author of "The Metallurgy of Gold" Second Edition, Enlarged. With 150 Illustrations. Crown 8vo, 10s. 6d. cloth. [*Just published.*

"A practical treatise, and a technical work which we are convinced will supply a long-felt want amongst practical men, and at the same time be of value to students and others indirectly connected with the industries."—*Mining Journal.*

"From first to last the book is thoroughly sound and reliable."—*Colliery Guardian.*

"For chemists, practical miners, assayers and investors alike, we do not know of any work on the subject so handy and yet so comprehensive."—*Glasgow Herald.*

Silver-Lead, Metallurgy of.

THE METALLURGY OF ARGENTIFEROUS LEAD: A Practical Treatise on the Smelting of Silver-Lead Ores and the Refining of Lead Bullion. Including Reports on various Smelting Establishments and Descriptions of Modern Furnaces and Plants in Europe and America. By M. EISSLER, M.E., Author of "The Metallurgy of Gold," &c. Crown 8vo. 400 pp., with numerous Illustrations, 12s. 6d. cloth. [*Just published.*

Metalliferous Minerals and Mining.

TREATISE ON METALLIFEROUS MINERALS AND MINING. By D. C. DAVIES, F.G.S., Mining Engineer, &c., Author of "A Treatise on Slate and Slate Quarrying." Illustrated with numerous Wood Engravings. Fourth Edition, carefully Revised. Crown 8vo, 12s. 6d. cloth.

"Neither the practical miner nor the general reader interested in mines can have a better book for his companion and his guide."—*Mining Journal.* [*Mining World.*

"We are doing our readers a service in calling their attention to this valuable work."—

"As a history of the present state of mining throughout the world this book has a real value, and it supplies an actual want."—*Athenæum.*

Earthy Minerals and Mining.

A TREATISE ON EARTHY & OTHER MINERALS AND MINING. By D. C. DAVIES, F.G.S. Uniform with, and forming a Companion Volume to, the same Author's "Metalliferous Minerals and Mining." With 76 Wood Engravings. Second Edition. Crown 8vo, 12s. 6d. cloth.

"We do not remember to have met with any English work on mining matters that contains the same amount of information packed in equally convenient form."—*Academy.*

"We should be inclined to rank it as among the very best of the handy technical and trades manuals which have recently appeared."—*British Quarterly Review.*

Mineral Surveying and Valuing.

THE MINERAL SURVEYOR AND VALUER'S COMPLETE
GUIDE, comprising a Treatise on Improved Mining Surveying and the Valuation of Mining Properties, with New Traverse Tables. By WM. LINTERN,
Mining and Civil Engineer. Third Edition, with an Appendix on "Magnetic
and Angular Surveying," with Records of the Peculiarities of Needle Disturbances. With Four Plates of Diagrams, Plans, &c. 12mo, 4s. cloth.

"Mr. Lintern's book forms a valuable and thoroughly trustworthy guide."—*Iron and Coal
Trades Review.*
"This new edition must be of the highest value to colliery surveyors, proprietors and managers."—*Colliery Guardian.*

Asbestos and its Uses.

ASBESTOS : Its Properties, Occurrence and Uses. With some
Account of the Mines of Italy and Canada. By ROBERT H. JONES. With
Eight Collotype Plates and other Illustrations. Crown 8vo, 12s. 6d. cloth.

"An interesting and invaluable work."—*Colliery Guardian.*
"We counsel our readers to get this exceedingly interesting work for themselves; they will
find in it much that is suggestive, and a great deal that is of immediate and practical usefulness."—
Builder.
"A valuable addition to the architect's and engineer's library."—*Building News.*

Underground Pumping Machinery.

MINE DRAINAGE. Being a Complete and Practical Treatise
on Direct-Acting Underground Steam Pumping Machinery, with a Description of a large number of the best known Engines, their General Utility and
the Special Sphere of their Action, the Mode of their Application, and
their merits compared with other forms of Pumping Machinery. By STEPHEN
MICHELL. 8vo, 15s. cloth.

"Will be highly esteemed by colliery owners and lessees, mining engineers, and students
generally who require to be acquainted with the best means of securing the drainage of mines. It
is a most valuable work, and stands almost alone in the literature of steam pumping machinery."—
Colliery Guardian.
"Much valuable information is given, so that the book is thoroughly worthy of an extensive
circulation amongst practical men and purchasers of machinery."—*Mining Journal.*

Mining Tools.

A MANUAL OF MINING TOOLS. For the Use of Mine
Managers, Agents, Students, &c. By WILLIAM MORGANS, Lecturer on Practical Mining at the Bristol School of Mines. 12mo, 2s. 6d. cloth limp.

ATLAS OF ENGRAVINGS to Illustrate the above, containing 235 Illustrations of Mining Tools, drawn to scale. 4to, 4s. 6d. cloth.

"Students in the science of mining, and overmen, captains, managers, and viewers may gain
practical knowledge and useful hints by the study of Mr. Morgans' manual."—*Colliery Guardian.*
"A valuable work, which will tend materially to improve our mining literature."—*Mining
Journal.*

Coal Mining.

COAL AND COAL MINING : A Rudimentary Treatise on. By
the late Sir WARINGTON W. SMYTH, M.A., F.R.S., &c., Chief Inspector of the
Mines of the Crown. Seventh Edition, Revised and Enlarged. With
numerous Illustrations. 12mo, 4s. cloth boards.

"As an outline is given of every known coal-field in this and other countries, as well as of the
principal methods of working, the book will doubtless interest a very large number of readers."—
Mining Journal.

Subterraneous Surveying.

SUBTERRANEOUS SURVEYING, Elementary and Practical
Treatise on, with and without the Magnetic Needle. By THOMAS FENWICK,
Surveyor of Mines, and THOMAS BAKER, C.E. Illust. 12mo, 3s. cloth boards.

Granite Quarrying.

GRANITES AND OUR GRANITE INDUSTRIES. By
GEORGE F. HARRIS, F.G.S., Membre de la Société Belge de Géologie, Lecturer on Economic Geology at the Birkbeck Institution, &c. With Illustrations. Crown 8vo, 2s. 6d. cloth.

"A clearly and well-written manual for persons engaged or interested in the granite industry."
—*Scotsman.*
"An interesting work, which will be deservedly esteemed."—*Colliery Guardian.*
"An exceedingly interesting and valuable monograph on a subject which has hitherto received
unaccountably little attention in the shape of systematic literary treatment."—*Scottish Leader.*

ELECTRICITY, ELECTRICAL ENGINEERING, etc.

Electrical Engineering.

THE ELECTRICAL ENGINEER'S POCKET-BOOK OF MODERN RULES, FORMULÆ, TABLES AND DATA. By H. R. KEMPE, M.Inst.E.E., A.M.Inst C.E., Technical Officer Postal Telegraphs, Author of "A Handbook of Electrical Testing," &c. With numerous Illustrations, royal 32mo, oblong, 5s. leather. [*Just published.*

"There is very little in the shape of formulæ or data which the electrician is likely to want in a hurry which cannot be found in its pages."—*Practical Engineer.*

"A very useful book of reference for daily use in practical electrical engineering and its various applications to the industries of the present day."—*Iron.*

"It is the best book of its kind."—*Electrical Engineer.*

"The Electrical Engineer's Pocket-Book is a good one."—*Electrician.*

"Strongly recommended to those engaged in the various electrical industries."—*Electrical Review.*

Electric Lighting.

ELECTRIC LIGHT FITTING: A Handbook for Working Electrical Engineers, embodying Practical Notes on Installation Management. By JOHN W. URQUHART, Electrician, Author of "Electric Light," &c. With numerous Illustrations, crown 8vo, 5s. cloth. [*Just published.*

"This volume deals with what may be termed the mechanics of electric lighting, and is addressed to men who are already engaged in the work or are training for it. The work traverses a great deal of ground, and may be read as a sequel to the same author's useful work on 'Electric Light.'"—*Electrician.*

"This is an attempt to state in the simplest language the precautions which should be adopted in instal ing the electric light, and to give information, for the guidance of those who have to run the plant when installed. The book is well worth the perusal of the workmen for whom it is written."—*Electrical Review.*

"Eminently practical and useful. . . . Ought to be in the hands of everyone in charge of an electric light plant."—*Electrical Engineer.*

"A really capital book, which we have no hesitation in recommending to the notice of working electricians and electrical engineers."—*Mechanical World.*

Electric Light.

ELECTRIC LIGHT: Its Production and Use. Embodying Plain Directions for the Treatment cf Dynamo-Electric Machines, Batteries, Accumulators, and Electric Lamps. By J. W. URQUHART, C.E., Author of "Electric Light Fitting," &c. Fourth Edition, Revised, with Large Additions and 145 Illustrations. Crown 8vo, 7s. 6d. cloth. [*Just published.*

"The book is by far the best that we have yet met with on the subject."—*Athenæum.*

"It is the only work at present available which gives, in language intelligible for the most part to the ordinary reader, a general but concise history of the means which have been adopted up to the present time in producing the electric light."—*Metropolitan.*

"The book contains a general account of the means adopted in producing the electric light, not only as obtained from voltaic or galvanic batteries, but treats at length of the dynamo-electric machine in several of its forms."—*Colliery Guardian.*

Construction of Dynamos.

DYNAMO CONSTRUCTION: A Practical Handbook for the Use of Engineer Constructors and Electricians in Charge. With Examples of leading English, American and Continental Dynamos and Motors. By J. W. URQUHART, Author of "Electric Light," &c. Crown 8vo, 7s. 6d. cloth. [*Just published.*

"The author has produced a book for which a demand has long existed. The subject is treated in a thoroughly practical manner."—*Mechanical World.*

Dynamic Electricity and Magnetism.

THE ELEMENTS OF DYNAMIC ELECTRICITY AND MAGNETISM. By PHILIP ATKINSON, A.M., Ph.D. Crown 8vo. 400 pp. With 120 Illustrations. 10s. 6d. cloth. [*Just publish. d.*

Text Book of Electricity.

THE STUDENT'S TEXT-BOOK OF ELECTRICITY. By HENRY M. NOAD, Ph.D., F.R.S., F.C.S. New Edition, carefully Revised. With an Introduction and Additional Chapters, by W. H. PREECE, M.I.C.E., Vice-President of the Society of Telegraph Engineers, &c. With 470 Illustrations. Crown 8vo, 12s. 6d. cloth.

"We can recommend Dr. Noad's book for clear style, great range of subject, a good index, and a plethora of woodcuts. Such collections as the present are indispensable."—*Athenæum.*

"An admirable text book for every student — beginner or advanced — of electricity."—*Engineering.*

Electric Lighting.

THE ELEMENTARY PRINCIPLES OF ELECTRIC LIGHT-
ING. By ALAN A. CAMPBELL SWINTON, Associate I.E.E. Second Edition,
Enlarged and Revised. With 16 Illustrations. Crown 8vo, 1s. 6d. cloth.
"Anyone who desires a short and thoroughly clear exposition of the elementary principles of
electric-lighting cannot do better than read this little work."—*Bradford Observer.*

Electricity.

A MANUAL OF ELECTRICITY: Including Galvanism, Mag-
netism, Dia-Magnetism, Electro-Dynamics, Magno-Electricity, and the Electric
Telegraph. By HENRY M. NOAD, Ph.D., F.R.S., F.C.S. Fourth Edition.
With 500 Woodcuts. 8vo, £1 4s. cloth.
"It is worthy of a place in the library of every public institution."—*Mining Journal.*

Dynamo Construction.

HOW TO MAKE A DYNAMO: A Practical Treatise for Amateurs.
Containing numerous Illustrations and Detailed Instructions for Constructing a Small Dynamo, to Produce the Electric Light. By ALFRED CROFTS.
Third Edition, Revised and Enlarged. Crown 8vo, 2s. cloth.
"The instructions given in this unpretentious little book are sufficiently clear and explicit to
enable any amateur mechanic possessed of average skill and the usual tools to be found in an
amateur's workshop, to build a practical dynamo machine."—*Electrician.*

NATURAL SCIENCE, etc.

Pneumatics and Acoustics.

PNEUMATICS: including Acoustics and the Phenomena of Wind
Currents, for the Use of Beginners. By CHARLES TOMLINSON, F.R.S.
F.C.S., &c. Fourth Edition, Enlarged. 12mo, 1s. 6d. cloth.
"Beginners in the study of this important application of science could not have a better manual."
—*Scotsman.* "A valuable and suitable text-book for students of Acoustics and the Pheno-
mena of Wind Currents."—*Schoolmaster.*

Conchology.

A MANUAL OF THE MOLLUSCA: Being a Treatise on Recent
and Fossil Shells. By S. P. WOODWARD, A.L.S., F.G.S., late Assistant
Palæontologist in the British Museum. With an Appendix on Recent and
Fossil Conchological Discoveries, by RALPH TATE, A.L.S., F.G.S. Illustrated
by A. N. WATERHOUSE and JOSEPH WILSON LOWRY. With 23 Plates and
upwards of 300 Woodcuts. Reprint of Fourth Ed., 1880. Cr. 8vo, 7s. 6d. cl.
"A most valuable storehouse of conchological and geological information."—*Science Gossip.*

Geology.

RUDIMENTARY TREATISE ON GEOLOGY, PHYSICAL
AND HISTORICAL. Consisting of "Physical Geology," which sets forth
the leading Principles of the Science; and "Historical Geology," which
treats of the Mineral and Organic Conditions of the Earth at each successive
epoch, especial reference being made to the British Series of Rocks. By
RALPH TATE, A.L.S., F.G.S., &c. With 250 Illustrations. 12mo, 5s. cloth.
"The fulness of the matter has elevated the book into a manual. Its information is exhaustive
and well arranged."—*School Board Chronicle.*

Geology and Genesis.

THE TWIN RECORDS OF CREATION; or, Geology and
Genesis: their Perfect Harmony and Wonderful Concord. By GEORGE W.
VICTOR LE VAUX. Numerous Illustrations. Fcap. 8vo, 5s. cloth.
"A valuable contribution to the evidences of Revelation, and disposes very conclusively of the
arguments of those who would set God's Works against God's Word."—*The Rock.*

The Constellations.

STAR GROUPS: A Student's Guide to the Constellations. By
J. ELLARD GORE, F.R.A.S., M.R.I.A., &c., Author of "The Scenery of the
Heavens." With 30 Maps. Small 4to, 5s. cloth, silvered. [Just published.

Astronomy.

ASTRONOMY. By the late Rev. ROBERT MAIN, M.A., F.R.S.,
formerly Radcliffe Observer at Oxford. Third Edition, Revised and Cor-
rected to the present time, by W. T. LYNN, B.A., F.R.A.S. 12mo, 2s. cloth.
"A sound and simple treatise, very carefully edited, and a capital book for beginners."—
Knowledge. [Nonal Times.
"Accurately brought down to the requirements of the present time by Mr. Lynn."—*Educa-*

DR. LARDNER'S COURSE OF NATURAL PHILOSOPHY.

THE HANDBOOK OF MECHANICS. Enlarged and almost re-written by BENJAMIN LOEWY, F.R.A.S. With 378 Illustrations. Post 8vo, 6s. cloth.

"The perspicuity of the original has been retained, and chapters which had become obsolete have been replaced by others of more modern character. The explanations throughout are studiously popular, and care has been taken to show the application of the various branches of physics to the industrial arts, and to the practical business of life."—*Mining Journal.*

"Mr. Loewy has carefully revised the book, and brought it up to modern requirements."—*Nature.*

"Natural philosophy has had few exponents more able or better skilled in the art of popu-larising the subject than Dr. Lardner; and Mr. Loewy is doing good service in fitting this treatise, and the others of the series, for use at the present time."—*Scotsman.*

THE HANDBOOK OF HYDROSTATICS AND PNEUMATICS. New Edition, Revised and Enlarged, by BENJAMIN LOEWY, F.R.A.S. With 236 Illustrations. Post 8vo, 5s. cloth.

"For those 'who desire to attain an accurate knowledge of physical science without the pro-found methods of mathematical investigation,' this work is not merely intended, but well adapted."—*Chemical News.*

"The volume before us has been carefully edited, augmented to nearly twice the bulk of the former edition, and all the most recent matter has been added. . . . It is a valuable text-book."—*Nature.*

"Candidates for pass examinations will find it, we think, specially suited to their requirements."—*English Mechanic.*

THE HANDBOOK OF HEAT. Edited and almost entirely re-written by BENJAMIN LOEWY, F.R.A.S., &c. 117 Illustrations. Post 8vo, 6s. cloth.

"The style is always clear and precise, and conveys instruction without leaving any cloudiness or lurking doubts behind."—*Engineering.*

"A most exhaustive book on the subject on which it treats, and is so arranged that it can be understood by all who desire to attain an accurate knowledge of physical science. . . . . Mr. Loewy has included all the latest discoveries in the varied laws and effects of heat."—*Standard.*

"A complete and handy text-book for the use of students and general readers."—*English Mechanic.*

THE HANDBOOK OF OPTICS. By DIONYSIUS LARDNER, D.C.L., formerly Professor of Natural Philosophy and Astronomy in University College, London. New Edition. Edited by T. OLVER HARDING, B.A. Lond., of University College, London. With 298 Illustrations. Small 8vo, 448 pages, 5s. cloth.

"Written by one of the ablest English scientific writers, beautifully and elaborately illustrated."*Mechanic's Magazine.*

THE HANDBOOK OF ELECTRICITY, MAGNETISM, AND ACOUSTICS. By Dr. LARDNER. Ninth Thousand. Edit. by GEORGE CAREY FOSTER, B.A., F.C.S. With 400 Illustrations. Small 8vo, 5s. cloth.

"The book could not have been entrusted to anyone better calculated to preserve the terse and lucid style of Lardner, while correcting his errors and bringing up his work to the present state of scientific knowledge."—*Popular Science Review.*

THE HANDBOOK OF ASTRONOMY. Forming a Companion to the "Handbook of Natural Philosophy." By DIONYSIUS LARDNER, D.C.L., formerly Professor of Natural Philosophy and Astronomy in University College, London. Fourth Edition. Revised and Edited by EDWIN DUNKIN, F.R.A.S., Royal Observatory, Greenwich. With 38 Plates and upwards of 100 Woodcuts. In One Vol., small 8vo, 550 pages, 9s. 6d. cloth.

"Probably no other book contains the same amount of information in so compendious and well-arranged a form—certainly none at the price at which this is offered to the public."—*Athenæum.*

"We can do no other than pronounce this work a most valuable manual of astronomy, and we strongly recommend it to all who wish to acquire a general—but at the same time correct—acquaint-ance with this sublime science."—*Quarterly Journal of Science.*

"One of the most deservedly popular books on the subject . . . We would recommend not only the student of the elementary principles of the science, but he who aims at mastering the higher and mathematical branches of astronomy, not to be without this work beside him."—*Practi-cal Magazine.*

Dr. Lardner's Electric Telegraph.

THE ELECTRIC TELEGRAPH. By Dr. LARDNER. Re-vised and Re-written by E. B. BRIGHT, F.R.A.S. 140 Illustrations. Small 8vo, 2s. 6d. cloth.

"One of the most readable books extant on the Electric Telegraph."—*English Mechanic.*

DR. LARDNER'S MUSEUM OF SCIENCE AND ART.

THE MUSEUM OF SCIENCE AND ART. Edited by DIONYSIUS LARDNER, D.C.L., formerly Professor of Natural Philosophy and Astronomy in University College, London. With upwards of 1,200 Engravings on Wood. In 6 Double Volumes, £1 1s., in a new and elegant cloth binding; or handsomely bound in half-morocco, 31s. 6d.

⁎⁎ OPINIONS OF THE PRESS.

"This series, besides affording popular but sound instruction on scientific subjects, with which the humblest man in the country ought to be acquainted, also undertakes that teaching of 'Common Things' which every well-wisher of his kind is anxious to promote. Many thousand copies of this serviceable publication have been printed, in the belief and hope that the desire for instruction and improvement widely prevails; and we have no fear that such enlightened faith will meet with disappointment."—*Times.*

"A cheap and interesting publication, alike informing and attractive. The papers combine subjects of importance and great scientific knowledge, considerable inductive powers, and a popular style of treatment."—*Spectator.*

"The 'Museum of Science and Art' is the most valuable contribution that has ever been made to the Scientific Instruction of every class of society."—Sir DAVID BREWSTER, in the *North British Review.*

"Whether we consider the liberality and beauty of the illustrations, the charm of the writing, or the durable interest of the matter, we must express our belief that there is hardly to be found among the new books one that would be welcomed by people of so many ages and classes as a valuable present."—*Examiner.*

⁎⁎ *Separate books formed from the above, suitable for Workmen's Libraries, Science Classes, etc.*

Common Things Explained. Containing Air, Earth, Fire, Water, Time, Man, the Eye, Locomotion, Colour, Clocks and Watches, &c. 233 Illustrations, cloth gilt, 5s.

The Microscope. Containing Optical Images, Magnifying Glasses, Origin and Description of the Microscope, Microscopic Objects, the Solar Microscope, Microscopic Drawing and Engraving, &c. 147 Illustrations, cloth gilt, 2s.

Popular Geology. Containing Earthquakes and Volcanoes, the Crust of the Earth, &c. 201 Illustrations, cloth gilt, 2s. 6d.

Popular Physics. Containing Magnitude and Minuteness, the Atmosphere, Meteoric Stones, Popular Fallacies, Weather Prognostics, the Thermometer, the Barometer, Sound, &c. 85 Illustrations, cloth gilt, 2s. 6d.

Steam and its Uses. Including the Steam Engine, the Locomotive, and Steam Navigation. 89 Illustrations, cloth gilt, 2s.

Popular Astronomy. Containing How to observe the Heavens—The Earth, Sun, Moon, Planets, Light, Comets, Eclipses, Astronomical Influences, &c. 182 Illustrations, 4s. 6d.

The Bee and White Ants: Their Manners and Habits. With Illustrations of Animal Instinct and Intelligence. 135 Illustrations, cloth gilt, 2s.

The Electric Telegraph Popularized. To render intelligible to all who can Read, irrespective of any previous Scientific Acquirements, the various forms of Telegraphy in Actual Operation. 100 Illustrations, cloth gilt, 1s. 6d.

Dr. Lardner's School Handbooks.

NATURAL PHILOSOPHY FOR SCHOOLS. By Dr. LARDNER. 328 Illustrations. Sixth Edition. One Vol., 3s. 6d. cloth.

"A very convenient class-book for junior students in private schools. It is intended to convey, in clear and precise terms, general notions of all the principal divisions of Physical Science."—*British Quarterly Review.*

ANIMAL PHYSIOLOGY FOR SCHOOLS. By Dr. LARDNER. With 190 Illustrations. Second Edition. One Vol., 3s. 6d. cloth.

"Clearly written, well arranged, and excellently illustrated."—*Gardener's Chronicle.*

COUNTING-HOUSE WORK, TABLES, etc.

Introduction to Business.
LESSONS IN COMMERCE. By Professor R. GAMBARO, of the Royal High Commercial School at Genoa. Edited and Revised by JAMES GAULT, Professor of Commerce and Commercial Law in King's College, London. Crown 8vo, price about 3s. 6d. [In the press.

Accounts for Manufacturers.
FACTORY ACCOUNTS: Their Principles and Practice. A Handbook for Accountants and Manufacturers, with Appendices on the Nomenclature of Machine Details; the Income Tax Acts; the Rating of Factories; Fire and Boiler Insurance; the Factory and Workshop Acts, &c., including also a Glossary of Terms and a large number of Specimen Rulings. By EMILE GARCKE and J. M. FELLS. Third Edition. Demy 8vo, 250 pages, price 6s. strongly bound.

"A very interesting description of the requirements of Factory Accounts. . . . the princip.e of assimilating the Factory Accounts to the general commercial books is one which we thoroughly agree with."—*Accountants' Journal.*

"There are few owners of Factories who would not derive great benefit from the perusal of this most admirable work."—*Local Government Chronicle.*

Foreign Commercial Correspondence.
THE FOREIGN COMMERCIAL CORRESPONDENT: Being Aids to Commercial Correspondence in Five Languages—English, French, German, Italian and Spanish. By CONRAD E. BAKER. Second Edition, Revised. Crown 8vo, 3s. cloth.

"Whoever wishes to correspond in all the languages mentioned by Mr. Baker cannot do better than study this work, the materials of which are excellent and conveniently arranged."—*Athenæum.*

"A careful examina i n has convinced us that it is unusually complete, well arranged and reliable. The book is a thoroughly good one."—*Schoolmaster.*

Intuitive Calculations.
THE COMPENDIOUS CALCULATOR; or, Easy and Concise Methods of Performing the various Arithmetical Operations required in Commercial and Business Transactions, together with Useful Tables. By D. O'GORMAN. Corrected by Professor J. R. YOUNG. Twenty-seventh Ed., Revised by C. NORRIS. Fcap. 8vo, 2s. 6d. cloth; or, 3s. 6d. half-bound.

"It would be difficult to exaggerate the usefulness of a book like this to everyone engaged in commerce or manufacturing industry."—*Knowledge.*

"Supplies special and rapid methods for all kinds of calculations. Of great utility to persons engaged in any kind of commercial transactions."—*Scotsman.*

Modern Metrical Units and Systems.
MODERN METROLOGY: A Manual of the Metrical Units and Systems of the Present Century. With an Appendix containing a proposed English System. By LEWIS D'A. JACKSON, A.M.Inst.C.E., Author of " Aid to Survey Practice," &c. Large crown 8vo, 12s. 6d. cloth.

"The author has brought together much valuable and interesting information. . . . We cannot but recommend the work."—*Nature.*

"For exhaustive tables of equivalent weights and measures of all sorts, and for clear demonstrations of the effects of the various systems that have been proposed or adopted, Mr. Jackson s treatise is without a rival."—*Academy.*

The Metric System and the British Standards.
A SERIES OF METRIC TABLES, in which the British Standard Measures and Weights are compared with those of the Metric System at present in Use on the Continent. By C. H. DOWLING, C.E. 8vo, 10s. 6d. strongly bound.

"Their accuracy has been certified by Professor Airy, the Astronomer-Royal."—*Builder.*

"Mr. Dowling's Tables are well put together as a ready-reckoner for the conversion of one system into the other."—*Athenæum.*

Iron and Metal Trades' Calculator.
THE IRON AND METAL TRADES' COMPANION. For expeditiously ascertaining the Value of any Goods bought or sold by Weight, from 1s. per cwt. to 112s. per cwt., and from one farthing per pound to one shilling per pound. Each Table extends from one pound to 100 tons. To which are appended Rules on Decimals, Square and Cube Root, Mensuration of Superficies and Solids, &c.; also Tables of Weights of Materials, and other Useful Memoranda. By THOS. DOWNIE. Strongly bound in leather, 396 pp., 9s.

"A most useful set of tables. . . . Nothing like them before existed."—*Building News.*

"Although specially adapted to the iron and metal trades, the tables will be found useful in every other business in which merchandise is bought and sold by weight."—*Railway News.*

Calculator for Numbers and Weights Combined.

THE NUMBER, WEIGHT AND FRACTIONAL CALCU-LATOR. Containing upwards of 250,000 Separate Calculations, showing at a glance the value at 422 different rates, ranging from $\frac{1}{16}$th of a Penny to 20s. each, or per cwt., and £20 per ton, of any number of articles consecutively, from 1 to 470.—Any number of cwts., qrs., and lbs., from 1 cwt. to 470 cwts.—Any number of tons, cwts., qrs., and lbs., from 1 to 1,000 tons. By WILLIAM CHADWICK, Public Accountant. Third Edition, Revised and Improved. 8vo, price 18s., strongly bound for Office wear and tear.

* *This work is specially adapted for the Apportionment of Mileage Charges for Railway Traffic.*

☞ *This comprehensive and entirely unique and original Calculator is adapted for the use of Accountants and Auditors, Railway Companies, Canal Companies, Shippers, Shipping Agents, General Carriers, etc.*

Ironfounders, Brassfounders, Metal Merchants, Iron Manufacturers, Ironmongers, Engineers, Machinists, Boiler Makers, Millwrights, Roofing, Bridge and Girder Makers, Colliery Proprietors, etc.

Timber Merchants, Builders, Contractors, Architects, Surveyors, Auctioneers Valuers, Brokers, Mill Owners and Manufacturers, Mill Furnishers, Merchants and General Wholesale Tradesmen.

* OPINIONS OF THE PRESS.

"The book contains the answers to questions, and not simply a set of ingenious puzzle methods of arriving at results. It is as easy of reference for any answer or any number of answers as a dictionary, and the references are even more quickly made. For making up accounts or estimates, the book must prove invaluable to all who have any considerable quantity of calculations involving price and measure in any combination to do."—*Engineer.*

"The most perfect work of the kind yet prepared."—*Glasgow Herald.*

Comprehensive Weight Calculator.

THE WEIGHT CALCULATOR. Being a Series of Tables upon a New and Comprehensive Plan, exhibiting at One Reference the exact Value of any Weight from 1 lb. to 15 tons, at 300 Progressive Rates, from 1d. to 168s. per cwt., and containing 186,000 Direct Answers, which, with their Combinations, consisting of a single addition (mostly to be performed at sight), will afford an aggregate of 10,266,000 Answers; the whole being calculated and designed to ensure correctness and promote despatch. By HENRY HARBEN, Accountant. Fourth Edition, carefully Corrected. Royal 8vo, strongly half-bound, £1 5s.

"A practical and useful work of reference for men of business generally; it is the best of the kind we have seen.'—*Ironmonger.*

"Of priceless value to business men. It is a necessary book in all mercantile offices."—*Sheffield Independent.*

Comprehensive Discount Guide.

THE DISCOUNT GUIDE. Comprising several Series of Tables for the use of Merchants, Manufacturers, Ironmongers, and others, by which may be ascertained the exact Profit arising from any mode of using Discounts, either in the Purchase or Sale of Goods, and the method of either Altering a Rate of Discount or Advancing a Price, so as to produce, by one operation, a sum that will realise any required profit after allowing one or more Discounts: to which are added Tables of Profit or Advance from 1¼ to 90 per cent., Tables of Discount from 1¼ to 98¾ per cent., and Tables of Commission, &c., from ¼ to 10 per cent. By HENRY HARBEN, Accountant, Author of "The Weight Calculator." New Edition, carefully Revised and Corrected. Demy 8vo, 544 pp. half-bound, £1 5s.

"A book such as this can only be appreciated by business men, to whom the saving of time means saving of money. We have the high authority of Professor J. R. Young that the tables throughout the work are constructed upon strictly accurate principles. The work is a mode of typographical clearness, and must prove of great value to merchants, manufacturers, and general traders."—*British Trade Journal.*

Iron Shipbuilders' and Merchants' Weight Tables.

IRON-PLATE WEIGHT TABLES: *For Iron Shipbuilders, Engineers and Iron Merchants.* Containing the Calculated Weights of upwards of 150,000 different sizes of Iron Plates, from 1 foot by 6 in. by ¼ in. to 10 feet by 5 feet by 1 in. Worked out on the basis of 40 lbs. to the square foot of Iron of 1 inch in thickness. Carefully compiled and thoroughly Revised by H. BURLINSON and W. H. SIMPSON. Oblong 4to, 25s. half-bound.

"This work will be found of great utility. The authors have had much practical experience of what is wanting in making estimates; and the use of the book will save much time in making elaborate calculations."—*English Mechanic.*

INDUSTRIAL AND USEFUL ARTS.

Soap-making.

THE ART OF SOAP-MAKING : *A Practical Handbook of the Manufacture of Hard and Soft Soaps, Toilet Soaps, etc.* Including many New Processes, and a Chapter on the Recovery of Glycerine from Waste Leys. By ALEXANDER WATT, Author of " Electro-Metallurgy Practically Treated," &c. With numerous Illustrations. Fourth Edition, Revised and Enlarged. Crown 8vo, 7s. 6d. cloth.

"The work will prove very useful, not merely to the technological student, but to the practical soap-boiler who wishes to understand the theory of his art."—*Chemical News.*

"Mr. Watt's book is a thoroughly practical treatise on an art which has almost no literature in our language. We congratulate the author on the success of his endeavour to fill a void in English technical literature."—*Nature.*

Paper Making.

THE ART OF PAPER MAKING : *A Practical Handbook of the Manufacture of Paper from Rags, Esparto, Straw and other Fibrous Materials.* Including the Manufacture of Pulp from Wood Fibre, with a Description of the Machinery and Appliances used. To which are added Details of Processes for Recovering Soda from Waste Liquors. By ALEXANDER WATT. With Illustrations. Crown 8vo, 7s. 6d. cloth.

"This book is succinct, lucid, thoroughly practical, and includes everything of interest to the modern paper-maker. It is the latest, most practical and most complete work on the paper-making art before the British public."—*Paper Record.*

"It may be regarded as the standard work on the subject. The book is full of valuable information. The ' Art of Paper-making,' is in every respect a model of a text-book, either for a technical class or for the private student."—*Paper and Printing Trades Journal.*

"Admirably adapted for general as well as ordinary technical reference, and as a handbook for students in technical education may be warmly commended."—*The Paper Maker's Monthly Journal.*

Leather Manufacture.

THE ART OF LEATHER MANUFACTURE. Being a Practical Handbook, in which the Operations of Tanning, Currying, and Leather Dressing are fully Described, the Principles of Tanning Explained and many Recent Processes introduced. By ALEXANDER WATT, Author of " Soap-Making," &c. With numerous Illustrations. Second Edition. Crown 8vo, 9s. cloth.

"A sound, comprehensive treatise on tanning and its accessories. This book is an eminently valuable production, which redounds to the credit of both author and publishers."—*Chemical Review.*

"This volume is technical without being tedious, comprehensive and complete without being prosy, and it bears on every page the impress of a master hand. We have never come across a better trade treatise, nor one that so thoroughly supplied an absolute want."—*Shoe and Leather Trades Chronicle.*

Boot and Shoe Making.

THE ART OF BOOT AND SHOE-MAKING. A Practical Handbook, including Measurement, Last-Fitting, Cutting-Out, Closing and Making, with a Description of the most approved Machinery employed. By JOHN B. LENO, late Editor of *St. Crispin*, and *The Boot and Shoe-Maker.* With numerous Illustrations. Third Edition. 12mo, 2s. cloth limp.

"This excellent treatise is by far the best work ever written on the subject. A new work, embracing all modern improvements, was much wanted. This want is now satisfied. The chapter on clicking, which shows how waste may be prevented, will save fifty times the price of the book."—*Scottish Leather Trader.*

Dentistry.

MECHANICAL DENTISTRY : *A Practical Treatise on the Construction of the various kinds of Artificial Dentures.* Comprising also Useful Formulæ, Tables and Receipts for Gold Plate, Clasps, Solders, &c. &c. By CHARLES HUNTER. Third Edition, Revised. With upwards of 100 Wood Engravings. Crown 8vo, 3s. 6d. cloth.

"The work is very practical."—*Monthly Review of Dental Surgery.*

"We can strongly recommend Mr. Hunter's treatise to all students preparing for the profession of dentistry, as well as to every mechanical dentist."—*Dublin Journal of Medical Science.*

Wood Engraving.

WOOD ENGRAVING: *A Practical and Easy Introduction to the Study of the Art.* By WILLIAM NORMAN BROWN. Second Edition. With numerous Illustrations. 12mo, 1s. 6d. cloth limp.

"The book is clear and complete, and will be useful to anyone wanting to understand the first elements of the beautiful art of wood engraving."—*Graphic.*

HANDYBOOKS FOR HANDICRAFTS. By PAUL N. HASLUCK.

Metal Turning.

THE METAL TURNER'S HANDYBOOK. *A Practical Manual for Workers at the Foot-Lathe:* Embracing Information on the Tools, Appliances and Processes employed in Metal Turning. By PAUL N. HAS-LUCK, Author of "Lathe-Work." With upwards of One Hundred Illustrations. Second Edition, Revised. Crown 8vo, 2s. cloth.
"Clearly and concisely written, excellent in every way."—*Mechanical World.*

Wood Turning.

THE WOOD TURNER'S HANDYBOOK. *A Practical Manual for Workers at the Lathe:* Embracing Information on the Tools, Appliances and Processes Employed in Wood Turning. By PAUL N. HASLUCK. With upwards of One Hundred Illustrations. Crown 8vo, 2s. cloth.
"We recommend the book to young turners and amateurs. A multitude of workmen have hitherto sought in vain for a manual of this special industry."—*Mechanical World.*

WOOD AND METAL TURNING. By P. N. HASLUCK. (Being the Two preceding Vols. bound together.) 300 pp., with upwards of 200 Illustrations, crown 8vo, 3s. 6d. cloth.

Watch Repairing.

THE WATCH JOBBER'S HANDYBOOK. *A Practical Manual on Cleaning, Repairing and Adjusting.* Embracing Information on the Tools, Materials, Appliances and Processes Employed in Watchwork. By PAUL N. HASLUCK. With upwards of One Hundred Illustrations. Cr. 8vo, 2s. cloth.
"All young persons connected with the trade should acquire and study this excellent, and at the same time, inexpensive work."—*Clerkenwell Chronicle.*

Clock Repairing.

THE CLOCK JOBBER'S HANDYBOOK : *A Practical Manual on Cleaning, Repairing and Adjusting.* Embracing Information on the Tools, Materials, Appliances and Processes Employed in Clockwork. By PAUL N. HASLUCK. With upwards of 100 Illustrations. Cr. 8vo, 2s. cloth.
"Of inestimable service to those commencing the trade."—*Coventry Standard.*

WATCH AND CLOCK JOBBING. By P. N. HASLUCK. (Being the Two preceding Vols. bound together.) 320 pp., with upwards of 200 Illustrations, crown 8vo, 3s. 6d. cloth.

Pattern Making.

THE PATTERN MAKER'S HANDYBOOK. A Practical Manual, embracing Information on the Tools, Materials and Appliances employed in Constructing Patterns for Founders. By PAUL N. HASLUCK. With One Hundred Illustrations. Crown 8vo, 2s. cloth.
"This handy volume contains sound information of considerable value to students and artificers."—*Hardware Trades Journal.*

Mechanical Manipulation.

THE MECHANIC'S WORKSHOP HANDYBOOK. *A Practical Manual on Mechanical Manipulation.* Embracing Information on various Handicraft Processes, with Useful Notes and Miscellaneous Memoranda. By PAUL N. HASLUCK. Crown 8vo, 2s. cloth.
"It is a book which should be found in every workshop, as it is one which will be continually referred to for a very great amount of standard information."—*Saturday Review.*

Model Engineering.

THE MODEL ENGINEER'S HANDYBOOK : *A Practical Manual on Model Steam Engines.* Embracing Information on the Tools, Materials and Processes Employed in their Construction. By PAUL N. HASLUCK. With upwards of 100 Illustrations. Crown 8vo, 2s. cloth.
"By carefully going through the work, amateurs may pick up an excellent notion of the construction of full-sized steam engines."—*Telegraphic Journal.*

Cabinet Making.

THE CABINET WORKER'S HANDYBOOK : A Practical Manual, embracing Information on the Tools, Materials, Appliances and Processes employed in Cabinet Work. By PAUL N. HASLUCK, Author of "Lathe Work," &c. With upwards of 100 Illustrations. Crown 8vo, 2s. cloth. [*Glasgow Herald.*
"Thoroughly practical throughout. The amateur worker in wood will find it most useful."—

Electrolysis of Gold, Silver, Copper, etc.

ELECTRO-DEPOSITION : A Practical Treatise on the Electrolysis of Gold, Silver, Copper, Nickel, and other Metals and Alloys. With descriptions of Voltaic Batteries, Magneto and Dynamo-Electric Machines, Thermopiles, and of the Materials and Processes used in every Department of the Art, and several Chapters on Electro-Metallurgy. By ALEXANDER WATT. Third Edition, Revised and Corrected. Crown 8vo, 9s. cloth.
"Eminently a book for the practical worker in electro-deposition. It contains practical descriptions of methods, processes and materials as actually pursued and used in the workshop."—*Engineer.*

Electro-Metallurgy.

ELECTRO-METALLURGY ; Practically Treated. By ALEXANDER WATT, Author of "Electro-Deposition," &c. Ninth Edition, Enlarged and Revised, with Additional Illustrations, and including the most recent Processes. 12mo, 4s. cloth boards.
"From this book both amateur and artisan may learn everything necessary for the successful prosecution of electroplating."—*Iron.*

Electroplating.

ELECTROPLATING : A Practical Handbook on the Deposition of Copper, Silver, Nickel, Gold, Aluminium, Brass, Platinum, &c. &c. With Descriptions of the Chemicals, Materials, Batteries and Dynamo Machines used in the Art. By J. W. URQUHART, C.E. Second Edition, with Additions. Numerous Illustrations. Crown 8vo. 5s. cloth.
" An excellent practical manual."—*Engineering.*
" An excellent work, giving the newest information."—*Horological Journal.*

Electrotyping.

ELECTROTYPING : The Reproduction and Multiplication of Printing Surfaces and Works of Art by the Electro-deposition of Metals. By J. W. URQUHART, C.E. Crown 8vo, 5s. cloth.
" The book is thoroughly practical. The reader is, therefore, conducted through the leading laws of electricity, then through the metals used by electrotypers, the apparatus, and the depositing processes, up to the final preparation of the work."—*Art Journal.*

Horology.

A TREATISE ON MODERN HOROLOGY, in Theory and Practice. Translated from the French of CLAUDIUS SAUNIER, by JULIEN TRIPPLIN, F.R.A.S., and EDWARD RIGG, M.A., Assayer in the Royal Mint. With 78 Woodcuts and 22 Coloured Plates. Second Edition. Royal 8vo, £2 2s. cloth ; £2 10s. half-calf.
" There is no horological work in the English language at all to be compared to this production of M. Saunier's for clearness and completeness. It is alike good as a guide for the student and as a reference for the experienced horologist and skilled workman."—*Horological Journal.*
" The latest, the most complete, and the most reliable of those literary productions to which continental watchmakers are indebted for the mechanical superiority over their English brethren —in fact, the Book of Books, is M. Saunier's 'Treatise.'"—*Watchmaker, Jeweller and Silversmith.*

Watchmaking.

THE WATCHMAKER'S HANDBOOK. A Workshop Companion for those engaged in Watchmaking and the Allied Mechanical Arts. From the French of CLAUDIUS SAUNIER. Enlarged by JULIEN TRIPPLIN, F.R.A.S., and EDWARD RIGG, M.A., Assayer in the Royal Mint. Woodcuts and Copper Plates. Third Edition, Revised. Crown 8vo, 9s. cloth.
" Each part is truly a treatise in itself. The arrangement is good and the language is clear and concise. It is an admirable guide for the young watchmaker."—*Engineering.*
" It is impossible to speak too highly of its excellence. It fulfils every requirement in a handbook intended for the use of a workman."—*Watch and Clockmaker.*
" This book contains an immense number of practical details bearing on the daily occupation of a watchmaker."—*Watchmaker and Metalworker* (Chicago).

Goldsmiths' Work.

THE GOLDSMITH'S HANDBOOK. By GEORGE E. GEE, Jeweller, &c. Third Edition, considerably Enlarged. 12mo, 3s. 6d. cl. bds.
"A good, sound educator, and will be accepted as an authority."—*Horological Journal.*

Silversmiths' Work.

THE SILVERSMITH'S HANDBOOK. By GEORGE E. GEE, Jeweller, &c. Second Edition, Revised, with numerous Illustrations. 12mo, 3s. 6d. cloth boards.
"Workers in the trade will speedily discover its merits when they sit down to study it."— *English Mechanic.*
₊ *The above two works together, strongly half-bound, price 7s.*

D

Bread and Biscuit Baking.

THE BREAD AND BISCUIT BAKER'S AND SUGAR-BOILER'S ASSISTANT. Including a large variety of Modern Recipes. With Remarks on the Art of Bread-making. By ROBERT WELLS, Practical Baker. Second Edition, with Additional Recipes. Crown 8vo, 2s. cloth.

" A large number of wrinkles for the ordinary cook, as well as the baker."—*Saturday Review.*

Confectionery.

THE PASTRYCOOK AND CONFECTIONER'S GUIDE. For Hotels, Restaurants and the Trade in general, adapted also for Family Use. By ROBERT WELLS, Author of " The Bread and Biscuit Baker's and Sugar Boiler's Assistant." Crown 8vo, 2s. cloth.

" We cannot speak too highly of this really excellent work. In these days of keen competition our readers cannot do better than purchase this book."—*Bakers' Times.*

Ornamental Confectionery.

ORNAMENTAL CONFECTIONERY : A Guide for Bakers, Confectioners and Pastrycooks; including a variety of Modern Recipes, and Remarks on Decorative and Coloured Work. With 129 Original Designs. By ROBERT WELLS. Crown 8vo, 5s. cloth.

" A valuable work, and should be in the hands of every baker and confectioner. The illustrative designs are alone worth treble the amount charged for the whole work."—*Bakers' Times.*

Flour Confectionery.

THE MODERN FLOUR CONFECTIONER. Wholesale and Retail.. Containing a large Collection of Recipes for Cheap Cakes, Biscuits, &c. With Remarks on the Ingredients used in their Manufacture, &c. By R. WELLS, Author of " Ornamental Confectionery," " The Bread and Biscuit Baker," " The Pastrycook's Guide," &c. Crown 8vo, 2s. cloth.

Laundry Work.

LAUNDRY MANAGEMENT. A Handbook for Use in Private and Public Laundries, Including Descriptive Accounts of Modern Machinery and Appliances for Laundry Work. By the EDITOR of " The Laundry Journal." With numerous Illustrations. Crown 8vo, 2s. 6d. cloth.

CHEMICAL MANUFACTURES & COMMERCE.

New Manual of Engineering Chemistry.

ENGINEERING CHEMISTRY : A Practical Treatise for the Use of Analytical Chemists, Engineers, Iron Masters, Iron Founders, Students, and others. Comprising Methods of Analysis and Valuation of the Principal Materials used in Engineering Work, with numerous Analyses, Examples, and Suggestions. By H. JOSHUA PHILLIPS, F.I.C., F.C.S., Analytical and Consulting Chemist to the Great Eastern Railway. Crown 8vo, 320 pp., with Illustrations, 10s. 6d. cloth.   [*Just published.*

" In this work the author has rendered no small service to a numerous body of practical men.   . . . The analytical methods may be pronounced most satisfactory, being as accurate as the despatch required of engineering chemists permits."—*Chemical News.*

Analysis and Valuation of Fuels.

FUELS: SOLID, LIQUID AND GASEOUS, Their Analysis and Valuation. For the Use of Chemists and Engineers. By H. J. PHILLIPS, F.C.S., Analytical and Consulting Chemist to the Great Eastern Railway. Crown 8vo, 3s. 6d. cloth.

" Ought to have its place in the laboratory of every metallurgical establishment, and wherever fuel is used on a large scale."—*Chemical News.*
" Cannot fail to be of wide interest, especially at the present time."—*Railway News.*

Alkali Trade, Manufacture of Sulphuric Acid, etc.

A MANUAL OF THE ALKALI TRADE, including the Manufacture of Sulphuric Acid, Sulphate of Soda, and Bleaching Powder. By JOHN LOMAS. 390 pages. With 232 Illustrations and Working Drawings. Second Edition. Royal 8vo, £1 10s. cloth.

"This book is written by a manufacturer for manufacturers. The working details of the most approved forms of apparatus are given, and these are accompanied by no less than 232 wood engravings, all of which may be used for the purposes of construction."—*Athenæum.*

The Blowpipe.

THE BLOWPIPE IN CHEMISTRY, MINERALOGY, AND GEOLOGY. Containing all known Methods of Anhydrous Analysis, Working Examples, and Instructions for Making Apparatus. By Lieut.-Col. W. A. Ross, R.A. With 120 Illustrations. New Edition. Crown 8vo, 5s. cloth.

"The student who goes through the course of experimentation here laid down will gain a better insight into inorganic chemistry and mineralogy than if he had 'got up' any of the best text-books of the day, and passed any number of examinations in their contents."—*Chemical News.*

Commercial Chemical Analysis.

THE COMMERCIAL HANDBOOK OF CHEMICAL ANALYSIS; or, Practical Instructions for the determination of the Intrinsic or Commercial Value of Substances used in Manufactures, Trades, and the Arts. By A. NORMANDY. New Edition by H. M. NOAD, F.R.S. Cr. 8vo, 12s. 6d. cl.

"Essential to the analysts appointed under the new Act. The most recent results are given, and the work is well edited and carefully written."—*Nature.*

Brewing.

A HANDBOOK FOR YOUNG BREWERS. By HERBERT EDWARDS WRIGHT, B.A. New Edition, much Enlarged. [*In the press.*

Dye-Wares and Colours.

THE MANUAL OF COLOURS AND DYE-WARES : Their Properties, Applications, Valuation, Impurities, and Sophistications. For the use of Dyers, Printers, Drysalters, Brokers, &c. By J. W. SLATER. Second Edition, Revised and greatly Enlarged. Crown 8vo, 7s. 6d. cloth.

"A complete encyclopædia of the *materia tinctoria.* The information given respecting each article is full and precise, and the methods of determining the value of articles such as these, so liable to sophistication, are given with clearness, and are practical as well as valuable."—*Chemist and Druggist.*

"There is no other work which covers precisely the same ground. To students preparing for examinations in dyeing and printing it will prove exceedingly useful."—*Chemical News.*

Pigments.

THE ARTIST'S MANUAL OF PIGMENTS. Showing their Composition, Conditions of Permanency, Non-Permanency, and Adulterations; Effects in Combination with Each Other and with Vehicles; and the most Reliable Tests of Purity. By H. C. STANDAGE Second Edition. Crown 8vo, 2s. 6d. cloth.

"This work is indeed *multum-in-parvo,* and we can, with good conscience, recommend it to all who come in contact with pigments, whether as makers, dealers or users."—*Chemical Review.*

Gauging. Tables and Rules for Revenue Officers, Brewers, etc.

A POCKET BOOK OF MENSURATION AND GAUGING : Containing Tables, Rules and Memoranda for Revenue Officers, Brewers, Spirit Merchants, &c. By J. B. MANT (Inland Revenue). Second Edition Revised. Oblong 18mo, 4s. leather, with elastic band.

"This handy and useful book is adapted to the requirements of the Inland Revenue Department, and will be a favourite book of reference."—*Civilian.*

"Should be in the hands of every practical brewer."—*Brewers' Journal.*

AGRICULTURE, FARMING, GARDENING, etc.

Youatt and Burn's Complete Grazier.

THE COMPLETE GRAZIER, and FARMER'S and CATTLE-BREEDER'S ASSISTANT. Including the Breeding, Rearing, and Feeding of Stock; Management of the Dairy, Culture and Management of Grass Land, and of Grain and Root Crops, &c. By W. YOUATT and R. SCOTT BURN. An entirely New Edition, partly Re-written and greatly Enlarged, by W. FREAM, B.Sc.Lond., LL.D. In medium 8vo, about 1,000 pp. [*In the press.*

Agricultural Facts and Figures.

NOTE-BOOK OF AGRICULTURAL FACTS AND FIGURES FOR FARMERS AND FARM STUDENTS. By PRIMROSE McCONNELL, late Professor of Agriculture, Glasgow Veterinary College. Third Edition. Royal 32mo, 4s. leather.

"The most complete and comprehensive Note-book for Farmers and Farm Students that we have seen. It literally teems with information, and we can cordially recommend it to all connected with agriculture."—*North British Agriculturist.*

Flour Manufacture, Milling, etc.

FLOUR MANUFACTURE : A Treatise on Milling Science and Practice. By FRIEDRICH KICK, Imperial Regierungsrath, Professor of Mechanical Technology in the Imperial German Polytechnic Institute, Prague. Translated from the Second Enlarged and Revised Edition with Supplement. By H. H. P. POWLES, A.M.I.C.E. Nearly 400 pp. Illustrated with 28 Folding Plates, and 167 Woodcuts. Royal 8vo, 25s. cloth.

"This valuable work is, and will remain, the standard authority on the science of milling. . . The miller who has read and digested this work will have laid the foundation, so to speak, of a successful career ; he will have acquired a number of general principles which he can proceed to apply. In this handsome volume we at last have the accepted text-book of modern milling in good, sound English, which has little, if any, trace of the German idiom."—*The Miller.*
"The appearance of this celebrated work in English is very opportune, and British millers will, we are sure, not be slow in availing themselves of its pages."—*Millers' Gazette.*

Small Farming.

SYSTEMATIC SMALL FARMING; or, The Lessons of my Farm. Being an Introduction to Modern Farm Practice for Small Farmers in the Culture of Crops; The Feeding of Cattle; The Management of the Dairy, Poultry and Pigs, &c. &c. By ROBERT SCOTT BURN, Author of "Outlines of Landed Estates' Management." Numerous Illusts., cr. 8vo, 6s. cloth.

"This is the completest book of its class we have seen, and one which every amateur farmer will read with pleasure and accept as a guide."—*Field.*
"The volume contains a vast amount of useful information. No branch of farming is left untouched, from the labour to be done to the results achieved. It may be safely recommended to all who think they will be in paradise when they buy or rent a three-acre farm."—*Glasgow Herald.*

Modern Farming.

OUTLINES OF MODERN FARMING. By R. SCOTT BURN. Soils, Manures, and Crops—Farming and Farming Economy—Cattle, Sheep, and Horses — Management of Dairy, Pigs and Poultry — Utilisation of Town-Sewage, Irrigation, &c. Sixth Edition. In One Vol., 1,250 pp., half-bound, profusely Illustrated, 12s.

"The aim of the author has been to make his work at once comprehensive and trustworthy, and in this aim he has succeeded to a degree which entitles him to much credit."—*Morning Advertiser.*   "No farmer should be without this book."—*Banbury Guardian.*

Agricultural Engineering.

FARM ENGINEERING, THE COMPLETE TEXT-BOOK OF. Comprising Draining and Embanking; Irrigation and Water Supply; Farm Roads, Fences, and Gates; Farm Buildings, their Arrangement and Construction, with Plans and Estimates; Barn Implements and Machines; Field Implements and Machines; Agricultural Surveying, Levelling, &c. By Prof. JOHN SCOTT, Editor of the "Farmers' Gazette," late Professor of Agriculture and Rural Economy at the Royal Agricultural College, Cirencester, &c. &c. In One Vol., 1,150 pages, half-bound, with over 600 Illustrations, 12s.

"Written with great care, as well as with knowledge and ability. The author has done his work well; we have found him a very trustworthy guide wherever we have tested his statements. The volume will be of great value to agricultural students."—*Mark Lane Express.*
"For a young agriculturist we know of no handy volume likely to be more usefully studied. —*Bell's Weekly Messenger.*

English Agriculture.

THE FIELDS OF GREAT BRITAIN : A Text-Book of Agriculture, adapted to the Syllabus of the Science and Art Department. For Elementary and Advanced Students. By HUGH CLEMENTS (Board of Trade). Second Ed., Revised, with Additions. 18mo, 2s. 6d. cl.

"A most comprehensive volume, giving a mass of information."—*Agricultural Economist.*
"It is a long time since we have seen a book which has pleased us more, or which contains such a vast and useful fund of knowledge."—*Educational Times.*

Tables for Farmers, etc.

TABLES, MEMORANDA, AND CALCULATED RESULTS for Farmers, Graziers, Agricultural Students, Surveyors, Land Agents Auctioneers, etc. With a New System of Farm Book-keeping. Selected and Arranged by SIDNEY FRANCIS. Second Edition, Revised. 272 pp., waistcoat-pocket size, 1s. 6d. limp leather.

"Weighing less than 1 oz., and occupying no more space than a match box, it contains a mass of facts and calculations which has never before, in such handy form, been obtainable. . Every operation on the farm is dealt with. The work may be taken as thoroughly accurate, the whole of the tables having been revised by Dr. Fream. We cordially recommend it."—*Bell's Weekly Messenger.*
"A marvellous little book. . . . The agriculturist who possesses himself of it will not be disappointed with his investment."—*The Farm.*

Farm and Estate Book-keeping.

BOOK-KEEPING FOR FARMERS & ESTATE OWNERS.
A Practical Treatise, presenting, in Three Plans, a System adapted for all
Classes of Farms. By JOHNSON M. WOODMAN, Chartered Accountant. Second
Edition, Revised. Cr. 8vo, 3s. 6d. cl. bds.; or 2s. 6d. cl. limp.
"The volume is a capital study of a most important subject."—*Agricultural Gazette.*
"Will be found of great assistance by those who intend to commence a system of book-keep-
ing, the author's examples being clear and explicit, and his explanations, while full and accurate,
being to a large extent free from technicalities."—*Live Stock Journal.*

Farm Account Book.

WOODMAN'S YEARLY FARM ACCOUNT BOOK. Giving
a Weekly Labour Account and Diary, and showing the Income and Expen-
diture under each Department of Crops, Live Stock, Dairy, &c. &c. With
Valuation, Profit and Loss Account, and Balance Sheet at the end of the
Year, and an Appendix of Forms. Ruled and Headed for Entering a Com-
plete Record of the Farming Operations. By JOHNSON M. WOODMAN,
Chartered Accountant. Folio, 7s. 6d. half-bound. [culture.
"Contains every requisite form for keeping farm accounts readily and accurately."—*Agri-*

Early Fruits, Flowers and Vegetables.

THE FORCING GARDEN; or, How to Grow Early Fruits,
Flowers, and Vegetables. With Plans and Estimates for Building Glass-
houses, Pits and Frames. By SAMUEL WOOD. Crown 8vo, 3s. 6d. cloth.
"A good book, and fairly fills a place that was in some degree vacant. The book is written with
great care, and contains a great deal of valuable teaching."—*Gardeners' Magazine.*
"Mr. Wood's book is an original and exhaustive answer to the question 'How to Grow Early
Fruits, Flowers and Vegetables!'"—*Land and Water.*

Good Gardening.

A PLAIN GUIDE TO GOOD GARDENING; or, How to Grow
Vegetables, Fruits, and Flowers. With Practical Notes on Soils, Manures,
Seeds, Planting, Laying-out of Gardens and Grounds, &c. By S. WOOD.
Fourth Edition, with numerous Illustrations. Crown 8vo, 3s. 6d. cloth.
"A very good book, and one to be highly recommended as a practical guide. The practical
directions are excellent."—*Athenæum.*
"May be recommended to young gardeners, cottagers, and specially to amateurs, for the
plain, simple, and trustworthy information it gives on common matters too often neglected."—
Gardeners' Chronicle.

Gainful Gardening.

MULTUM-IN-PARVO GARDENING; or, How to make One
Acre of Land produce £620 a-year by the Cultivation of Fruits and Vegetables;
also, How to Grow Flowers in Three Glass Houses, so as to realise £176 per
annum clear Profit. By S. WOOD. Fifth Edition. Crown 8vo, 1s. sewed.
"We are bound to recommend it as not only suited to the case of the amateur and gentleman's
gardener, but to the market grower."—*Gardeners' Magazine.*

Gardening for Ladies.

THE LADIES' MULTUM-IN-PARVO FLOWER GARDEN,
and Amateurs' Complete Guide. By S. WOOD. With Illusts. Cr. 8vo, 3s. 6d. cl.
"This volume contains a good deal of sound, common sense instruction."—*Florist.*
"Full of shrewd hints and useful instructions, based on a lifetime of experience."—*Scotsman.*

Receipts for Gardeners.

GARDEN RECEIPTS. By C. W. QUIN. 12mo, 1s. 6d. cloth.
"A useful and handy book, containing a good deal of valuable information."—*Athenæum.*

Market Gardening.

MARKET AND KITCHEN GARDENING. By Contributors
to "The Garden." Compiled by C. W. SHAW, late Editor of "Gardening
Illustrated." 12mo, 3s. 6d. cloth boards.
"The most valuable compendium of kitchen and market-garden work published."—*Farmer.*

Cottage Gardening.

COTTAGE GARDENING; or, Flowers, Fruits, and Vegetables for
Small Gardens. By E. HOBDAY. 12mo, 1s. 6d. cloth limp.

Potato Culture.

POTATOES: How to Grow and Show Them. A Practical Guide
to the Cultivation and General Treatment of the Potato. By JAMES PINK.
Second Edition. Crown 8vo, 2s. cloth.

LAND AND ESTATE MANAGEMENT, LAW, etc.

Hudson's Land Valuer's Pocket-Book.

THE LAND VALUER'S BEST ASSISTANT: Being Tables on a very much Improved Plan, for Calculating the Value of Estates. With Tables for reducing Scotch, Irish, and Provincial Customary Acres to Statute Measure, &c. By R. HUDSON, C.E. New Edition. Royal 32mo, leather, elastic band, 4s.

"This new edition includes tables for ascertaining the value of leases for any term of years ; and for showing how to lay out plots of ground of certain acres in forms, square, round, &c., with valuable rules for ascertaining the probable worth of standing timber to any amount; and is of incalculable value to the country gentleman and professional man."—*Farmers' Journal.*

Ewart's Land Improver's Pocket-Book.

THE LAND IMPROVER'S POCKET-BOOK OF FORMULÆ, TABLES and MEMORANDA required in any Computation relating to the Permanent Improvement of Landed Property. By JOHN EWART, Land Surveyor and Agricultural Engineer. Second Edition, Revised. Royal 32mo, oblong, leather, gilt edges, with elastic band, 4s.

"A compendious and handy little volume."—*Spectator.*

Complete Agricultural Surveyor's Pocket-Book.

THE LAND VALUER'S AND LAND IMPROVER'S COMPLETE POCKET-BOOK. Consisting of the above Two Works bound together. Leather, gilt edges, with strap, 7s. 6d.

"Hudson's book is the best ready-reckoner on matters relating to the valuation of land and crops, and its combination with Mr. Ewart's work greatly enhances the value and usefulness of the latter-mentioned. . . . It is most useful as a manual for reference."—*North of England Farmer.*

Auctioneer's Assistant.

THE APPRAISER, AUCTIONEER, BROKER, HOUSE AND ESTATE AGENT AND VALUER'S POCKET ASSISTANT, for the Valuation for Purchase, Sale, or Renewal of Leases, Annuities and Reversions, and of property generally; with Prices for Inventories, &c. By JOHN WHEELER, Valuer, &c. Fifth Edition, re-written and greatly extended by C. NORRIS, Surveyor, Valuer, &c. Royal 32mo, 5s. cloth.

"A neat and concise book of reference, containing an admirable and clearly-arranged list of prices for inventories, and a very practical guide to determine the value of furniture, &c."—*Standard.*

"Contains a large quantity of varied and useful information as to the valuation for purchase, sale, or renewal of leases, annuities and reversions, and of property generally, with prices for inventories, and a guide to determine the value of interior fittings and other effects."—*Builder.*

Auctioneering.

AUCTIONEERS: THEIR DUTIES AND LIABILITIES. A Manual of Instruction and Counsel for the Young Auctioneer. By ROBERT SQUIBBS, Auctioneer. Second Edition, Revised and partly Re-written. Demy 8vo, 12s. 6d. cloth.

"The position and duties of auctioneers treated compendiously and clearly."—*Builder.*

"Every auctioneer ought to possess a copy of this excellent work."—*Ironmonger.*

"Of great value to the profession. . . . We readily welcome this book from the fact that it treats the subject in a manner somewhat new to the profession."—*Estates Gazette.*

Legal Guide for Pawnbrokers.

THE PAWNBROKERS', FACTORS' AND MERCHANTS' GUIDE TO THE LAW OF LOANS AND PLEDGES. With the Statutes and a Digest of Cases on Rights and Liabilities, Civil and Criminal, as to Loans and Pledges of Goods, Debentures, Mercantile and other Securities. By H. C. FOLKARD, Esq., Barrister-at-Law, Author of "The Law of Slander and Libel," &c. With Additions and Corrections. Fcap. 8vo, 3s. 6d. cloth.

"This work contains simply everything that requires to be known concerning the department of the law of which it treats. We can safely commend the book as unique and very nearly perfect.'—*Iron.*

"The task undertaken by Mr. Folkard has been very satisfactorily performed. . . . Such explanations as are needful have been supplied with great clearness and with due regard to brevity.'
City Press.

Law of Patents.

PATENTS FOR INVENTIONS, AND HOW TO PROCURE THEM. Compiled for the Use of Inventors, Patentees and others. By G. G. M. HARDINGHAM, Assoc.Mem.Inst.C.E., &c. Demy 8vo, cloth, price 2s. 6d.

Metropolitan Rating Appeals.

REPORTS OF APPEALS HEARD BEFORE THE COURT OF GENERAL ASSESSMENT SESSIONS, from the Year 1871 to 1885. By EDWARD RYDE and ARTHUR LYON RYDE. Fourth Edition, brought down to the Present Date, with an Introduction to the Valuation (Metropolis) Act, 1869, and an Appendix by WALTER C. RYDE, of the Inner Temple, Barrister-at-Law. 8vo, 16s. cloth.

"A useful work, occupying a place mid-way between a handbook for a lawyer and a guide to the surveyor. It is compiled by a gentleman eminent in his profession as a land agent, whose specialty, it is acknowledged, lies i the direction of assessing property for rating purposes."—*Land Agents' Record.*

"It is an indispensable work of reference for all engaged in assessment business."—*Journal of Gas Lighting.*

House Property.

HANDBOOK OF HOUSE PROPERTY. A Popular and Practical Guide to the Purchase, Mortgage, Tenancy, and Compulsory Sale of Houses and Land, including the Law of Dilapidations and Fixtures; with Examples of all kinds of Valuations, Useful Information on Building, and Suggestive Elucidations of Fine Art. By E. L. TARBUCK, Architect and Surveyor. Fourth Edition, Enlarged. 12mo, 5s. cloth.

"The advice is thoroughly practical."—*Law Journal.*
"For all who have dealings with house property, this is an indispensable guide."—*Decoration.*
"Carefully brought up to date, and much improved by the addition of a division on fine art."
"A well-written and thoughtful work."—*Land Agent's Record.*

Inwood's Estate Tables.

TABLES FOR THE PURCHASING OF ESTATES, *Freehold, Copyhold, or Leasehold; Annuities, Advowsons, etc.,* and for the Renewing of Leases held under Cathedral Churches, Colleges, or other Corporate bodies, for Terms of Years certain, and for Lives; also for Valuing Reversionary Estates, Deferred Annuities, Next Presentations, &c.; together with SMART'S Five Tables of Compound Interest, and an Extension of the same to Lower and Intermediate Rates. By W. INWOOD. 23rd Edition, with considerable Additions, and new and valuable Tables of Logarithms for the more Difficult Computations of the Interest of Money, Discount, Annuities, &c., by M. FEDOR THOMAN, of the Société Crédit Mobilier of Paris. Crown 8vo, 8s. cloth.

"Those interested in the purchase and sale of estates, and in the adjustment of compensation cases, as well as in transactions in annuities, life insurances, &c., will find the present edition of eminent service."—*Engineering.*

"'Inwood's Tables' still maintain a most enviable reputation. The new issue has been enriched by large additional contributions by M. Fedor Theman, whose carefully arranged Tables cannot fail to be of the utmost utility."—*Mining Journal.*

Agricultural and Tenant-Right Valuation.

THE AGRICULTURAL AND TENANT-RIGHT-VALUER'S ASSISTANT. A Practical Handbook on Measuring and Estimating the Contents, Weights and Values of Agricultural Produce and Timber, the Values of Estates and Agricultural Labour, Forms of Tenant-Right-Valuations, Scales of Compensation under the Agricultural Holdings Act, 1883, &c. &c. By TOM BRIGHT, Agricultural Surveyor. Crown 8vo, 3s. 6d. cloth.

"Full of tables and examples in connection with the valuation of tenant-right, estates, labour, contents, and weights of timber, and farm produce of all kinds."—*Agricultural Gazette.*

"An eminently practical handbook, full of practical tables and data of undoubted interest and value to surveyors and auctioneers in preparing valuations of all kinds."—*Farmer.*

Plantations and Underwoods.

POLE PLANTATIONS AND UNDERWOODS: A Practical Handbook on Estimating the Cost of Forming, Renovating, Improving and Grubbing Plantations and Underwoods, their Valuation for Purposes of Transfer, Rental, Sale or Assessment. By TOM BRIGHT, F.S.Sc., Author of "The Agricultural and Tenant-Right-Valuer's Assistant," &c. Crown 8vo, 3s. 6d. cloth. [*Just published.*]

"Will be found very useful to those who are actually engaged in managing wood."—*Bell's Weekly Messenger.*

"To valuers, foresters and agents it will be a welcome aid."—*North British Agriculturist.*

"Well calculated to assist the valuer in the discharge of his duties, and of undoubted interest and use both to surveyors and auctioneers in preparing valuations of all kinds."—*Kent Herald.*

Berva the brat

A Complete Epitome of the Laws of this Country.

EVERY MAN'S OWN LAWYER: A Handy-Book of the *Principles of Law and Equity.* By A BARRISTER. Twenty-ninth Edition. Revised and Enlarged. Including the Legislation of 1891, and including careful digests of *The Tithe Act*, 1891; the *Mortmain and Charitable Uses Act*, 1891; the *Charitable Trusts (Recovery) Act*, 1891; the *Forged Transfers Act*, 1891; the *Custody of Children Act*, 1891; the *Slander of Women Act*, 1891; the *Public Health (London) Act*, 1891; the *Stamp Act*, 1891; the *Savings Bank Act*, 1891; the *Elementary Education* ("*Free Education*") *Act*, 1891; the *County Councils (Elections) Act*, 1891; and the *Land Registry (Middlesex Deeds) Act*, 1891; while other new Acts have been duly noted. Crown 8vo, 688 pp., price 6s. 8d. (saved at every consultation!), strongly bound in cloth.   [*Just published.*

*** THE BOOK WILL BE FOUND TO COMPRISE (AMONGST OTHER MATTER)—

THE RIGHTS AND WRONGS OF INDIVIDUALS—LANDLORD AND TENANT—VENDORS AND PURCHASERS—PARTNERS AND AGENTS—COMPANIES AND ASSOCIATIONS—MASTERS, SERVANTS AND WORKMEN—LEASES AND MORTGAGES—CHURCH AND CLERGY, RITUAL —LIBEL AND SLANDER—CONTRACTS AND AGREEMENTS—BONDS AND BILLS OF SALE— CHEQUES, BILLS AND NOTES—RAILWAY AND SHIPPING LAW—BANKRUPTCY AND IN- SURANCE—BORROWERS, LENDERS AND SURETIES—CRIMINAL LAW—PARLIAMENTARY ELECTIONS—COUNTY COUNCILS—MUNICIPAL CORPORATIONS—PARISH LAW, CHURCH- WARDENS, ETC.—PUBLIC HEALTH AND NUISANCES—FRIENDLY AND BUILDING SOCIETIES—COPYRIGHT AND PATENTS—TRADE MARKS AND DESIGNS—HUSBAND AND WIFE, DIVORCE, ETC.—TRUSTEES AND EXECUTORS—INTESTACY, LAW OF—GUARDIAN AND WARD, INFANTS, ETC.—GAME LAWS AND SPORTING—HORSES, HORSE-DEALING AND DOGS—INNKEEPERS, LICENSING, ETC.—FORMS OF WILLS, AGREEMENTS, ETC. ETC.

NOTE.—*The object of this work is to enable those who consult it to help them- selves to the law; and thereby to dispense, as far as possible, with professional assistance and advice. There are many wrongs and grievances which persons sub- mit to from time to time through not knowing how or where to apply for redress; and many persons have as great a dread of a lawyer's office as of a lion's den. With this book at hand it is believed that many a* SIX-AND-EIGHTPENCE *may be saved; many a wrong redressed; many a right reclaimed; many a law suit avoided; and many an evil abated. The work has established itself as the standard legal adviser of all classes, and also made a reputation for itself as a useful book of reference for lawyers residing at a distance from law libraries, who are glad to have at hand a work em- bodying recent decisions and enactments.*

*** OPINIONS OF THE PRESS.

"It is a complete code of English Law, written in plain language, which all can understand. . . Should be in the hands of every business man, and all who wish to abolish lawyers' bills."— *Weekly Times.*

"A useful and concise epitome of the law, compiled with considerable care."—*Law Magazine.*

"A complete digest of the most useful facts which constitute English law."—*Globe.*

"Admirably done, admirably arranged, and admirably cheap."—*Leeds Mercury.*

'A concise, cheap and complete epitome of the English law So plainly written that he who runs may read, and he who reads may understand."—*Figaro.*

"A dictionary of legal facts well put together. The book is a very useful one."—*Spectator.*

"The latest edition of this popular book ought to be in every business establishment, and on every library table."—*Sheffield Post.*

Private Bill Legislation and Provisional Orders.

HANDBOOK FOR THE USE OF SOLICITORS AND EN- GINEERS Engaged in Promoting Private Acts of Parliament and Provi- sional Orders, for the Authorization of Railways, Tramways, Works for the Supply of Gas and Water, and other undertakings of a like character. By L. LIVINGSTON MACASSEY, of the Middle Temple, Barrister-at-Law, M.Inst.C.E.; Author of "Hints on Water Supply." 8vo, 950 pp., 25s. cloth.

"The volume is a desideratum on a subject which can be only acquired by practical experi- ence, and the order of procedure in Private Bill Legislation and Provisional Orders is followed. The author's suggestions and notes will be found of great value to engineers and others profession- ally engaged in this class of practice."—*Building News.*

"The author's double experience as an engineer and barrister has eminently qualified him for the task, and enabled him to approach the subject alike from an engineering and legal point of view. The volume will be found a great help both to engineers and lawyers engaged in promoting Private Acts of Parliament and Provisional Orders."—*Local Government Chronicle.*

OGDEN, SMALE AND CO. LIMITED, PRINTERS, GREAT SAFFRON HILL, E.C

WEALE'S RUDIMENTARY SCIENTIFIC SERIES.

** The volumes of this Series are freely Illustrated with Woodcuts, or otherwise, where requisite. Throughout the following List it must be understood that the books are bound in limp cloth, unless otherwise stated; *but the volumes marked with a ‡ may also be had strongly bound in cloth boards for 6d. extra.*

N.B.—In ordering from this List it is recommended, as a means of facilitating business and obviating error, to quote the numbers affixed to the volumes, as well as the titles and prices.

CIVIL ENGINEERING, SURVEYING, ETC.

No.

31. *WELLS AND WELL-SINKING.* By JOHN GEO. SWINDELL, A.R.I.B.A., and G. R. BURNELL, C.E. Revised Edition. With a New Appendix on the Qualities of Water. Illustrated. 2s.

35. *THE BLASTING AND QUARRYING OF STONE,* for Building and other Purposes. By Gen. Sir J. BURGOYNE, Bart. 1s. 6d.

43. *TUBULAR, AND OTHER IRON GIRDER BRIDGES,* particularly describing the Britannia and Conway Tubular Bridges. By G. DRYSDALE DEMPSEY, C.E. Fourth Edition. 2s.

44. *FOUNDATIONS AND CONCRETE WORKS,* with Practical Remarks on Footings, Sand, Concrete, Béton, Pile-driving, Caissons, and Cofferdams, &c. By E. DOBSON. Seventh Edition. 1s. 6d.

60. *LAND AND ENGINEERING SURVEYING.* By T. BAKER, C.E. Fifteenth Edition, revised by Professor J. R. YOUNG. 2s.‡

80*. *EMBANKING LANDS FROM THE SEA.* With examples and Particulars of actual Embankments, &c. By J. WIGGINS, F.G.S. 2s.

81. *WATER WORKS,* for the Supply of Cities and Towns. With a Description of the Principal Geological Formations of England as influencing Supplies of Water, &c. By S. HUGHES, C.E. New Edition. 4s.‡

118. *CIVIL ENGINEERING IN NORTH AMERICA,* a Sketch of. By DAVID STEVENSON, F.R.S.E., &c. Plates and Diagrams. 3s.

167. *IRON BRIDGES, GIRDERS, ROOFS, AND OTHER WORKS.* By FRANCIS CAMPIN, C.E. 2s. 6d.‡

197. *ROADS AND STREETS.* By H. LAW, C.E., revised and enlarged by D. K. CLARK, C.E., including pavements of Stone, Wood, Asphalte, &c. 4s. 6d.‡

203. *SANITARY WORK IN THE SMALLER TOWNS AND IN VILLAGES.* By C. SLAGG, A.M.I.C.E. Revised Edition. 3s.‡

212. *GAS-WORKS, THEIR CONSTRUCTION AND ARRANGEMENT;* and the Manufacture and Distribution of Coal Gas. Originally written by SAMUEL HUGHES, C.E. Re-written and enlarged by WILLIAM RICHARDS, C.E. Eighth Edition, with important additions. 5s. 6d.‡

213. *PIONEER ENGINEERING.* A Treatise on the Engineering Operations connected with the Settlement of Waste Lands in New Countries. By EDWARD DOBSON, Assoc. Inst. C.E. 4s. 6d.‡

216. *MATERIALS AND CONSTRUCTION;* A Theoretical and Practical Treatise on the Strains, Designing, and Erection of Works of Construction. By FRANCIS CAMPIN, C.E. Second Edition, revised. 3s.‡

219. *CIVIL ENGINEERING.* By HENRY LAW, M.Inst. C.E. Including HYDRAULIC ENGINEERING by GEO. R. BURNELL, M.Inst. C.E. Seventh Edition, revised, with large additions by D. KINNEAR CLARK, M.Inst. C.E. 6s. 6d., Cloth boards, 7s. 6d.

268. *THE DRAINAGE OF LANDS, TOWNS, & BUILDINGS.* By G. D. DEMPSEY, C.E. Revised, with large Additions on Recent Practice in Drainage Engineering, by D. KINNEAR CLARK, M.I.C.E. Second Edition, Corrected. 4s. 6d.‡

☞ *The ‡ indicates that these vols. may be had strongly bound at 6d. extra.*

LONDON : CROSBY LOCKWOOD AND SON,

MECHANICAL ENGINEERING, ETC.

33. *CRANES,* the Construction of, and other Machinery for Raising Heavy Bodies. By JOSEPH GLYNN, F.R.S. Illustrated. 1s. 6d.
34. *THE STEAM ENGINE.* By Dr. LARDNER. Illustrated. 1s. 6d.
59. *STEAM BOILERS:* their Construction and Management. By R. ARMSTRONG, C.E. Illustrated. 1s. 6d.
82. *THE POWER OF WATER,* as applied to drive Flour Mills, and to give motion to Turbines, &c. By JOSEPH GLYNN, F.R.S. 2s.‡
98. *PRACTICAL MECHANISM,* the Elements of; and Machine Tools. By T. BAKER, C.E. With Additions by J. NASMYTH, C.E. 2s. 6d.‡
139. *THE STEAM ENGINE,* a Treatise on the Mathematical Theory of, with Rules and Examples for Practical Men. By T. BAKER, C.E. 1s. 6d.
164. *MODERN WORKSHOP PRACTICE,* as applied to Steam ;Engines, Bridges, Ship-building, Cranes, &c. By J. G. WINTON. Fourth Edition, much enlarged and carefully revised. 3s. 6d.‡ [*Just published.*
165. *IRON AND HEAT,* exhibiting the Principles concerned in the Construction of Iron Beams, Pillars, and Girders. By J. ARMOUR. 2s. 6d.‡
166. *POWER IN MOTION:* Horse-Power, Toothed-Wheel Gearing, Long and Short Driving Bands, and Angular Forces. By J. ARMOUR. 2s.‡
171. *THE WORKMAN'S MANUAL OF ENGINEERING* DRAWING. By J. MAXTON. 6th Edn. With 7 Plates and 350 Cuts. 3s. 6d.‡
190. *STEAM AND THE STEAM ENGINE,* Stationary and Portable. Being an Extension of the Elementary Treatise on the Steam Engine of MR. JOHN SEWELL. By D. K. CLARK, M.I.C.E. 3s. 6d.‡
200. *FUEL,* its Combustion and Economy. By C. W. WILLIAMS. With Recent Practice in the Combustion and Economy of Fuel—Coal, Coke Wood, Peat, Petroleum, &c.—by D. K. CLARK, M.I.C.E. 3s. 6d.‡
202. *LOCOMOTIVE ENGINES.* By G. D. DEMPSEY, C.E.; with large additions by D. KINNEAR CLARK, M.I.C.E. 3s.‡
211. *THE BOILERMAKER'S ASSISTANT* in Drawing, Templating, and Calculating Boiler and Tank Work. By JOHN COURTNEY. Practical Boiler Maker. Edited by D. K. CLARK, C.E. 100 Illustrations. 2s,
217. *SEWING MACHINERY:* Its Construction, History, &c., with full Technical Directions for Adjusting, &c. By J. W. URQUHART, C.E. 2s.‡
223. *MECHANICAL ENGINEERING.* Comprising Metallurgy, Moulding, Casting, Forging, Tools, Workshop Machinery, Manufacture of the Steam Engine, &c. By FRANCIS CAMPIN, C.E. Second Edition. 2s, 6d.‡
236. *DETAILS OF MACHINERY.* Comprising Instructions for the Execution of various Works in Iron. By FRANCIS CAMPIN,'C.E. 3s.‡
237. *THE SMITHY AND FORGE;* including the Farrier's Art and Coach Smithing. By W. J. E. CRANE. Illustrated. 2s. 6d.‡
238. *THE SHEET-METAL WORKER'S GUIDE;* a Practical Handbook for Tinsmiths, Coppersmiths, Zincworkers, &c. With 94 Diagrams and Working Patterns. By W. J. E. CRANE. Second Edition, revised. 1s. 5d.
251. *STEAM AND MACHINERY MANAGEMENT:* with Hints on Construction and Selection. By M. POWIS BALE, M.I.M.E. 2s. 6d.‡
254. *THE BOILERMAKER'S READY-RECKONER.* By J. COURTNEY. Edited by D. K. CLARK, C.E. 4s., limp; 5s., half-bound.
255. *LOCOMOTIVE ENGINE-DRIVING.* A Practical Manual for Engineers in charge of Locomotive Engines. By MICHAEL REYNOLDS, M.S.E Eighth Edition. 3s. 6d., limp; 4s. 6d. cloth boards.
256. *STATIONARY ENGINE-DRIVING.* A Practical Manual Engineers in charge of Stationary Engines. By MICHAEL REYNOLDS, M.S.E. Third Edition. 3s. 6d. limp; 4s. 6d. cloth boards,
260. *IRON BRIDGES OF MODERATE SPAN:* their Construction and Erection. By HAMILTON W. PENDRED, C.E. 2s.

The ‡ indicates that these vols. may be had strongly bound at 6d. extra.

MINING, METALLURGY, ETC.

ARCHITECTURE, BUILDING, ETC.

The ‡ indicates that these vols. may be had strongly bound at 6d. extra.

LONDON : CROSBY LOCKWOOD AND SON,

Architecture, Building, etc., *continued.*

116. *THE ACOUSTICS OF PUBLIC BUILDINGS;* or, The Principles of the Science of Sound applied to the purposes of the Architect and Builder. By T. Roger Smith, M.R.I.B.A., Architect. Illustrated. 1s. 6d.

127. *ARCHITECTURAL MODELLING IN PAPER,* the Art of. By T. A. Richardson, Architect. Illustrated. 1s. 6d.

128. *VITRUVIUS—THE ARCHITECTURE OF MARCUS VITRUVIUS POLLO.* In Ten Books. Translated from the Latin by Joseph Gwilt, F.S.A., F.R.A.S. With 23 Plates. 5s.

130. *GRECIAN ARCHITECTURE,* An Inquiry into the Principles of Beauty in; with an Historical View of the Rise and Progress of the Art in Greece. By the Earl of Aberdeen. 1s.

• *The two preceding Works in One handsome Vol., half bound, entitled "Ancient Architecture," price 6s.*

132. *THE ERECTION OF DWELLING-HOUSES.* Illustrated by a Perspective View, Plans, Elevations, and Sections of a pair of Semi-detached Villas, with the Specification, Quantities, and Estimates, &c. By S. H. Brooks. New Edition, with Plates. 2s. 6d.‡

156. *QUANTITIES & MEASUREMENTS* in Bricklayers', Masons', Plasterers', Plumbers', Painters', Paperhangers', Gilders', Smiths', Carpenters' and Joiners' Work. By A. C. Beaton, Surveyor. Ninth Edition. 1s. 6d.

175. *LOCKWOOD'S BUILDER'S PRICE BOOK FOR* 1892. A Comprehensive Handbook of the Latest Prices and Data for Builders, Architects, Engineers, and Contractors. Re-constructed, Re-written, and further Enlarged. By Francis T. W. Miller, A.R.I.B.A. 700 pages. 3s. 6d.; cloth boards, 4s. [*Just Published.*

182. *CARPENTRY AND JOINERY*—The Elementary Principles of Carpentry. Chiefly composed from the Standard Work of Thomas Tredgold, C.E. With a TREATISE ON JOINERY by E. Wyndham Tarn, M.A. Fifth Edition, Revised. 3s. 6d.‡

182*. *CARPENTRY AND JOINERY.* ATLAS of 35 Plates to accompany the above. With Descriptive Letterpress. 4to. 6s.

185. *THE COMPLETE MEASURER;* the Measurement of Boards, Glass, &c.; Unequal-sided, Square-sided, Octagonal-sided, Round Timber and Stone, and Standing Timber, &c. By Richard Horton. Fifth Edition. 4s.; strongly bound in leather, 5s.

187. *HINTS TO YOUNG ARCHITECTS.* By G. Wightwick. New Edition. By G. H. Guillaume. Illustrated. 3s. 6d.‡

188. *HOUSE PAINTING, GRAINING, MARBLING, AND SIGN WRITING:* with a Course of Elementary Drawing for House-Painters, Sign-Writers, &c., and a Collection of Useful Receipts. By Ellis A. Davidson. Sixth Edition. With Coloured Plates. 5s. cloth limp; 6s. cloth boards.

189. *THE RUDIMENTS OF PRACTICAL BRICKLAYING.* In Six Sections: General Principles; Arch Drawing, Cutting, and Setting; Pointing; Paving. Tiling, Materials; Slating and Plastering; Practical Geometry, Mensuration, &c. By Adam Hammond. Seventh Edition. 1s. 6d.

191. *PLUMBING.* A Text-Book to the Practice of the Art or Craft of the Plumber. With Chapters upon House Drainage and Ventilation. Sixth Edition. With 380 Illustrations. By W. P. Buchan. 3s. 6d.‡

192. *THE TIMBER IMPORTER'S, TIMBER MERCHANT'S,* and BUILDER'S STANDARD GUIDE. By R. E. Grandy. 2s.

206. *A BOOK ON BUILDING, Civil and Ecclesiastical,* including Church Restoration. With the Theory of Domes and the Great Pyramid, &c. By Sir Edmund Beckett, Bart., LL.D., Q.C., F.R.A.S. 4s. 6d.‡

226. *THE JOINTS MADE AND USED BY BUILDERS* in the Construction of various kinds of Engineering and Architectural Works. By Wyvill J. Christy, Architect. With upwards of 160 Engravings on Wood. 3s.‡

228. *THE CONSTRUCTION OF ROOFS OF WOOD AND IRON.* By E. Wyndham Tarn, M.A., Architect. Second Edition, revised. 1s. 6d.

☞ *The ‡ indicates that these vols. may be had strongly bound at 6d. extra.*

Architecture, Building, etc., *continued.*

229. *ELEMENTARY DECORATION:* as applied to the Interior and Exterior Decoration of Dwelling-Houses, &c. By J. W. FACEY. 2s.

257. *PRACTICAL HOUSE DECORATION.* A Guide to the Art of Ornamental Painting. By JAMES W. FACEY. 2s. 6d.

*** *The two preceding Works, in One handsome Vol., half-bound, entitled "*HOUSE DECORATION, ELEMENTARY AND PRACTICAL,*" price 5s.*

230. *HANDRAILING.* Showing New and Simple Methods for finding the Pitch of the Plank, Drawing the Moulds, Bevelling, Jointing-up, and Squaring the Wreath. By GEORGE COLLINGS. Second Edition, Revised, including A TREATISE ON STAIRBUILDING. Plates and Diagrams. 2s. 6d.

247. *BUILDING ESTATES:* a Rudimentary Treatise on the Development, Sale, Purchase, and General Management of Building Land. By FOWLER MAITLAND, Surveyor. Second Edition, revised. 2s.

248. *PORTLAND CEMENT FOR USERS.* By HENRY FAIJA, Assoc. M. Inst. C.E. Third Edition, corrected. Illustrated. 2s.

252. *BRICKWORK:* a Practical Treatise, embodying the General and Higher Principles of Bricklaying, Cutting and Setting, &c. By F. WALKER. Third Edition, Revised and Enlarged. 1s. 6d.

23. *THE PRACTICAL BRICK AND TILE BOOK.* Comprising:
189. BRICK AND TILE MAKING, by E. DOBSON, A.I.C.E.; PRACTICAL BRICKLAY-
265. ING, by A. HAMMOND; BRICKCUTTING AND SETTING, by A. HAMMOND. 534 pp. with 270 Illustrations. 6s. Strongly half-bound.

253. *THE TIMBER MERCHANT'S, SAW-MILLER'S, AND IMPORTER'S FREIGHT-BOOK AND ASSISTANT.* By WM. RICHARDSON. With a Chapter on Speeds of Saw-Mill Machinery, &c. By M. Powis BALE, A.M.Inst.C.E. 3s.‡

258. *CIRCULAR WORK IN CARPENTRY AND JOINERY.* A Practical Treatise on Circular Work of Single and Double Curvature. By GEORGE COLLINGS. Second Edition, 2s. 6d.

259. *GAS FITTING:* A Practical Handbook treating of every Description of Gas Laying and Fitting. By JOHN BLACK. With 122 Illustrations. 2s. 6d.‡

261. *SHORING AND ITS APPLICATION:* A Handbook for the Use of Students. By GEORGE H. BLAGROVE. 1s. 6d.

265. *THE ART OF PRACTICAL BRICK CUTTING & SETTING.* By ADAM HAMMOND. With 90 Engravings. 1s. 6d.   [*Just published.*

267. *THE SCIENCE OF BUILDING:* An Elementary Treatise on the Principles of Construction. Adapted to the Requirements of Architectural Students. By E. WYNDHAM TARN, M.A. Lond. Third Edition, Revised and Enlarged. With 59 Wood Engravings. 3s. 6d.‡

271. *VENTILATION:* a Text-book to the Practice of the Art of Ventilating Buildings, with a Supplementary Chapter upon Air Testing. By WILLIAM PATON BUCHAN, R.P., Sanitary and Ventilating Engineer, Author of "Plumbing," &c. 3s. 6d.‡   [*Just published.*

SHIPBUILDING, NAVIGATION, MARINE ENGINEERING, ETC.

51. *NAVAL ARCHITECTURE.* An Exposition of the Elementary Principles of the Science, and their Practical Application to Naval Construction. By J. PEAKE. Fifth Edition, with Plates and Diagrams. 3s. 6d.‡

53*. *SHIPS FOR OCEAN & RIVER SERVICE*, Elementary and Practical Principles of the Construction of. By H. A. SOMMERFELDT. 1s. 6d.

53**. *AN ATLAS OF ENGRAVINGS* to Illustrate the above. Twelve large folding plates. Royal 4to, cloth. 7s. 6d.

54. *MASTING, MAST-MAKING, AND RIGGING OF SHIPS*, Also Tables of Spars, Rigging, Blocks; Chain, Wire, and Hemp Ropes, &c., relative to every class of vessels. By ROBERT KIPPING, N.A. 2s.

☞ *The ‡ indicates that these vols. may be had strongly bound at 6d. extra.*

LONDON : CROSBY LOCKWOOD AND SON,

Shipbuilding, Navigation, Marine Engineering, etc., *cont.*

54*. *IRON SHIP-BUILDING.* With Practical Examples and Details.
By JOHN GRANTHAM, C.E. Fifth Edition. 4s.

55. *THE SAILOR'S SEA BOOK:* a Rudimentary Treatise on
Navigation. By JAMES GREENWOOD, B.A. With numerous Woodcuts and
Coloured Plates. New and enlarged edition. By W. H. ROSSER. 2s. 6d.‡

80. *MARINE ENGINES AND STEAM VESSELS.* By ROBERT
MURRAY, C.E. Eighth Edition, thoroughly Revised, with Additions by the
Author and by GEORGE CARLISLE, C.E. 4s. 6d. limp; 5s. cloth boards.

83*bis*. *THE FORMS OF SHIPS AND BOATS.* By W. BLAND.
Eighth Edition, Revised, with numerous Illustrations and Models. 1s. 6d.

99. *NAVIGATION AND NAUTICAL ASTRONOMY*, in Theory
and Practice. By Prof. J. R. YOUNG. New Edition. 2s. 6d.

106. *SHIPS' ANCHORS*, a Treatise on. By G. COTSELL, N.A. 1s. 6d.

149. *SAILS AND SAIL-MAKING*. With Draughting, and the Centre
of Effort of the Sails; Weights and Sizes of Ropes; Masting, Rigging,
and Sails of Steam Vessels, &c. 12th Edition. By R. KIPPING, N.A., 2s. 6d.‡

155. *ENGINEER'S GUIDE TO THE ROYAL & MERCANTILE*
NAVIES. By a PRACTICAL ENGINEER. Revised by D. F. M'CARTHY. 3s.

55
&
204.

PRACTICAL NAVIGATION. Consisting of The Sailor's
Sea-Book. By JAMES GREENWOOD and W. H. ROSSER. Together with
the requisite Mathematical and Nautical Tables for the Working of the
Problems. By H. LAW, C.E., and Prof. J. R. YOUNG. 7s. Half-bound.

AGRICULTURE, GARDENING, ETC.

61*. *A COMPLETE READY RECKONER FOR THE ADMEA-*
SUREMENT OF LAND, &c. By A. ARMAN. Third Edition, revised
and extended by C. NORRIS, Surveyor, Valuer, &c. 2s.

131. *MILLER'S, CORN MERCHANT'S, AND FARMER'S*
READY RECKONER. Second Edition, with a Price List of Modern
Flour-Mill Machinery, by W. S. HUTTON, C.E. 2s.

140. *SOILS, MANURES, AND CROPS.* (Vol. 1. OUTLINES OF
MODERN FARMING.) By R. SCOTT BURN. Woodcuts. 2s.

141. *FARMING & FARMING ECONOMY*, Notes, Historical and
Practical, on. (Vol. 2. OUTLINES OF MODERN FARMING.) By R. SCOTT BURN. 3s.

142. *STOCK; CATTLE, SHEEP, AND HORSES.* (Vol. 3.
OUTLINES OF MODERN FARMING.) By R. SCOTT BURN. Woodcuts. 2s. 6d.

145. *DAIRY, PIGS, AND POULTRY*, Management of the. By
R. SCOTT BURN. (Vol. 4. OUTLINES OF MODERN FARMING.) 2s.

146. *UTILIZATION OF SEWAGE, IRRIGATION, AND*
RECLAMATION OF WASTE LAND. (Vol. 5. OUTLINES OF MODERN
FARMING.) By R. SCOTT BURN. Woodcuts. 2s. 6d.

₊ *Nos.* 140-1-2-5-6, *in One Vol., handsomely half-bound, entitled* "OUTLINES OF
MODERN FARMING." By ROBERT SCOTT BURN. *Price* 12s.

177. *FRUIT TREES*, The Scientific and Profitable Culture of. From
the French of DU BREUIL. Revised by GEO. GLENNY. 187 Woodcuts. 3s. 6d.‡

198. *SHEEP; THE HISTORY, STRUCTURE, ECONOMY, AND*
DISEASES OF. By W. C. SPOONER, M.R.V.C., &c. Fifth Edition,
enlarged, including Specimens of New and Improved Breeds. 3s. 6d.‡

201. *KITCHEN GARDENING MADE EASY.* By GEORGE M. F.
GLENNY. Illustrated. 1s. 6d.‡

207. *OUTLINES OF FARM MANAGEMENT, and the Organi-*
zation of Farm Labour. By R. SCOTT BURN. 2s. 6d.‡

208. *OUTLINES OF LANDED ESTATES MANAGEMENT.*
By R. SCOTT BURN. 2s. 6d.

₊ *Nos.* 207 & 208 *in One Vol., handsomely half-bound, entitled* "OUTLINES OF
LANDED ESTATES AND FARM MANAGEMENT." By R. SCOTT BURN. *Price* 6s.

☞ *The ‡ indicates that these vols. may be had strongly bound at 6d. extra.*

Agriculture, Gardening, etc., *continued.*

209. *THE TREE PLANTER AND PLANT PROPAGATOR.* A Practical Manual on the Propagation of Forest Trees, Fruit Trees, Flowering Shrubs, Flowering Plants, &c. By SAMUEL WOOD. 2s.

210. *THE TREE PRUNER.* A Practical Manual on the Pruning of Fruit Trees, including also their Training and Renovation; also the Pruning of Shrubs, Climbers, and Flowering Plants. By SAMUEL WOOD. 1s. 6d.

⁎⁎ Nos. 209 & 210 *in One Vol., handsomely half-bound, entitled* "THE TREE PLANTER, PROPAGATOR, AND PRUNER." By SAMUEL WOOD. *Price* 3s. 6d.

218. *THE HAY AND STRAW MEASURER :* Being New Tables for the Use of Auctioneers, Valuers, Farmers, Hay and Straw Dealers, &c. By JOHN STEELE. Fifth Edition. 2s.

222. *SUBURBAN FARMING.* The Laying-out and Cultivation of Farms, adapted to the Produce of Milk, Butter, and Cheese, Eggs, Poultry, and Pigs. By Prof. JOHN DONALDSON and R. SCOTT BURN. 3s. 6d.‡

231. *THE ART OF GRAFTING AND BUDDING.* By CHARLES BALTET. With Illustrations. 2s. 6d.‡

232. *COTTAGE GARDENING;* or, Flowers, Fruits, and Vegetables for Small Gardens. By E. HOBDAY. 1s. 6d.

233. *GARDEN RECEIPTS.* Edited by CHARLES W. QUIN. 1s. 6d.

234. *MARKET AND KITCHEN GARDENING.* By C. W. SHAW, late Editor of "Gardening Illustrated." 3s.‡

239. *DRAINING AND EMBANKING.* A Practical Treatise, embodying the most recent experience in the Application of Improved Methods. By JOHN SCOTT, late Professor of Agriculture and Rural Economy at the Royal Agricultural College, Cirencester. With 68 Illustrations. 1s. 6d.

240. *IRRIGATION AND WATER SUPPLY.* A Treatise on Water Meadows, Sewage Irrigation, and Warping; the Construction of Wells, Ponds, and Reservoirs, &c. By Prof. JOHN SCOTT. With 34 Illus. 1s. 6d.

241. *FARM ROADS, FENCES, AND GATES.* A Practical Treatise on the Roads, Tramways, and Waterways of the Farm; the Principles of Enclosures; and the different kinds of Fences, Gates, and Stiles. By Professor JOHN SCOTT. With 75 Illustrations. 1s. 6d.

242. *FARM BUILDINGS.* A Practical Treatise on the Buildings necessary for various kinds of Farms, their Arrangement and Construction, with Plans and Estimates. By Prof. JOHN SCOTT. With 105 Illus. 2s.

243. *BARN IMPLEMENTS AND MACHINES.* A Practical Treatise on the Application of Power to the Operations of Agriculture; and on various Machines used in the Threshing-barn, in the Stock-yard, and in the Dairy, &c. By Prof. J. SCOTT. With 123 Illustrations. 2s.

244. *FIELD IMPLEMENTS AND MACHINES.* A Practical Treatise on the Varieties now in use, with Principles and Details of Construction, their Points of Excellence, and Management. By Professor JOHN SCOTT. With 138 Illustrations. 2s.

245. *AGRICULTURAL SURVEYING.* A Practical Treatise on Land Surveying, Levelling, and Setting-out; and on Measuring and Estimating Quantities, Weights, and Values of Materials, Produce, Stock, &c. By Prof. JOHN SCOTT. With 62 Illustrations. 1s. 6d.

⁎⁎ Nos. 239 *to* 245 *in One Vol., handsomely half-bound, entitled* "THE COMPLETE TEXT-BOOK OF FARM ENGINEERING." By Professor JOHN SCOTT. *Price* 12s.

250. *MEAT PRODUCTION.* A Manual for Producers, Distributors, &c. By JOHN EWART. 2s. 6d.‡

266. *BOOK-KEEPING FOR FARMERS & ESTATE OWNERS.* By J. M. WOODMAN, Chartered Accountant. 2s. 6d. cloth limp; 3s. 6d. cloth boards.

☞ *The* ‡ *indicates that these vols. may be had strongly bound at* 6d. *extra.*

LONDON : CROSBY LOCKWOOD AND SON,

MATHEMATICS, ARITHMETIC, ETC.

32. *MATHEMATICAL INSTRUMENTS*, a Treatise on; Their Construction, Adjustment, Testing, and Use concisely Explained. By J. F. HEATHER, M.A. Fourteenth Edition, revised, with additions, by A. T. WALMISLEY, M.I.C.E., Fellow of the Surveyors' Institution. Original Edition, in 1 vol., Illustrated. 2s.‡

. *In ordering the above, be careful to say, "Original Edition" (No. 32), to distinguish it from the Enlarged Edition in 3 vols. (Nos. 168-9-70.)*

76. *DESCRIPTIVE GEOMETRY*, an Elementary Treatise on; with a Theory of Shadows and of Perspective, extracted from the French of G. MONGE. To which is added, a description of the Principles and Practice of Isometrical Projection. By J. F. HEATHER, M.A. With 14 Plates. 2s.

178. *PRACTICAL PLANE GEOMETRY:* giving the Simplest Modes of Constructing Figures contained in one Plane and Geometrical Construction of the Ground. By J. F. HEATHER, M.A. With 215 Woodcuts. 2s.

83. *COMMERCIAL BOOK-KEEPING*. With Commercial Phrases and Forms in English, French, Italian, and German. By JAMES HADDON, M.A., Arithmetical Master of King's College School, London. 1s. 6d.

84. *ARITHMETIC*, a Rudimentary Treatise on: with full Explanations of its Theoretical Principles, and numerous Examples for Practice. By Professor J. R. YOUNG. Eleventh Edition. 1s. 6d.

84*. A KEY to the above, containing Solutions in full to the Exercises, together with Comments, Explanations, and Improved Processes, for the Use of Teachers and Unassisted Learners. By J. R. YOUNG. 1s. 6d.

85. *EQUATIONAL ARITHMETIC*, applied to Questions of Interest, Annuities, Life Assurance, and General Commerce; with various Tables by which all Calculations may be greatly facilitated. By W. HIPSLEY. 2s.

86. *ALGEBRA*, the Elements of. By JAMES HADDON, M.A. With Appendix, containing miscellaneous Investigations, and a Collection of Problems in various parts of Algebra. 2s.

86*. A KEY AND COMPANION to the above Book, forming an extensive repository of Solved Examples and Problems in Illustration of the various Expedients necessary in Algebraical Operations. By J. R. YOUNG. 1s. 6d.

88. *EUCLID*, THE ELEMENTS OF: with many additional Propositions
89. and Explanatory Notes: to which is prefixed, an Introductory Essay Logic. By HENRY LAW, C.E. 2s. 6d.‡

. *Sold also separately, viz. :—*

88. EUCLID, The First Three Books. By HENRY LAW, C.E. 1s. 6d.
89. EUCLID, Books 4, 5, 6, 11, 12. By HENRY LAW, C.E. 1s. 6d.

90. *ANALYTICAL GEOMETRY AND CONIC SECTIONS,* By JAMES HANN. A New Edition, by Professor J. R. YOUNG. 2s.‡

91. *PLANE TRIGONOMETRY*, the Elements of. By JAMES HANN, formerly Mathematical Master of King's College, London. 1s. 6d.

92. *SPHERICAL TRIGONOMETRY*, the Elements of. By JAMES HANN. Revised by CHARLES H. DOWLING, C.E.
. *Or with "The Elements of Plane Trigonometry," in One Volume, 2s. 6d.*

98. *MENSURATION AND MEASURING*. With the Mensuration and Levelling of Land for the Purposes of Modern Engineering. By T. BAKER, C.E. New Edition by E. NUGENT, C.E. Illustrated. 1s. 6d.

101. *DIFFERENTIAL CALCULUS*, Elements of the. By W. S. B. WOOLHOUSE, F.R.A.S., &c. 1s. 6d.

102. *INTEGRAL CALCULUS*, Rudimentary Treatise on the. By HOMERSHAM COX, B.A. Illustrated. 1s.

136. *ARITHMETIC*, Rudimentary, for the Use of Schools and Self-Instruction. By JAMES HADDON, M.A. Revised by A. ARMAN. 1s. 6d.

137. A KEY TO HADDON'S RUDIMENTARY ARITHMETIC. By A. ARMAN. 1s. 6d.

☞ *The ‡ indicates that these vols. may be had strongly bound at 6d. extra.*

7, STATIONERS' HALL COURT, LUDGATE HILL, E.C.

Mathematics, Arithmetic, etc., *continued.*

168. *DRAWING AND MEASURING INSTRUMENTS.* Including—I. Instruments employed in Geometrical and Mechanical Drawing, and in the Construction, Copying, and Measurement of Maps and Plans. II. Instruments used for the purposes of Accurate Measurement, and for Arithmetical Computations. By J. F. HEATHER, M.A. Illustrated. 1s. 6d.

169. *OPTICAL INSTRUMENTS.* Including (more especially) Telescopes, Microscopes, and Apparatus for producing copies of Maps and Plans by Photography. By J. F. HEATHER, M.A. Illustrated. 1s. 6d.

170. *SURVEYING AND ASTRONOMICAL INSTRUMENTS.* Including—I. Instruments Used for Determining the Geometrical Features of a portion of Ground. II. Instruments Employed in Astronomical Observations. By J. F. HEATHER, M.A. Illustrated. 1s. 6d.

⁎ *The above three volumes form an enlargement of the Author's original work "Mathematical Instruments." (See No. 32 in the Series.)*

168.⎫ *MATHEMATICAL INSTRUMENTS.* By J. F. HEATHER,
169.⎬ M.A. Enlarged Edition, for the most part entirely re-written. The 3 Parts as
170.⎭ above, in One thick Volume. With numerous Illustrations. 4s. 6d.‡

158. *THE SLIDE RULE, AND HOW TO USE IT;* containing full, easy, and simple Instructions to perform all Business Calculations with unexampled rapidity and accuracy. By CHARLES HOARE, C.E. Sixth Edition. With a Slide Rule in tuck of cover. 2s. 6d.‡

196. *THEORY OF COMPOUND INTEREST AND ANNUITIES;* with Tables of Logarithms for the more Difficult Computations of Interest, Discount, Annuities, &c. By FÉDOR THOMAN. Fourth Edition. 4s.‡

199. *THE COMPENDIOUS CALCULATOR;* or, Easy and Concise Methods of Performing the various Arithmetical Operations required in Commercial and Business Transactions; together with Useful Tables. By D. O'GORMAN. Twenty-seventh Edition, carefully revised by C. NORRIS. 2s. 6d., cloth limp; 3s. 6d., strongly half-bound in leather.

204. *MATHEMATICAL TABLES,* for Trigonometrical, Astronomical, and Nautical Calculations; to which is prefixed a Treatise on Logarithms. By HENRY LAW, C.E. Together with a Series of Tables for Navigation and Nautical Astronomy. By Prof. J. R. YOUNG. New Edition. 4s.

204⁎. *LOGARITHMS.* With Mathematical Tables for Trigonometrical, Astronomical, and Nautical Calculations. By HENRY LAW, M.Inst.C.E. New and Revised Edition. (Forming part of the above Work). 3s.

221. *MEASURES, WEIGHTS, AND MONEYS OF ALL NATIONS,* and an Analysis of the Christian, Hebrew, and Mahometan Calendars. By W. S. B. WOOLHOUSE, F.R.A.S., F.S.S. Seventh Edition, 2s. 6d.‡

227. *MATHEMATICS AS APPLIED TO THE CONSTRUCTIVE ARTS.* Illustrating the various processes of Mathematical Investigation, by means of Arithmetical and Simple Algebraical Equations and Practical Examples. By FRANCIS CAMPIN, C.E. Second Edition. 3s.‡

PHYSICAL SCIENCE, NATURAL PHILOSOPHY, ETC.

1. *CHEMISTRY.* By Professor GEORGE FOWNES, F.R.S. With an Appendix on the Application of Chemistry to Agriculture. 1s.

2. *NATURAL PHILOSOPHY,* Introduction to the Study of. By C. TOMLINSON. Woodcuts. 1s. 6d.

6. *MECHANICS,* Rudimentary Treatise on. By CHARLES TOMLINSON. Illustrated. 1s. 6d.

7. *ELECTRICITY;* showing the General Principles of Electrical Science, and the purposes to which it has been applied. By Sir W. SNOW HARRIS, F.R.S., &c. With Additions by R. SABINE, C.E., F.S.A. 1s. 6d.

7⁎. *GALVANISM.* By Sir W. SNOW HARRIS. New Edition by ROBERT SABINE, C.E., F.S.A. 1s. 6d.

8. *MAGNETISM;* being a concise Exposition of the General Principles of Magnetical Science. By Sir W. SNOW HARRIS. New Edition, revised by H. M. NOAD, Ph.D. With 165 Woodcuts. 3s. 6d.‡

The ‡ *indicates that these vols. may be had strongly bound at 6d. extra.*

Physical Science, Natural Philosophy, etc., *continued.*

11. *THE ELECTRIC TELEGRAPH;* its History and Progress; with Descriptions of some of the Apparatus. By R. Sabine, C.E., F.S.A. 3s.

12. *PNEUMATICS,* including Acoustics and the Phenomena of Wind Currents, for the Use of Beginners. By Charles Tomlinson, F.R.S. Fourth Edition, enlarged. Illustrated. 1s. 6d.

72. *MANUAL OF THE MOLLUSCA;* a Treatise on Recent and Fossil Shells. By Dr. S. P. Woodward, A.L.S. Fourth Edition. With Plates and 300 Woodcuts. 7s. 6d., cloth.

96. *ASTRONOMY.* By the late Rev. Robert Main, M.A. Third Edition, by William Thynne Lynn, B.A., F.R.A.S. 2s.

97. *STATICS AND DYNAMICS,* the Principles and Practice of; embracing also a clear development of Hydrostatics, Hydrodynamics, and Central Forces. By T. Baker, C.E. Fourth Edition. 1s. 6d.

173. *PHYSICAL GEOLOGY,* partly based on Major-General Portlock's "Rudiments of Geology." By Ralph Tate, A.L.S., &c. Woodcuts. 2s.

174. *HISTORICAL GEOLOGY,* partly based on Major-General Portlock's "Rudiments." By Ralph Tate, A.L.S., &c. Woodcuts. 2s. 6d.

173 & 174. *RUDIMENTARY TREATISE ON GEOLOGY,* Physical and Historical. Partly based on Major-General Portlock's "Rudiments of Geology." By Ralph Tate, A.L.S., F.G.S., &c. In One Volume. 4s. 6d.‡

183 & 184. *ANIMAL PHYSICS,* Handbook of. By Dr. Lardner, D.C.L., formerly Professor of Natural Philosophy and Astronomy in University College, Lond. With 520 Illustrations. In One Vol. 7s. 6d., cloth boards.
*** *Sold also in Two Parts, as follows:—*

183. Animal Physics. By Dr. Lardner. Part I., Chapters I.—VII. 4s.

184. Animal Physics. By Dr. Lardner. Part II., Chapters VIII.—XVIII. 3s.

269. *LIGHT:* an Introduction to the Science of Optics, for the Use of Students of Architecture, Engineering, and other Applied Sciences. By E. Wyndham Tarn, M.A. 1s. 6d. *[Just published.*

FINE ARTS.

20. *PERSPECTIVE FOR BEGINNERS.* Adapted to Young Students and Amateurs in Architecture, Painting, &c. By George Pyne. 2s.

40. *GLASS STAINING, AND THE ART OF PAINTING ON GLASS.* From the German of Dr. Gessert and Emanuel Otto Fromberg. With an Appendix on The Art of Enamelling. 2s. 6d.

69. *MUSIC,* A Rudimentary and Practical Treatise on. With numerous Examples. By Charles Child Spencer. 2s. 6d.

71. *PIANOFORTE,* The Art of Playing the. With numerous Exercises & Lessons from the Best Masters. By Charles Child Spencer. 1s. 6d.

69-71. *MUSIC & THE PIANOFORTE.* In one vol. Half bound, 5s.

181. *PAINTING POPULARLY EXPLAINED,* including Fresco, Oil, Mosaic, Water Colour, Water-Glass, Tempera, Encaustic, Miniature, Painting on Ivory, Vellum, Pottery, Enamel, Glass, &c. With Historical Sketches of the Progress of the Art by Thomas John Gullick, assisted John Times, F.S.A. Sixth Edition, revised and enlarged. 5s.‡

186. *A GRAMMAR OF COLOURING,* applied to Decorative Painting and the Arts. By George Field. New Edition, enlarged and adapted to the Use of the Ornamental Painter and Designer. By Ellis A. Davidson. With two new Coloured Diagrams, &c. 3s.‡

246. *A DICTIONARY OF PAINTERS, AND HANDBOOK FOR PICTURE AMATEURS;* including Methods of Painting, Cleaning, Relining and Restoring, Schools of Painting, &c. With Notes on the Copyists and Imitators of each Master. By Philippe Daryl. 2s. 6d.‡

The ‡ indicates that these vols. may be had strongly bound at 6d. extra.

INDUSTRIAL AND USEFUL ARTS.

23. *BRICKS AND TILES,* Rudimentary Treatise on the Manufacture of. By E. DOBSON, M.R.I.B.A. Illustrated, 3s.‡

67. *CLOCKS, WATCHES, AND BELLS,* a Rudimentary Treatise on. By Sir EDMUND BECKETT, LL.D., Q.C. Seventh Edition, revised and enlarged. 4s. 6d. limp; 5s. 6d. cloth boards.

83**. *CONSTRUCTION OF DOOR LOCKS.* Compiled from the Papers of A. C. HOBBS, and Edited by CHARLES TOMLINSON, F.R.S. 2s. 6d.

162. *THE BRASS FOUNDER'S MANUAL;* Instructions for Modelling, Pattern-Making, Moulding, Turning, Filing, Burnishing, Bronzing, &c. With copious Receipts, &c. By WALTER GRAHAM. 2s.‡

205. *THE ART OF LETTER PAINTING MADE EASY.* By J. G. BADENOCH. Illustrated with 12 full-page Engravings of Examples. 1s. 6d.

215. *THE GOLDSMITH'S HANDBOOK,* containing full Instructions for the Alloying and Working of Gold. By GEORGE E. GEE, 3s.‡

225. *THE SILVERSMITH'S HANDBOOK,* containing full Instructions for the Alloying and Working of Silver. By GEORGE E. GEE, 3s.‡

* *The two preceding Works, in One handsome Vol., half-bound, entitled "THE GOLDSMITH'S & SILVERSMITH'S COMPLETE HANDBOOK," 7s.*

249. *THE HALL-MARKING OF JEWELLERY PRACTICALLY CONSIDERED.* By GEORGE E. GEE. 3s.‡

224. *COACH BUILDING,* A Practical Treatise, Historical and Descriptive. By J. W. HURGESS. 2s. 6d.‡

235. *PRACTICAL ORGAN BUILDING.* By W. E. DICKSON, M.A., Precentor of Ely Cathedral. Illustrated. 2s. 6d.‡

262. *THE ART OF BOOT AND SHOEMAKING.* By JOHN BEDFORD LENO. Numerous Illustrations. Third Edition. 2s.

263. *MECHANICAL DENTISTRY:* A Practical Treatise on the Construction of the Various Kinds of Artificial Dentures, with Formulæ, Tables, Receipts, &c. By CHARLES HUNTER. Third Edition. 3s.‡

270. *WOOD ENGRAVING:* A Practical and Easy Introduction to the Study of the Art. By W. N. BROWN. 1s. 6d.

MISCELLANEOUS VOLUMES.

36. *A DICTIONARY OF TERMS used in ARCHITECTURE, BUILDING, ENGINEERING, MINING, METALLURGY, ARCHÆOLOGY, the FINE ARTS, &c.* By JOHN WEALE. Sixth Edition. Revised by ROBERT HUNT, F.R.S. Illustrated. 5s. limp; 6s. cloth boards.

50. *THE LAW OF CONTRACTS FOR WORKS AND SERVICES.* By DAVID GIBBONS. Third Edition, enlarged. 3s.‡

112. *MANUAL OF DOMESTIC MEDICINE.* By R. GOODING, B.A., M.D. A Family Guide in all Cases of Accident and Emergency 2s.

112*. *MANAGEMENT OF HEALTH.* A Manual of Home and Personal Hygiene. By the Rev. JAMES BAIRD, B.A. 1s.

150. *LOGIC,* Pure and Applied. By S. H. EMMENS. 1s. 6d.

153. *SELECTIONS FROM LOCKE'S ESSAYS ON THE HUMAN UNDERSTANDING.* With Notes by S. H. EMMENS. 2s.

154. *GENERAL HINTS TO EMIGRANTS.* 2s.

157. *THE EMIGRANT'S GUIDE TO NATAL.* By ROBERT JAMES MANN, F.R.A.S., F.M.S. Second Edition. Map. 2s.

193. *HANDBOOK OF FIELD FORTIFICATION.* By Major W. W. KNOLLYS, F.R.G.S. With 163 Woodcuts. 3s.†

194. *THE HOUSE MANAGER:* Being a Guide to Housekeeping. Practical Cookery, Pickling and Preserving, Household Work, Dairy Management, &c. By AN OLD HOUSEKEEPER. 3s. 6d.‡

194, *HOUSE BOOK (The).* Comprising:—I. THE HOUSE MANAGER.
112 & By an OLD HOUSEKEEPER. II. DOMESTIC MEDICINE. By R. GOODING, M.D.
112*. III. MANAGEMENT OF HEALTH. By J. BAIRD. In One Vol., half-bound, 6s.

The ‡ indicates that these vols may be had strongly bound at 6d. extra.

LONDON : CROSBY LOCKWOOD AND SON,

EDUCATIONAL AND CLASSICAL SERIES.

HISTORY.

1. **England, Outlines of the History of;** more especially with reference to the Origin and Progress of the English Constitution. By WILLIAM DOUGLAS HAMILTON, F.S.A., of Her Majesty's Public Record Office. 4th Edition, revised. 5s.; cloth boards, 6s.

5. **Greece, Outlines of the History of;** in connection with the Rise of the Arts and Civilization in Europe. By W. DOUGLAS HAMILTON, of University College, London, and EDWARD LEVIEN, M.A., of Balliol College, Oxford. 2s. 6d.; cloth boards, 3s. 6d.

7. **Rome, Outlines of the History of:** from the Earliest Period to the Christian Era and the Commencement of the Decline of the Empire. By EDWARD LEVIEN, of Balliol College, Oxford. Map, 2s. 6d.; cl. bds. 3s. 6d.

9. **Chronology of History, Art, Literature, and Progress,** from the Creation of the World to the Present Time. The Continuation by W. D. HAMILTON, F.S.A. 3s.; cloth boards, 3s. 6d.

50. **Dates and Events in English History,** for the use of Candidates in Public and Private Examinations. By the Rev. E. RAND. 1s.

ENGLISH LANGUAGE AND MISCELLANEOUS.

11. **Grammar of the English Tongue,** Spoken and Written. With an Introduction to the Study of Comparative Philology. By HYDE CLARKE, D.C.L. Fifth Edition. 1s. 6d.

12. **Dictionary of the English Language,** as Spoken and Written. Containing above 100,000 Words. By HYDE CLARKE, D.C.L. 3s. 6d.; cloth boards, 4s. 6d.; complete with the GRAMMAR, cloth bds., 5s. 6d.

48. **Composition and Punctuation,** familiarly Explained for those who have neglected the Study of Grammar. By JUSTIN BRENAN. 18th Edition. 1s. 6d.

49. **Derivative Spelling-Book:** Giving the Origin of Every Word from the Greek, Latin, Saxon, German, Teutonic, Dutch, French, Spanish, and other Languages; with their present Acceptation and Pronunciation. By J. ROWBOTHAM, F.R.A.S. Improved Edition. 1s. 6d.

51. **The Art of Extempore Speaking:** Hints for the Pulpit, the Senate, and the Bar. By M. BAUTAIN, Vicar-General and Professor at the Sorbonne. Translated from the French. 8th Edition, carefully corrected. 2s. 6d.

54. **Analytical Chemistry,** Qualitative and Quantitative, a Course of. To which is prefixed, a Brief Treatise upon Modern Chemical Nomenclature and Notation. By WM. W. PINK and GEORGE E. WEBSTER. 2s.

THE SCHOOL MANAGERS' SERIES OF READING BOOKS,

Edited by the Rev. A. R. GRANT, Rector of Hitcham, and Honorary Canon of Ely; formerly H.M. Inspector of Schools.

INTRODUCTORY PRIMER, 3d.

	s. d.		s. d.
FIRST STANDARD	. 0 6	FOURTH STANDARD . . .	1 2
SECOND ,,	. 0 10	FIFTH ,, . . .	1 6
THIRD ,,	. 1 0	SIXTH ,, . . .	1 6

LESSONS FROM THE BIBLE. Part I. Old Testament. 1s.

LESSONS FROM THE BIBLE. Part II. New Testament, to which is added THE GEOGRAPHY OF THE BIBLE, for very young Children. By Rev. C. THORNTON FORSTER. 1s. 2d. *_* Or the Two Parts in One Volume. 2s.

FRENCH.

24. **French Grammar.** With Complete and Concise Rules on the Genders of French Nouns. By G. L. STRAUSS, Ph.D. 1s. 6d.
25. **French-English Dictionary.** Comprising a large number of New Terms used in Engineering, Mining, &c. By ALFRED ELWES. 1s. 6d.
26. **English-French Dictionary.** By ALFRED ELWES. 2s.
25,26. **French Dictionary** (as above). Complete, in One Vol., 3s.; cloth boards, 3s. 6d. *₊* Or with the GRAMMAR, cloth boards, 4s. 6d.
47. **French and English Phrase Book:** containing Introductory Lessons, with Translations, several Vocabularies of Words, a Collection of suitable Phrases, and Easy Familiar Dialogues. 1s. 6d.

GERMAN.

39. **German Grammar.** Adapted for English Students, from Heyse's Theoretical and Practical Grammar, by Dr. G. L. STRAUSS. 1s. 6d.
40. **German Reader:** A Series of Extracts, carefully culled from the most approved Authors of Germany; with Notes, Philological and Explanatory. By G. L. STRAUSS, Ph.D. 1s.
41-43. **German Triglot Dictionary.** By N. E. S. A. HAMILTON. In Three Parts. Part I. German-French-English. Part II. English-German-French. Part III. French-German-English. 3s., or cloth boards, 4s.
41-43 **German Triglot Dictionary** (as above), together with German & 39. Grammar (No. 39), in One Volume, cloth boards, 5s.

ITALIAN.

27. **Italian Grammar,** arranged in Twenty Lessons, with a Course of Exercises. By ALFRED ELWES. 1s. 6d.
28. **Italian Triglot Dictionary,** wherein the Genders of all the Italian and French Nouns are carefully noted down. By ALFRED ELWES. Vol. 1. Italian-English-French. 2s. 6d.
30. **Italian Triglot Dictionary.** By A. ELWES. Vol. 2. English-French-Italian. 2s. 6d.
32. **Italian Triglot Dictionary.** By ALFRED ELWES. Vol. 3. French-Italian-English. 2s. 6d.
28,30, **Italian Triglot Dictionary** (as above). In One Vol., 7s. 6d. 32. Cloth boards.

SPANISH AND PORTUGUESE.

34. **Spanish Grammar,** in a Simple and Practical Form. With a Course of Exercises. By ALFRED ELWES. 1s. 6d.
35. **Spanish-English and English-Spanish Dictionary.** Including a large number of Technical Terms used in Mining, Engineering, &c. with the proper Accents and the Gender of every Noun. By ALFRED ELWES 4s.; cloth boards, 5s. *₊* Or with the GRAMMAR, cloth boards, 6s.
55. **Portuguese Grammar,** in a Simple and Practical Form. With a Course of Exercises. By ALFRED ELWES. 1s. 6d.
56. **Portuguese-English and English-Portuguese Dictionary.** Including a large number of Technical Terms used in Mining, Engineering, &c., with the proper Accents and the Gender of every Noun. By ALFRED ELWES. Second Edition, Revised, 5s.; cloth boards, 6s. *₊* Or with the GRAMMAR, cloth boards, 7s.

HEBREW.

46*. **Hebrew Grammar.** By Dr. BRESSLAU. 1s. 6d.
44. **Hebrew and English Dictionary,** Biblical and Rabbinical; containing the Hebrew and Chaldee Roots of the Old Testament Post-Rabbinical Writings. By Dr. BRESSLAU. 6s.
46. **English and Hebrew Dictionary.** By Dr. BRESSLAU. 3s.
44,46. **Hebrew Dictionary** (as above), in Two Vols., complete, with 46*. the GRAMMAR, cloth boards, 12s.

LATIN.

19. **Latin Grammar.** Containing the Inflections and Elementary Principles of Translation and Construction. By the Rev. THOMAS GOODWIN, M.A., Head Master of the Greenwich Proprietary School. 1s. 6d.

20. **Latin-English Dictionary.** By the Rev. THOMAS GOODWIN, M.A. 2s.

22. **English-Latin Dictionary;** together with an Appendix of French and Italian Words which have their origin from the Latin. By the Rev. THOMAS GOODWIN, M.A. 1s. 6d.

20,22. **Latin Dictionary** (as above). Complete in One Vol., 3s. 6d. cloth boards, 4s. 6d. *₊* Or with the GRAMMAR, cloth boards, 5s. 6d.

LATIN CLASSICS. With Explanatory Notes in English.

1. **Latin Delectus.** Containing Extracts from Classical Authors, with Genealogical Vocabularies and Explanatory Notes, by H. YOUNG. 1s. 6d;

2. **Cæsaris** Commentarii de Bello Gallico. Notes, and a Geographical Register for the Use of Schools, by H. YOUNG. 2s.

3. **Cornelius Nepos.** With Notes. By H. YOUNG. 1s.

4. **Virgilii** Maronis Bucolica et Georgica. With Notes on the Bucolics by W. RUSHTON, M.A., and on the Georgics by H. YOUNG. 1s. 6d.

5. **Virgilii** Maronis Æneis. With Notes, Critical and Explanatory, by H. YOUNG. New Edition, revised and improved With copious Additional Notes by Rev. T. H. L. LEARY, D.C.L., formerly Scholar of Brasenose College, Oxford. 3s.
5* —— Part 1. Books i.—vi., 1s. 6d.
5** —— Part 2. Books vii.—xii., 2s.

6. **Horace;** Odes, Epode, and Carmen Sæculare. Notes by H. YOUNG. 1s. 6d.

7. **Horace;** Satires, Epistles, and Ars Poetica. Notes by W. BROWN-RIGG SMITH, M.A., F.R.G.S. 1s. 6d.

8. **Sallustii** Crispi Catalina et Bellum Jugurthinum. Notes, Critical and Explanatory, by W. M. DONNE, B.A., Trin. Coll., Cam. 1s. 6d.

9. **Terentii** Andria et Heautontimorumenos. With Notes, Critical and Explanatory, by the Rev. JAMES DAVIES, M.A. 1s. 6d.

10. **Terentii** Adelphi, Hecyra, Phormio. Edited, with Notes, Critical and Explanatory, by the Rev. JAMES DAVIES, M.A. 2s.

11. **Terentii** Eunuchus, Comœdia. Notes, by Rev. J. DAVIES, M.A. 1s. 6d.

12. **Ciceronis** Oratio pro Sexto Roscio Amerino. Edited, with an Introduction, Analysis, and Notes, Explanatory and Critical, by the Rev JAMES DAVIES, M.A. 1s. 6d.

13. **Ciceronis** Orationes in Catilinam, Verrem, et pro Archia. With Introduction, Analysis, and Notes, Explanatory and Critical, by Rev. T. H. L. LEARY, D.C.L. formerly Scholar of Brasenose College, Oxford. 1s. 6d.

14. **Ciceronis** Cato Major, Lælius, Brutus, sive de Senectute, de Amicitia, de Claris Oratoribus Dialogi. With Notes by W. BROWNRIGG SMITH M.A., F.R.G.S. 2s.

16. **Livy:** History of Rome. Notes by H. YOUNG and W. B. SMITH, M.A. Part 1. Books i., ii., 1s. 6d.
16*. —— Part 2. Books iii., iv., v., 1s. 6d.
17. —— Part 3. Books xxi., xxii., 1s. 6d.

19. **Latin Verse Selections,** from Catullus, Tibullus, Propertius, and Ovid. Notes by W. B. DONNE, M.A., Trinity College, Cambridge. 2s.

20. **Latin Prose Selections,** from Varro, Columella, Vitruvius, Seneca, Quintilian, Florus, Velleius Paterculus, Valerius Maximus Suetonius, Apuleius, &c. Notes by W. B. DONNE, M.A. 2s.

21. **Juvenalis** Satiræ. With Prolegomena and Notes by T. H. S. ESCOTT, B.A., Lecturer on Logic at King's College, London. 2s.

GREEK.

14. Greek Grammar, in accordance with the Principles and Philological Researches of the most eminent Scholars of our own day. By HANS
CLAUDE HAMILTON. 1s. 6d.

15,17. Greek Lexicon. Containing all the Words in General Use, with
their Significations, Inflections, and Doubtful Quantities. By HENRY R.
HAMILTON. Vol. 1. Greek-English, 2s. 6d.; Vol. 2. English-Greek, 2s. Or
the Two Vols. in One, 4s. 6d.: cloth boards, 5s.

14,15. Greek Lexicon (as above). Complete, with the GRAMMAR, in
17. One Vol., cloth boards, 6s.

GREEK CLASSICS. With Explanatory Notes in English.

1. Greek Delectus. Containing Extracts from Classical Authors,
with Genealogical Vocabularies and Explanatory Notes, by H. YOUNG. New
Edition, with an improved and enlarged Supplementary Vocabulary, by JOHN
HUTCHISON, M.A., of the High School, Glasgow. 1s. 6d.

2, 3. Xenophon's Anabasis; or, The Retreat of the Ten Thousand.
Notes and a Geographical Register, by H. YOUNG. Part 1. Books i. to iii.,
1s. Part 2. Books iv. to vii., 1s.

4. Lucian's Select Dialogues. The Text carefully revised, with
Grammatical and Explanatory Notes, by H. YOUNG. 1s. 6d.

5-12. Homer, The Works of. According to the Text of BAEUMLEIN.
With Notes, Critical and Explanatory, drawn from the best and latest
Authorities, with Preliminary Observations and Appendices, by T. H. L.
LEARY, M.A., D.C.L.

THE ILIAD: Part 1. Books i. to vi., 1s. 6d. | Part 3. Books xiii. to xviii., 1s. 6d
               Part 2. Books vii. to xii., 1s. 6d. | Part 4. Books xix. to xxiv., 1s. 6d.
THE ODYSSEY: Part 1. Books i. to vi., 1s. 6d. | Part 3. Books xiii. to xviii., 1s. 6d.
               Part 2. Books vii. to xii., 1s. 6d. | Part 4. Books xix. to xxiv., and
                                                     Hymns, 2s.

13. Plato's Dialogues: The Apology of Socrates, the Crito, and
the Phædo. From the Text of C. F. HERMANN. Edited with Notes, Critical
and Explanatory, by the Rev. JAMES DAVIES, M.A. 2s.

14-17. Herodotus, The History of, chiefly after the Text of GAISFORD.
With Preliminary Observations and Appendices, and Notes, Critical and
Explanatory, by T. H. L. LEARY, M.A., D.C.L.
               Part 1. Books i., ii. (The Clio and Euterpe), 2s.
               Part 2. Books iii., iv. (The Thalia and Melpomene), 2s.
               Part 3. Books v.-vii. (The Terpsichore, Erato, and Polymnia), 2s.
               Part 4. Books viii., ix. (The Urania and Calliope) and Index, 1s. 6d.

18. Sophocles: Œdipus Tyrannus. Notes by H. YOUNG. 1s.

20. Sophocles: Antigone. From the Text of DINDORF. Notes,
Critical and Explanatory, by the Rev. JOHN MILNER, B.A. 2s.

23. Euripides: Hecuba and Medea. Chiefly from the Text of DIN-
DORF. With Notes, Critical and Explanatory, by W. BROWNRIGG SMITH,
M.A., F.R.G.S. 1s. 6d.

26. Euripides: Alcestis. Chiefly from the Text of DINDORF. With
Notes, Critical and Explanatory, by JOHN MILNER, B.A. 1s. 6d.

30. Æschylus: Prometheus Vinctus: The Prometheus Bound. From
the Text of DINDORF. Edited, with English Notes, Critical and Explanatory,
by the Rev. JAMES DAVIES, M.A. 1s.

32. Æschylus: Septem Contra Thebes: The Seven against Thebes.
From the Text of DINDORF. Edited, with English Notes, Critical and Explanatory, by the Rev. JAMES DAVIES, M.A. 1s.

40. Aristophanes: Acharnians. Chiefly from the Text of C. H.
WEISE. With Notes, by C. S. T. TOWNSHEND, M.A. 1s. 6d.

41. Thucydides: History of the Peloponnesian War. Notes by H.
YOUNG. Book 1. 1s. 6d.

42. Xenophon's Panegyric on Agesilaus. Notes and Introduction by LL. F. W. JEWITT. 1s. 6d.

43. Demosthenes. The Oration on the Crown and the Philippics.
With English Notes. By Rev. T. H. L. LEARY, D.C.L., formerly Scholar of
Brasenose College, Oxford. 1s. 6d.

CROSBY LOCKWOOD AND SON, 7, STATIONERS' HALL COURT, E.C.

www.ingramcontent.com/pod-product-compliance
Lightning Source LLC
Chambersburg PA
CBHW030950110726
47900CB00004B/1213